, Brett, Robbie
cially Linda

ook is in the
on-Wesley Series in Computer Science

sulting Editor
ael A. Harrison

nsoring Editor
es T. DeWolf

brary of Congress Cataloging in Publication Data

edgewick, Robert, 1946-
　　Algorithms.

　　1. Algorithms. I. Title.
QA76.6.S435　1983
ISBN 0-201-06672-6　　　　519.4　　　　82-11672

Reproduced by Addison-Wesley from camera-ready copy supplied by the author.

ISBN 0-201-06672-6
BCDEFGHIJ-HA-89876543

To Adam
and esp

ALGORIT

ROBERT SE

This b
Addis

Cons
Mich

Spo
Jan

Li

S

ADDISON-WESLEY PUBLISHING COM
Reading, Massachusetts • Menlo Park, Cali
London • Amsterdam • Don Mills, Ontario • Sy

Preface

This book is intended to survey the most important algorithms in use on computers today and to teach fundamental techniques to the growing number of people who are interested in becoming serious computer users. It is appropriate for use as a textbook for a second, third or fourth course in computer science: after students have acquired some programming skills and familiarity with computer systems, but before they have specialized courses in advanced areas of computer science or computer applications. Additionally, the book may be useful as a reference for those who already have some familiarity with the material, since it contains a number of computer implementations of useful algorithms.

The book consists of forty chapters which are grouped into seven major parts: mathematical algorithms, sorting, searching, string processing, geometric algorithms, graph algorithms and advanced topics. A major goal in the development of this book has been to bring together the fundamental methods from these diverse areas, in order to provide access to the best methods that we know for solving problems by computer for as many people as possible. The treatment of sorting, searching and string processing (which are not likely to be covered in other courses) is somewhat more complete than the treatment of mathematical algorithms (which are likely to be covered in more depth in applied mathematics or engineering courses), geometric algorithms (which are likely to be covered in more depth in advanced computer science courses). Some of the chapters involve introduction treatment of advanced material. It is hoped that the descriptions here can provide students with some understanding of the basic properties of fundamental algorithms such as the FFT or the simplex method, while at the same time preparing them to better appreciate the methods when they learn them in advanced courses.

The orientation of the book is towards algorithms that are likely to be of practical use. The emphasis is on teaching students the tools of their trade to the point that they can confidently implement, run and debug useful algorithms. Full implementations of the methods discussed (in an actual programming language) are included in the text, along with descriptions of the operations of these programs on a consistent set of examples. Though not emphasized, connections to theoretical computer science and the analysis of algorithms are not ignored. When appropriate, analytic results are discussed to illustrate why certain algorithms are preferred. When interesting, the relationship of the practical algorithms being discussed to purely theoretical results is described. More information of the orientation and coverage of the material in the book may be found in the Introduction which follows.

One or two previous courses in computer science are recommended for students to be able to appreciate the material in this book: one course in programming in a high-level language such as Pascal, and perhaps another course which teaches fundamental concepts of programming systems. In short, students should be conversant with a modern programming language and have a comfortable understanding of the basic features of modern computer systems. There is some mathematical material which requires knowledge of calculus, but this is isolated within a few chapters and could be skipped.

There is a great deal of flexibility in the way that the material in the book can be taught. To a large extent, the individual chapters in the book can each be read independently of the others. The material can be adapted for use for various courses by selecting perhaps thirty of the forty chapters. An elementary course on "data structures and algorithms" might omit some of the mathematical algorithms and some of the advanced graph algorithms and other advanced topics, then emphasize the ways in which various data structures are used in the implementation. An intermediate course on "design and analysis of algorithms" might omit some of the more practically-oriented sections, then emphasize the identification and study of the ways in which good algorithms achieve good asymptotic performance. A course on "software tools" might omit the mathematical and advanced algorithmic material, then emphasize means by which the implementations given here can be integrated for use into large programs or systems. Some supplementary material might be required for each of these examples to reflect their particular orientation (on elementary data structures for "data structures and algorithms," on mathematical analysis for "design and analysis of algorithms," and on software engineering techniques for "software tools"); in this book, the emphasis is on the algorithms themselves.

At Brown University, we've used preliminary versions of this book in our third course in computer science, which is prerequisite to all later courses. Typically, about one-hundred students take the course, perhaps half of which are majors. Our experience has been that the breadth of coverage of material in this book provides an "introduction to computer science" for our majors which can later be expanded upon in later courses on analysis of algorithms, systems programming and theoretical computer science, while at the same time providing all the students with a large set of techniques that they can immediately put to good use.

The programming language used throughout the book is Pascal. The advantage of using Pascal is that it is widely available and widely known; the disadvantage is that it lacks many features needed by sophisticated algorithms. The programs are easily translatable to other modern programming languages, since relatively few Pascal constructs are used. Some of the programs can be simplified by using more advanced language features (some not available in Pascal), but this is true less often than one might think. A goal of this book is to present the algorithms in as simple and direct form as possible.

The programs are not intended to be read by themselves, but as part of the surrounding text. This style was chosen as an alternative, for example, to having inline comments. Consistency in style is used whenever possible, so that programs which are similar, look similar. There are 400 exercises, ten following each chapter, which generally divide into one of two types. Most of the exercises are intended to test students' understanding of material in the text, and ask students to work through an example or apply concepts described in the text. A few of the exercises at the end of each chapter involve implementing and putting together some of the algorithms, perhaps running empirical studies to learn their properties.

Acknowledgments

Many people, too numerous to mention here, have provided me with helpful feedback on earlier drafts of this book. In particular, students and teaching assistants at Brown have suffered through preliminary versions of the material in this book over the past three years. Thanks are due to Trina Avery, Tom Freeman and Janet Incerpi, all of whom carefully read the last two drafts of the book. Janet provided extensive detailed comments and suggestions which helped me fix innumerable technical errors and omissions; Tom ran and checked the programs; and Trina's copy editing helped me make the text clearer and more nearly correct.

Much of what I've written in this book I've learned from the teaching and writings of Don Knuth, my thesis advisor at Stanford. Though Don had no direct influence at all on this work, his presence may be felt in the book, for it was he who put the study of algorithms on a scientific footing that makes a work such as this possible.

Special thanks are due to Janet Incerpi who initially converted the book into TEX format, added the thousands of changes I made after the "last draft," guided the files through various systems to produce printed pages and even wrote the scan conversion routine for TEX that we used to produce draft manuscripts, among many other things.

The text for the book was typeset at the American Mathematical Society; the drawings were done with pen-and-ink by Linda Sedgewick; and the final assembly and printing were done by Addison-Wesley under the guidance of Jim DeWolf. The help of all the people involved is gratefully acknowledged.

Finally, I am very thankful for the support of Brown University and INRIA where I did most of the work on the book, and the Institute for Defense Analyses and the Xerox Palo Alto Research Center, where I did some work on the book while visiting.

Robert Sedgewick
Marly-le-Roi, France
February, 1983

Contents

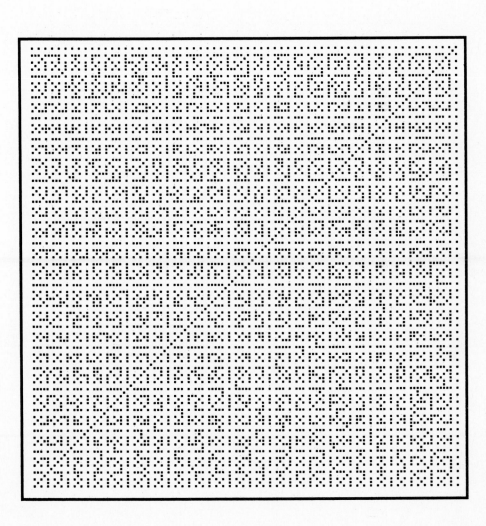

Introduction

The objective of this book is to study a broad variety of important and useful *algorithms*: methods for solving problems which are suited for computer implementation. We'll deal with many different areas of application, always trying to concentrate on "fundamental" algorithms which are important to know and interesting to study. Because of the large number of areas and algorithms to be covered, we won't have room to study many of the methods in great depth. However, we will try to spend enough time on each algorithm to understand its essential characteristics and to respect its subtleties. In short, our goal is to learn a large number of the most important algorithms used on computers today, well enough to be able to use and appreciate them.

To learn an algorithm well, one must implement it. Accordingly, the best strategy for understanding the programs presented in this book is to implement and test them, experiment with variants, and try them out on real problems. We will use the Pascal programming language to discuss and implement most of the algorithms; since, however, we use a relatively small subset of the language, our programs are easily translatable to most modern programming languages.

Readers of this book are expected to have at least a year's experience in programming in high- and low-level languages. Also, they should have some familiarity with elementary algorithms on simple data structures such as arrays, stacks, queues, and trees. (We'll review some of this material but within the context of their use to solve particular problems.) Some elementary acquaintance with machine organization and computer architecture is also assumed. A few of the applications areas that we'll deal with will require knowledge of elementary calculus. We'll also be using some very basic material involving linear algebra, geometry, and discrete mathematics, but previous knowledge of these topics is not necessary.

3

This book is divided into forty chapters which are organized into seven major parts. The chapters are written so that they can be read independently, to as great extent as possible. Generally, the first chapter of each part gives the basic definitions and the "ground rules" for the chapters in that part; otherwise specific references make it clear when material from an earlier chapter is required.

Algorithms

When one writes a computer program, one is generally implementing a method of solving a problem which has been previously devised. This method is often independent of the particular computer to be used: it's likely to be equally appropriate for many computers. In any case, it is the method, not the computer program itself, which must be studied to learn how the problem is being attacked. The term *algorithm* is universally used in computer science to describe problem-solving methods suitable for implementation as computer programs. Algorithms are the "stuff" of computer science: they are central objects of study in many, if not most, areas of the field.

Most algorithms of interest involve complicated methods of organizing the data involved in the computation. Objects created in this way are called *data structures*, and they are also central objects of study in computer science. Thus algorithms and data structures go hand in hand: in this book we will take the view that data structures exist as the byproducts or endproducts of algorithms, and thus need to be studied in order to understand the algorithms. Simple algorithms can give rise to complicated data structures and, conversely, complicated algorithms can use simple data structures.

When a very large computer program is to be developed, a great deal of effort must go into understanding and defining the problem to be solved, managing its complexity, and decomposing it into smaller subtasks which can be easily implemented. It is often true that many of the algorithms required after the decomposition are trivial to implement. However, in most cases there are a few algorithms the choice of which is critical since most of the system resources will be spent running those algorithms. In this book, we will study a variety of fundamental algorithms basic to large programs in many applications areas.

The sharing of programs in computer systems is becoming more widespread, so that while it is true that a serious computer user will *use* a large fraction of the algorithms in this book, he may need to *implement* only a somewhat smaller fraction of them. However, implementing simple versions of basic algorithms helps us to understand them better and thus use advanced versions more effectively in the future. Also, mechanisms for sharing software on many computer systems often make it difficult to tailor standard programs to perform effectively on specific tasks, so that the opportunity to reimplement basic algorithms frequently arises.

Computer programs are often overoptimized. It may be worthwhile to take pains to ensure that an implementation is the most efficient possible only if an algorithm is to be used for a very large task or is to be used many times. In most situations, a careful, relatively simple implementation will suffice: the programmer can have some confidence that it will work, and it is likely to run only five or ten times slower than the best possible version, which means that it may run for perhaps an extra fraction of a second. By contrast, the proper choice of algorithm in the first place can make a difference of a factor of a hundred or a thousand or more, which translates to minutes, hours, days or more in running time. In this book, we will concentrate on the simplest reasonable implementations of the best algorithms.

Often several different algorithms (or implementations) are available to solve the same problem. The choice of the very best algorithm for a particular task can be a very complicated process, often involving sophisticated mathematical analysis. The branch of computer science where such questions are studied is called *analysis of algorithms*. Many of the algorithms that we will study have been shown to have very good performance through analysis, while others are simply known to work well through experience. We will not dwell on comparative performance issues: our goal is to learn some reasonable algorithms for important tasks. But we will try to be aware of roughly how well these algorithms might be expected to perform.

Outline of Topics

Below are brief descriptions of the major parts of the book, which give some of the specific topics covered as well as some indication of the general orientation towards the material described. This set of topics is intended to allow us to cover as many fundamental algorithms as possible. Some of the areas covered are "core" computer science areas which we'll study in some depth to learn basic algorithms of wide applicability. We'll also touch on other disciplines and advanced fields of study within computer science (such as numerical analysis, operations research, compiler construction, and the theory of algorithms): in these cases our treatment will serve as an introduction to these fields of study through examination of some basic methods.

MATHEMATICAL ALGORITHMS include fundamental methods from arithmetic and numerical analysis. We study methods for addition and multiplication of integers, polynomials, and matrices as well as algorithms for solving a variety of mathematical problems which arise in many contexts: random number generation, solution of simultaneous equations, data fitting, and integration. The emphasis is on algorithmic aspects of the methods, not the mathematical basis. Of course we can't do justice to advanced topics with this kind of treatment, but the simple methods given here may serve to introduce the reader to some advanced fields of study.

SORTING methods for rearranging files into order are covered in some depth, due to their fundamental importance. A variety of methods are developed, described, and compared. Algorithms for several related problems are treated, including priority queues, selection, and merging. Some of these algorithms are used as the basis for other algorithms later in the book.

SEARCHING methods for finding things in files are also of fundamental importance. We discuss basic and advanced methods for searching using trees and digital key transformations, including binary search trees, balanced trees, hashing, digital search trees and tries, and methods appropriate for very large files. These methods are related to each other and similarities to sorting methods are discussed.

STRING PROCESSING algorithms include a range of methods for dealing with (long) sequences of characters. String searching leads to pattern matching which leads to parsing. File compression techniques and cryptology are also considered. Again, an introduction to advanced topics is given through treatment of some elementary problems which are important in their own right.

GEOMETRIC ALGORITHMS comprise a collection of methods for solving problems involving points and lines (and other simple geometric objects) which have only recently come into use. We consider algorithms for finding the convex hull of a set of points, for finding intersections among geometric objects, for solving closest point problems, and for multidimensional searching. Many of these methods nicely complement more elementary sorting and searching methods.

GRAPH ALGORITHMS are useful for a variety of difficult and important problems. A general strategy for searching in graphs is developed and applied to fundamental connectivity problems, including shortest-path, minimal spanning tree, network flow, and matching. Again, this is merely an introduction to quite an advanced field of study, but several useful and interesting algorithms are considered.

ADVANCED TOPICS are discussed for the purpose of relating the material in the book to several other advanced fields of study. Special-purpose hardware, dynamic programming, linear programming, exhaustive search, and NP-completeness are surveyed from an elementary viewpoint to give the reader some appreciation for the interesting advanced fields of study that are suggested by the elementary problems confronted in this book.

The study of algorithms is interesting because it is a new field (almost all of the algorithms we will study are less than twenty-five years old) with a rich tradition (a few algorithms have been known for thousands of years). New discoveries are constantly being made, and few algorithms are completely understood. In this book we will consider intricate, complicated, and difficult algorithms as well as elegant, simple, and easy algorithms. Our challenge is to understand the former and appreciate the latter in the context of many different potential application areas. In doing so, we will explore a variety of useful tools and develop a way of "algorithmic thinking" that will serve us well in computational challenges to come.

1. Preview

To introduce the general approach that we'll be taking to studying algorithms, we'll examine a classic elementary problem: "Reduce a given fraction to lowest terms." We want to write 2/3, not 4/6, 200/300, or 178468/267702. Solving this problem is equivalent to finding the *greatest common divisor* (gcd) of the numerator and the denominator: the largest integer which divides them both. A fraction is reduced to lowest terms by dividing both numerator and denominator by their greatest common divisor.

Pascal

A concise description of the Pascal language is given in the Wirth and Jensen *Pascal User Manual and Report* that serves as the definition for the language. Our purpose here is not to repeat information from that book but rather to examine the implementation of a few simple algorithms which illustrate some of the basic features of the language and the style that we'll be using.

Pascal has a rigorous high-level syntax which allows easy identification of the main features of the program. The variables (**var**) and functions (**function**) used by the program are declared first, followed by the body of the program. (Other major program parts, not used in the program below which are declared before the program body are **const**ants and **type**s.) Functions have the same format as the main program except that they return a value, which is set by assigning something to the function name within the body of the function. (Functions that return no value are called **procedure**s.)

The built-in function *readln* reads a line from the input and assigns the values found to the variables given as arguments; *writeln* is similar. A standard built-in predicate, *eof*, is set to *true* when there is no more input. (Input and output within a line are possible with *read*, *write*, and *eoln*.) The declaration of *input* and *output* in the program statement indicates that the program is using the "standard" input and output streams.

To begin, we'll consider a Pascal program which is essentially a translation of the definition of the concept of the greatest common divisor into a programming language.

```
program example(input, output);
var x, y: integer;
function gcd(u, v: integer): integer;
  var t: integer;
  begin
  if u<v then t:=u else t:=v;
  while (u mod t<>0) or (v mod t<>0) do t:=t−1;
  gcd:=t
  end;
begin
while not eof do
  begin
  readln(x, y);
  writeln(x, y, gcd(abs(x), abs(y)));
  end
end.
```

The body of the program above is trivial: it reads two numbers from the input, then writes them and their greatest common divisor on the output. The *gcd* function implements a "brute-force" method: start at the smaller of the two inputs and test every integer (decreasing by one until 1 is reached) until an integer is found that divides both of the inputs. The built-in function *abs* is used to ensure that *gcd* is called with positive arguments. (The **mod** function is used to test whether two numbers divide: u **mod** v is the remainder when u is divided by v, so a result of 0 indicates that v divides u.)

Many other similar examples are given in the *Pascal User Manual and Report*. The reader is encouraged to scan the manual, implement and test some simple programs and then read the manual carefully to become reasonably comfortable with most of the features of Pascal.

Euclid's Algorithm

A much more efficient method for finding the greatest common divisor than that above was discovered by Euclid over two thousand years ago. Euclid's method is based on the fact that if u is greater than v then the greatest common divisor of u and v is the same as the greatest common divisor of v and $u − v$. Applying this rule successively, we can continue to subtract off multiples of v from u until we get a number less than v. But this number is exactly the same as the remainder left after dividing u by v, which is what the **mod** function computes: the greatest common divisor of u and v is the same as the greatest common divisor of v and $u \bmod v$. If $u \bmod v$ is 0, then v divides u exactly and is itself their greatest common divisor, so we are done.

This mathematical description explains how to compute the greatest common divisor of two numbers by computing the greatest common divisor of two smaller numbers. We can implement this method directly in Pascal simply by having the *gcd* function call itself with smaller arguments:

```
function gcd(u, v:integer): integer;
  begin
  if v=0 then gcd:=u
         else  gcd:=gcd(v,  u mod v)
  end;
```

(Note that if u is less than v, then u **mod** v is just u, and the recursive call just exchanges u and v so things work as described the next time around.) If the two inputs are 461952 and 116298, then the following table shows the values of u and v each time *gcd* is invoked:

$$(461952, 116298)$$
$$(116298, 113058)$$
$$(113058, 3240)$$
$$(3240, 2898)$$
$$(2898, 342)$$
$$(342, 162)$$
$$(162, 18)$$
$$(18, 0)$$

It turns out that this algorithm always uses a relatively small number of steps: we'll discuss that fact in some more detail below.

Recursion

A fundamental technique in the design of efficient algorithms is *recursion*: solving a problem by solving smaller versions of the same problem, as in the program above. We'll see this general approach used throughout this book, and we will encounter recursion many times. It is important, therefore, for us to take a close look at the features of the above elementary recursive program.

An essential feature is that a recursive program must have a *termination condition*. It can't always call itself, there must be some way for it to do something else. This seems an obvious point when stated, but it's probably the most common mistake in recursive programming. For similar reasons, one shouldn't make a recursive call for a larger problem, since that might lead to a loop in which the program attempts to solve larger and larger problems.

Not all programming environments support a general-purpose recursion facility because of intrinsic difficulties involved. Furthermore, when recursion is provided and used, it can be a source of unacceptable inefficiency. For these reasons, we often consider ways of removing recursion. This is quite easy to do when there is only one recursive call involved, as in the function above. We simply replace the recursive call with a *goto* to the beginning, after inserting some assignment statements to reset the values of the parameters as directed by the recursive call. After cleaning up the program left by these mechanical transformations, we have the following implementation of Euclid's algorithm:

```
function gcd(u, v:integer):integer;
  var t: integer;
  begin
  while v<>0 do
    begin t:= u mod v;  u:=v;  v:=t end;
  gcd:=u
  end;
```

Recursion removal is much more complicated when there is more than one recursive call. The algorithm produced is sometimes not recognizable, and indeed is very often useful as a different way of looking at a fundamental algorithm. Removing recursion almost always gives a more efficient implementation. We'll see many examples of this later on in the book.

Analysis of Algorithms

In this short chapter we've already seen three different algorithms for the same problem; for most problems there are many different available algorithms. How is one to choose the best implementation from all those available?

This is actually a well developed area of study in computer science. Frequently, we'll have occasion to call on research results describing the performance of fundamental algorithms. However, comparing algorithms can be challenging indeed, and certain general guidelines will be useful.

Usually the problems that we solve have a natural "size" (usually the amount of data to be processed; in the above example the magnitude of the numbers) which we'll normally call N. We would like to know the resources used (most often the amount of time taken) as a function of N. We're interested in the *average case*, the amount of time a program might be expected to take on "typical" input data, and in the *worst case*, the amount of time a program would take on the worst possible input configuration.

Many of the algorithms in this book are very well understood, to the point that accurate mathematical formulas are known for the average- and worst-case running time. Such formulas are developed first by carefully studying the program, to find the running time in terms of fundamental mathematical quantities and then doing a mathematical analysis of the quantities involved.

For some algorithms, it is easy to figure out the running time. For example, the brute-force algorithm above obviously requires $\min(u, v) - \gcd(u, v)$ iterations of the **while** loop, and this quantity dominates the running time if the inputs are not small, since all the other statements are executed either 0 or 1 times. For other algorithms, a substantial amount of analysis is involved. For example, the running time of the recursive Euclidean algorithm obviously depends on the "overhead" required for each recursive call (which can be determined only through detailed knowledge of the programming environment being used) as well as the number of such calls made (which can be determined only through extremely sophisticated mathematical analysis).

Several important factors go into this analysis which are somewhat outside a given programmer's domain of influence. First, Pascal programs are translated into machine code for a given computer, and it can be a challenging task to figure out exactly how long even one Pascal statement might take to execute (especially in an environment where resources are being shared, so that even the same program could have varying performance characteristics). Second, many programs are extremely sensitive to their input data, and performance might fluctuate wildly depending on the input. The average case might be a mathematical fiction that is not representative of the actual data on which the program is being used, and the worst case might be a bizarre construction that would never occur in practice. Third, many programs of interest are not well understood, and specific mathematical results may not be available. Finally, it is often the case that programs are not comparable at all: one runs much more efficiently on one particular kind of input, the other runs efficiently under other circumstances.

With these caveats in mind, we'll use rough estimates for the running time of our programs for purposes of classification, secure in the knowledge that a fuller analysis can be done for important programs when necessary. Such rough estimates are quite often easy to obtain via the old programming saw "90% of the time is spent in 10% of the code." (This has been quoted in the past for many different values of "90%.")

The first step in getting a rough estimate of the running time of a program is to identify the *inner loop*. Which instructions in the program are executed most often? Generally, it is only a few instructions, nested deep within the control structure of a program, that absorb all of the machine cycles. It is always worthwhile for the programmer to be aware of the inner loop, just to be sure that unnecessary expensive instructions are not put there.

Second, some analysis is necessary to estimate how many times the inner loop is iterated. It would be beyond the scope of this book to describe the mathematical mechanisms which are used in such analyses, but fortunately the running times many programs fall into one of a few distinct classes. When possible, we'll give a rough description of the analysis of the programs, but it will often be necessary merely to refer to the literature. (Specific references are given at the end of each major section of the book.) For example, the results of a sophisticated mathematical argument show that the number of recursive steps in Euclid's algorithm when u is chosen at random less than v is approximately $((12\ln 2)/\pi^2)\ln v$. Often, the results of a mathematical analysis are not exact, but approximate in a precise technical sense: the result might be an expression consisting of a sequence of decreasing terms. Just as we are most concerned with the inner loop of a program, we are most concerned with the *leading term* (the largest term) of a mathematical expression.

As mentioned above, most algorithms have a primary parameter N, usually the number of data items to be processed, which affects the running time most significantly. The parameter N might be the degree of a polynomial, the size of a file to be sorted or searched, the number of nodes in a graph, etc. Virtually all of the algorithms in this book have running time proportional to one of the following functions:

1   Most instructions of most programs are executed once or at most only a few times. If all the instructions of a program have this property, we say that its running time is *constant*. This is obviously the situation to strive for in algorithm design.

$\log N$ When the running time of a program is *logarithmic,* the program gets slightly slower as N grows.This running time commonly occurs in programs which solve a big problem by transforming it into a smaller problem by cutting the size by some constant fraction. For our range of interest, the running time can be considered to be less than a "large" constant. The base of the logarithm changes the constant, but not by much: when N is a thousand, $\log N$ is 3 if the base is 10, 10 if the base is 2; when N is a million, $\log N$ is twice as great. Whenever N doubles, $\log N$ increases by a constant, but $\log N$ doesn't double until N increases to N^2.

N When the running time of a program is *linear,* it generally is the case that a small amount of processing is done on each input element. When N is a million, then so is the running time. Whenever N

doubles, then so does the running time. This is the optimal situation for an algorithm that must process N inputs (or produce N outputs).

$N \log N$ This running time arises in algorithms which solve a problem by breaking it up into smaller subproblems, solving them independently, and then combining the solutions. For lack of a better adjective (*linearithmic?*), we'll say that the running time of such an algorithm is "$N \log N$." When N is a million, $N \log N$ is perhaps twenty million. When N doubles, the running time more than doubles (but not much more).

N^2 When the running time of an algorithm is *quadratic,* it is practical for use only on relatively small problems. Quadratic running times typically arise in algorithms which process all pairs of data items (perhaps in a double nested loop). When N is a thousand, the running time is a million. Whenever N doubles, the running time increases fourfold.

N^3 Similarly, an algorithm which processes triples of data items (perhaps in a triple-nested loop) has a *cubic* running time and is practical for use only on small problems. When N is a hundred, the running time is a million. Whenever N doubles, the running time increases eightfold.

2^N Few algorithms with *exponential* running time are likely to be appropriate for practical use, though such algorithms arise naturally as "brute-force" solutions to problems. When N is twenty, the running time is a million. Whenever N doubles, the running time squares!

The running time of a particular program is likely to be some constant times one of these terms (the "leading term") plus some smaller terms. The values of the constant coefficient and the terms included depends on the results of the analysis and on implementation details. Roughly, the coefficient of the leading term has to do with the number of instructions in the inner loop: at any level of algorithm design it's prudent to limit the number of such instructions. For large N the effect of the leading term dominates; for small N or for carefully engineered algorithms, more terms may contribute and comparisions of algorithms are more difficult. In most cases, we'll simply refer to the running time of programs as "linear," "$N \log N$, " "cubic," etc., with the implicit understanding that more detailed analysis or empirical studies must be done in cases where efficiency is very important.

A few other functions do arise. For example, an algorithm with N^2 inputs that has a running time that is cubic in N is more properly classed as an $N^{3/2}$ algorithm. Also some algorithms have two stages of subproblem decomposition, which leads to a running time proportional to $N(\log N)^2$. Both of these functions should be considered to be much closer to $N \log N$ than to N^2 for large N.

One further note on the "log" function. As mentioned above, the base of the logarithm changes things only by a constant factor. Since we usually deal with analytic results only to within a constant factor, it doesn't matter much what the base is, so we refer to "$\log N$," etc. On the other hand, it is sometimes the case that concepts can be explained more clearly when some specific base is used. In mathematics, the *natural logarithm* (base $e = 2.718281828\ldots$) arises so frequently that a special abbreviation is commonly used: $\log_e N \equiv \ln N$. In computer science, the *binary logarithm* (base 2) arises so frequently that the abbreviation $\log_2 N \equiv \lg N$ is commonly used. For example, $\lg N$ rounded up to the nearest integer is the number of bits required to represent N in binary.

Implementing Algorithms

The algorithms that we will discuss in this book are quite well understood, but for the most part we'll avoid excessively detailed comparisons. Our goal will be to try to identify those algorithms which are likely to perform best for a given type of input in a given application.

The most common mistake made in the selection of an algorithm is to ignore performance characteristics. Faster algorithms are often more complicated, and implementors are often willing to accept a slower algorithm to avoid having to deal with added complexity. But it is often the case that a faster algorithm is really not much more complicated, and dealing with slight added complexity is a small price to pay to avoid dealing with a slow algorithm. Users of a surprising number of computer systems lose substantial time waiting for simple quadratic algorithms to finish when only slightly more complicated $N \log N$ algorithms are available which could run in a fraction the time.

The second most common mistake made in the selection of an algorithm is to pay too much attention to performance characteristics. An $N \log N$ algorithm might be only slightly more complicated than a quadratic algorithm for the same problem, but a better $N \log N$ algorithm might give rise to a substantial increase in complexity (and might actually be faster only for very large values of N). Also, many programs are really run only a few times: the time required to implement and debug an optimized algorithm might be substantially more than the time required simply to run a slightly slower one.

The programs in this book use only basic features of Pascal, rather than taking advantage of more advanced capabilities that are available in Pascal and other programming environments. Our purpose is to study algorithms, not systems programming nor advanced features of programming languages.

It is hoped that the essential features of the algorithms are best exposed through simple direct implementations in a near-universal language. For the same reason, the programming style is somewhat terse, using short variable names and few comments, so that the control structures stand out. The "documentation" of the algorithms is the accompanying text. It is expected that readers who use these programs in actual applications will flesh them out somewhat in adapting them for a particular use.

□

Exercises

1. Solve our initial problem by writing a Pascal program to reduce a given fraction x/y to lowest terms.

2. Check what values your Pascal system computes for u **mod** v when u and v are not necessarily positive. Which versions of the *gcd* work properly when one or both of the arugments are 0?

3. Would our original *gcd* program ever be faster than the nonrecursive version of Euclid's algorithm?

4. Give the values of u and v each time the recursive *gcd* is invoked after the initial call *gcd*(*12345, 56789*).

5. Exactly how many Pascal statements are executed in each of the three *gcd* implementations for the call in the previous exercise?

6. Would it be more efficient to test for $u>v$ in the recursive implementation of Euclid's algorithm?

7. Write a recursive program to compute the largest integer less than $\log_2 N$ based on the fact that the value of this function for N **div** *2* is one greater than for N if $N > 1$.

8. Write an iterative program for the problem in the previous exercise. Also, write a program that does the computation using Pascal library subroutines. If possible on your computer system, compare the performance of these three programs.

9. Write a program to compute the greatest common divisor of *three* integers u, v, and w.

10. For what values of N is $10N \lg N > 2N^2$? (Thus a quadratic algorithm is not necessarily slower than an $N \log N$ one.)

SOURCES for background material

A reader interested in learning more about Pascal will find a large number of introductory textbooks available, for example, the ones by Clancy and Cooper or Holt and Hune. Someone with experience programming in other languages can learn Pascal effectively directly from the manual by Wirth and Jensen. Of course, the most important thing to do to learn about the language is to implement and debug as many programs as possible.

Many introductory Pascal textbooks contain some material on data structures. Though it doesn't use Pascal, an important reference for further information on basic data structures is volume one of D.E. Knuth's series on *The Art of Computer Programming*. Not only does this book provide encyclopedic coverage, but also it and later books in the series are primary references for much of the material that we'll be covering in this book. For example, anyone interested in learning more about Euclid's algorithm will find about fifty pages devoted to it in Knuth's volume two.

Another reason to study Knuth's volume one is that it covers in detail the mathematical techniques needed for the analysis of algorithms. A reader with little mathematical background should be warned that a substantial amount of discrete mathematics is required to properly analyze many algorithms; a mathematically inclined reader will find much of this material ably summarized in Knuth's first book and applied to many of the methods we'll be studying in later books.

M. Clancy and D. Cooper, *Oh! Pascal*, W. W. Norton & Company, New York, 1982.

R. Holt and J. P.Hume, *Programming Standard Pascal*, Reston (Prentice-Hall), Reston, Virginia, 1980.

D. E. Knuth, *The Art of Computer Programming. Volume 1: Fundamental Algorithms*, Addison-Wesley, Reading, MA, 1968.

D. E. Knuth, *The Art of Computer Programming. Volume 2: Seminumerical Algorithms*, Addison-Wesley, Reading, MA, Second edition, 1981.

K. Jensen and N. Wirth, *Pascal User Manual and Report*, Springer-Verlag, New York, 1974.

MATHEMATICAL ALGORITHMS

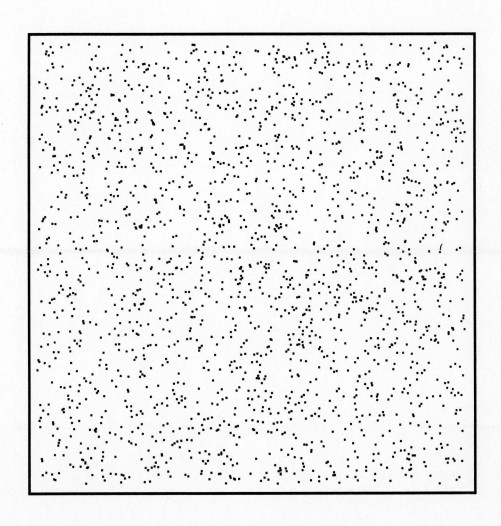

MATHEMATICAL ALGORITHMS

2. Arithmetic

Algorithms for doing elementary arithmetic operations such as addition, multiplication, and division have a very long history, dating back to the origins of algorithm studies in the work of the Arabic mathematician al-Khowârizmî, with roots going even further back to the Greeks and the Babylonians.

Though the situation is beginning to change, the *raison d'être* of many computer systems is their capability for doing fast, accurate numerical calculations. Computers have built-in capabilities to perform arithmetic on integers and floating-point representations of real numbers; for example, Pascal allows numbers to be of type *integer* or *real*, with all of the normal arithmetic operations defined on both types. Algorithms come into play when the operations must be performed on more complicated mathematical objects, such as polynomials or matrices.

In this section, we'll look at Pascal implementations of some simple algorithms for addition and multiplication of polynomials and matrices. The algorithms themselves are well-known and straightforward; we'll be examining sophisticated algorithms for these problems in Chapter 4. Our main purpose in this section is to get used to treating these mathematical objects as objects for manipulation by Pascal programs. This translation from abstract data to something which can be processed by a computer is fundamental in algorithm design. We'll see many examples throughout this book in which a proper representation can lead to an efficient algorithm and vice versa. In this chapter, we'll use two fundamental ways of structuring data, the *array* and the *linked list*. These data structures are used by many of the algorithms in this book; in later sections we'll study some more advanced data structures.

Polynomials

Suppose that we wish to write a program that adds two polynomials: we would like it to perform calculations like

$$(1 + 2x - 3x^3) + (2 - x) = 3 + x - 3x^3.$$

In general, suppose we wish our program to be able to compute $r(x) = p(x) + q(x)$, where p and q are polynomials with N coefficients. The following program is a straightforward implementation of polynomial addition:

```
program polyadd(input, output);
const maxN=100;
var p, q, r: array [0..maxN] of real;
    N, i: integer;
begin
  readln(N);
  for i:=0 to N−1 do read(p[i]);
  for i:=0 to N−1 do read(q[i]);
  for i:=0 to N−1 do r[i]:=p[i]+q[i];
  for i:=0 to N−1 do write(r[i]);
  writeln
end.
```

In this program, the polynomial $p(x) = p_0 + p_1 x + \cdots + p_{N-1} x^{N-1}$ is represented by the **array** $p[0..N-1]$ with $p[j] \equiv p_j$, etc. A polynomial of degree $N-1$ is defined by N coefficients. The input is assumed to be N, followed by the p coefficients, followed by the q coefficients. In Pascal, we must decide ahead of time how large N might get; this program will handle polynomials up to degree *100*. Obviously, *maxN* should be set to the maximum degree anticipated. This is inconvenient if the program is to be used at different times for various sizes from a wide range: many programming environments allow "dynamic arrays" which, in this case, could be set to the size N. We'll see another technique for handling this situation below.

The program above shows that addition is quite trivial once this representation for polynomials has been chosen; other operations are also easily coded. For example, to multiply we can replace the third **for** loop by

```
for i:=0 to 2*(N−1) do r[i]:=0;
for i:=0 to N−1 do
  for j:=0 to N−1 do
    r[i+j]:=r[i+j]+p[i]*q[j];
```

Also, the declaration of *r* has to be suitably changed to accomodate twice as many coefficients for the product. Each of the *N* coefficients of *p* is multiplied by each of the *N* coefficients of *q*, so this is clearly a quadratic algorithm.

An advantage of representing a polynomial by an array containing its coefficients is that it's easy to reference any coefficient directly; a disadvantage is that space may have to be saved for more numbers than necessary. For example, the program above couldn't reasonably be used to multiply

$$(1 + x^{10000})(1 + 2x^{10000}) = 1 + 3x^{10000} + 2x^{20000},$$

even though the input involves only four coefficients and the output only three.

An alternate way to represent a polynomial is to use a *linked list*. This involves storing items in noncontiguous memory locations, with each item containing the address of the next. The Pascal mechanisms for linked lists are somewhat more complicated than for arrays. For example, the following program computes the sum of two polynomials using a linked list representation (the bodies of the *readlist* and *add* functions and the *writelist* procedure are given in the text following):

```
program polyadd(input, output);
type  link = ↑node;
      node = record c: real; next: link end;
var N: integer;  z: link;
function readlist(N: integer): link;
procedure writelist(r: link);
function add(p, q: link): link;
begin
  readln(N);  new(z);
  writelist(add(readlist(N), readlist(N)))
end.
```

The polynomials are represented by linked lists which are built by the *readlist* procedure. The format of these is described in the **type** statement: the lists are made up of *node*s, each node containing a coefficient and a *link* to the next node on the list. If we have a *link* to the first *node* on a list, then we can examine the coefficients in order, by following *link*s. The last node on each list contains a *link* to a special dummy node called *z*: if we reach *z* when scanning through a list, we know we're at the end. (It is possible to get by without such dummy nodes, but they do make certain manipulations on the lists somewhat simpler.) The **type** statement only describes the formats of the nodes; nodes can be created only when the builtin procedure *new* is called. For example, the call *new(z)* creates a new node, putting a pointer to it in *z*. (The other nodes on the lists processed by this program are created in the *readlist* and *add* routines.)

The procedure to write out what's on a list is the simplest. It simply steps through the list, writing out the value of the coefficient in each node encountered, until *z* is found:

```
procedure writelist(r: link);
  begin
  while r<>z do
    begin write(r↑.c); r:=r↑.next end;
  writeln
  end;
```

The output of this program will be indistinguishable from that of the program above which uses the simple array representation.

Building a list involves first calling *new* to create a node, then filling in the coefficient, and then linking the node to the end of the partial list built so far. The following function reads in *N* coefficients, assuming the same format as before, and constructs the linked list which represents to corresponding polynomial:

```
function readlist(N: integer): link;
  var i: integer; t: link;
  begin
  t:=z;
  for i:=0 to N−1 do
    begin new(t↑.next); t:=t↑.next; read(t↑.c) end;
  t↑.next:=z; readlist:=z↑.next; z↑.next:=z
  end;
```

The dummy node *z* is used here to hold the link which points to the first node on the list while the list is being constructed. After this list is built, *z* is set to link to itself. This ensures that once we reach the end of a list, we stay there. Another convention which is sometimes convenient, would be to leave *z* pointing to the beginning, to provide a way to get from the back to the front.

Finally, the program which adds two polynomials constructs a new list in a manner similar to *readlist*, calculating the coefficients for the result by stepping through the argument lists and adding together corresponding coefficients:

```
function add(p, q: link): link;
    var t: link;
    begin
    t:=z;
    repeat
        new(t↑.next); t:=t↑.next;
        t↑.c:=p↑.c+q↑.c;
        p:=p↑.next; q:=q↑.next
    until (p=z) and (q=z);
    t↑.next:=z; add:=z↑.next
    end;
```

Employing linked lists in this way, we use only as many nodes as are required by our program. As N gets larger, we simply make more calls on *new*. By itself, this might not be reason enough to use linked lists for this program, because it does seem quite clumsy compared to the array implementation above. For example, it uses twice as much space, since a link must be stored along with each coefficient. However, as suggested by the example above, we can take advantage of the possibility that many of the coefficients may be zero. We can have list nodes represent only the nonzero terms of the polynomial by also including the degree of the term represented within the list node, so that each list node contains values of c and j to represent cx^j. It is then convenient to separate out the function of creating a node and adding it to a list, as follows:

```
type  link = ↑node;
      node = record c: real; j: integer; next: link end;
function listadd(t: link; c: real; j: integer): link;
    begin
    new(t↑.next); t:=t↑.next;
    t↑.c:=c; t↑.j:=j;
    listadd:=t;
    end;
```

The *listadd* function creates a new node, gives it the specified fields, and links it into a list after node t. Now the *readlist* routine can be changed either to accept the same input format as above (and create list nodes only for nonzero coefficients) or to input the coefficient and exponent directly for terms with nonzero coefficient. Of course, the *writelist* function also has to be changed suitably. To make it possible to process the polynomials in an organized way, the list nodes should be kept in increasing order of degree of the term represented.

Now the *add* function becomes more interesting, since it has to perform an addition only for terms whose degrees match, and then make sure that no term with coefficient 0 is output:

```
function add(p, q: link): link;
  begin
  t:=z;  z↑.j:=N+1;
  repeat
    if (p↑.j=q↑.j) and (p↑.c+q↑.c<>0.0) then
      begin
      t:=listadd(t, p↑.c+q↑.c, p↑.j);
      p:=p↑.next;  q:=q↑.next
      end
    else if p↑.j<q↑.j then
      begin t:=listadd(t, p↑.c, p↑.j);  p:=p↑.next end
    else if q↑.j<p↑.j then
      begin t:=listadd(t, q↑.c, q↑.j);  q:=q↑.next end;
  until (p=z) and (q=z);
  t↑.next:=z;  add:=z↑.next
  end;
```

These complications are worthwhile for processing "sparse" polynomials with many zero coefficients, but the array representation is better if there are only a few terms with zero coefficients. Similar savings are available for other operations on polynomials, for example multiplication.

Matrices

We can proceed in a similar manner to implement basic operations on two-dimensional matrices, though the programs become more complicated. Suppose that we want to compute the sum of the two matrices

$$\begin{pmatrix} 1 & 3 & -4 \\ 1 & 1 & -2 \\ -1 & -2 & 5 \end{pmatrix} + \begin{pmatrix} 8 & 3 & 0 \\ 3 & 10 & 2 \\ 0 & 2 & 6 \end{pmatrix} = \begin{pmatrix} 9 & 6 & -4 \\ 4 & 11 & 0 \\ -1 & 0 & 11 \end{pmatrix}.$$

This is term-by-term addition, just as for polynomials, so the addition program is a straightforward extension of our program for polynomials:

```
program matrixadd(input, output);
const maxN=10;
var p, q, r: array [0..maxN, 0..maxN] of real;
    N, i, j: integer;
begin
  readln(N);
  for i:=0 to N−1 do for j:=0 to N−1 do read(p[i,j]);
  for i:=0 to N−1 do for j:=0 to N−1 do read(q[i,j]);
  for i:=0 to N−1 do for j:=0 to N−1 do r[i,j]:=p[i,j]+q[i,j];
  for i:=0 to N−1 do for j:=0 to N do
    if j=N then writeln else write(r[i,j]);
end.
```

Matrix multiplication is a more complicated operation. For our example, we have

$$\begin{pmatrix} 1 & 3 & -4 \\ 1 & 1 & -2 \\ -1 & -2 & 5 \end{pmatrix}\begin{pmatrix} 8 & 3 & 0 \\ 3 & 10 & 2 \\ 0 & 2 & 6 \end{pmatrix} = \begin{pmatrix} 17 & 25 & -18 \\ 11 & 9 & -10 \\ -14 & -13 & 26 \end{pmatrix}.$$

Element $r[i,j]$ is the *dot product* of the ith row of p with the jth column of q. The dot product is simply the sum of the N term-by-term multiplications $p[i,1]*q[1,j]+p[i,2]*q[2,j]+\cdots p[i,N-1]*q[N-1,j]$ as in the following program:

```
for i:=0 to N−1 do
  for j:=0 to N−1 do
    begin
    t:=0.0;
    for k:=0 to N−1 do t:=t+p[i,k]*q[k,j];
    r[i,j]:=t
    end;
```

Each of the N^2 elements in the result matrix is computed with N multiplications, so about N^3 operations are required to multiply two N by N matrices together. (As noted in the previous chapter, this is not really a cubic algorithm, since the number of data items in this case is about N^2, not N.)

As with polynomials, *sparse* matrices (those with many zero elements) can be processed in a much more efficient manner using a linked list representation. To keep the two-dimensional structure intact, each nonzero matrix element is represented by a list node containing a value and two links: one pointing to the next nonzero element in the same row and the other pointing to the next nonzero element in the same column. Implementing addition for sparse matrices represented in this way is similar to our implementation for sparse polynomials, but is complicated by the fact that each node appears on two lists.

Data Structures

Even if there are no terms with zero coefficients in a polynomial or no zero elements in a matrix, an advantage of the linked list representation is that we don't need to know in advance how big the objects that we'll be processing are. This is a significant advantage that makes linked structures preferable in many situations. On the other hand, the links themselves can consume a significant part of the available space, a disadvantage in some situations. Also, access to individual elements in linked structures is much more restricted than in arrays.

We'll see examples of the use of these data structures in various algorithms, and we'll see more complicated data structures that involve more constraints on the elements in an array or more pointers in a linked representation. For example, multidimensional arrays can be defined which use multiple indices to access individual items. Similarly, we'll encounter many "multidimensional" linked structures with more than one pointer per node. The tradeoffs between competing structures are usually complicated, and different structures turn out to be appropriate for different situations.

When possible it is wise to think of the data and the specific operations to be performed on it as an *abstract data structure* which can be realized in several ways. For example, the abstract data structure for polynomials in the examples above is the set of coefficients: a user providing input to one of the programs above need not know whether a linked list or an array is being used. Modern programming systems have sophisticated mechanisms which make it possible to change representations easily, even in large, tightly integrated systems.

Exercises

1. Another way to represent polynomials is to write them in the form $r_0(x - r_1)(x - r_2)\ldots(x - r_N)$. How would you multiply two polynomials in this representation?

2. How would you *add* two polynomials represented as in Exercise 1?

3. Write a Pascal program that multiplies two polynomials, using a linked list representation with a list node for each term.

4. Write a Pascal program that multiplies sparse polynomials, using a linked list representation with no nodes for terms with 0 coefficients.

5. Write a Pascal function that returns the value of the element in the ith row and jth column of a sparse matrix, assuming that the matrix is represented using a linked list representation with no nodes for 0 entries.

6. Write a Pascal procedure that sets the value of the element in the ith row and jth column of a sparse matrix to v, assuming that the matrix is represented using a linked list representation with no nodes for 0 entries.

7. What is the running time of matrix multiplication in terms of the number of data items?

8. Does the running time of the polynomial addition programs for nonsparse input depend on the value of any of the coefficients?

9. Run an experiment to determine which of the polynomial addition programs runs fastest on your computer system, for relatively large N.

10. Give a counterexample to the assertion that the user of an abstract data structure need not know what representation is being used.

3. Random Numbers

Our next set of algorithms will be methods for using a computer to generate random numbers. We will find many uses for random numbers later on; let's begin by trying to get a better idea of exactly what they are.

Often, in conversation, people use the term *random* when they really mean *arbitrary*. When one asks for an *arbitrary* number, one is saying that one doesn't really care what number one gets: almost any number will do. By contrast, a *random* number is a precisely defined mathematical concept: every number should be equally likely to occur. A random number will satisfy someone who needs an arbitrary number, but not the other way around.

For "every number to be equally likely to occur" to make sense, we must restrict the numbers to be used to some finite domain. You can't have a random integer, only a random integer in some range; you can't have a random real number, only a random fraction in some range to some fixed precision.

It is almost always the case that not just one random number, but a *sequence* of random numbers is needed (otherwise an arbitrary number might do). Here's where the mathematics comes in: it's possible to prove many facts about properties of sequences of random numbers. For example, we can expect to see each value about the same number of times in a very long sequence of random numbers from a small domain. Random sequences model many natural situations, and a great deal is known about their properties. To be consistent with current usage, we'll refer to numbers from random sequences as random numbers.

There's no way to produce true random numbers on a computer (or any deterministic device). Once the program is written, the numbers that it will produce can be deduced, so how could they be random? The best we can hope to do is to write programs which produce sequences of numbers having many of the same properties as random numbers. Such numbers are commonly called *pseudo-random* numbers: they're not really random, but they can be useful as approximations to random numbers, in much the same way that floating-point numbers are useful as approximations to real numbers. (Sometimes it's convenient to make a further distinction: in some situations, a few properties of random numbers are of crucial interest while others are irrelevant. In such situations, one can generate *quasi-random* numbers, which are sure to have the properties of interest but are unlikely to have other properties of random numbers. For some applications, quasi-random numbers are provably preferable to pseudo-random numbers.)

It's easy to see that approximating the property "each number is equally likely to occur" in a long sequence is not enough. For example, each number in the range [1,100] appears once in the sequence $(1,2,\ldots,100)$, but that sequence is unlikely to be useful as an approximation to a random sequence. In fact, in a random sequence of length 100 of numbers in the range [1,100], it is likely that a few numbers will appear more than once and a few will not appear at all. If this doesn't happen in a sequence of pseudo-random numbers, then there is something wrong with the random number generator. Many sophisticated tests based on specific observations like this have been devised for random number generators, testing whether a long sequence of pseudo-random numbers has some property that random numbers would. The random number generators that we will study do very well in such tests.

We have been (and will be) talking exclusively about *uniform* random numbers, with each value equally likely. It is also common to deal with random numbers which obey some other distribution in which some values are more likely than others. Pseudo-random numbers with non-uniform distributions are usually obtained by performing some operations on uniformly distributed ones. Most of the applications that we will be studying use uniform random numbers.

Applications

Later in the book we will meet many applications in which random numbers will be useful. A few of them are outlined here. One obvious application is in *cryptography*, where the major goal is to encode a message so that it can't be read by anyone but the intended recipient. As we will see in Chapter 23, one way to do this is to make the message look random using a pseudo-random sequence to encode the message, in such a way that the recipient can use the same pseudo-random sequence to decode it.

Another area in which random numbers have been widely used is in *simulation*. A typical simulation involves a large program which models some aspect of the real world: random numbers are natural for the input to such programs. Even if true random numbers are not needed, simulations typically need many arbitrary numbers for input, and these are conveniently provided by a random number generator.

When a very large amount of data is to be analyzed, it is sometimes sufficient to process only a very small amount of the data, chosen according to random *sampling*. Such applications are widespread, the most prominent being national political opinion polls.

Often it is necessary to make a choice when all factors under consideration seem to be equal. The national draft lottery of the 70's or the mechanisms used on college campuses to decide which students get the choice dormitory rooms are examples of using random numbers for *decision making*. In this way, the responsibility for the decision is given to "fate" (or the computer).

Readers of this book will find themselves using random numbers extensively for simulation: to provide random or arbitrary inputs to programs. Also, we will see examples of algorithms which gain efficiency by using random numbers to do sampling or to aid in decision making.

Linear Congruential Method

The most well-known method for generating random numbers, which has been used almost exclusively since it was introduced by D. Lehmer in 1951, is the so-called *linear congruential* method. If $a[1]$ contains some arbitrary number, then the following statement fills up an array with N random numbers using this method:

---

for $i:=2$ **to** N **do**
  $a[i]:=(a[i-1]*b+1)$ **mod** m

That is, to get a new random number, take the previous one, multiply it by a constant b, add 1 and take the remainder when divided by a second constant m. The result is always an integer between 0 and $m-1$. This is attractive for use on computers because the **mod** function is usually trivial to implement: if we ignore overflow on the arithmetic operations, then most computer hardware will throw away the bits that overflowed and thus effectively perform a **mod** operation with m equal to one more than the largest integer that can be represented in the computer word.

Simple as it may seem, the linear congruential random number generator has been the subject of volumes of detailed and difficult mathematical analysis. This work gives us some guidance in choosing the constants b and m. Some "common-sense" principles apply, but in this case common sense isn't enough to ensure good random numbers. First, m should be large: it can be the computer word size, as mentioned above, but it needn't be quite that large if that's inconvenient (see the implementation below). It will normally be convenient to make m a power of 10 or 2. Second, b shouldn't be too large or too small: a safe choice is to use a number with one digit less than m. Third, *b* should be an arbitrary constant with no particular pattern in its digits, *except that* it should end with $\cdots x21$, with x even: this last requirement is admittedly peculiar, but it prevents the occurrence of some bad cases that have been uncovered by the mathematical analysis.

The rules described above were developed by D.E.Knuth, whose textbook covers the subject in some detail. Knuth shows that these choices will make the linear congruential method produce good random numbers which pass several sophisticated statistical tests. The most serious potential problem, which can become quickly apparent, is that the generator could get caught in a cycle and produce numbers it has already produced much sooner than it should. For example, the choice *b=19*, *m=381*, with *a[1]=0*, produces the sequence 0,1,20,0,1,20,..., a not-very-random sequence of integers between 0 and 380.

Any initial value can be used to get the random number generator started with no particular effect except of course that different initial values will give rise to different random sequences. Often, it is not necessary to store the whole sequence as in the program above. Rather, we simply maintain a global variable *a*, initialized with some value, then updated by the computation $a:=(a*b+1)$ **mod** m.

In Pascal (and many other programming languages) we're still one step away from a working implementation because we're not allowed to ignore overflow: it's defined to be an error condition that can lead to unpredictable results. Suppose that we have a computer with a 32-bit word, and we choose *m=100000000*, *b=3141582*, and, initially, *a=1234567*. All of these values are comfortably less than the largest integer that can be represented, but the first $a*b+1$ operation causes overflow. The part of the product that causes the overflow is not relevant to our computation, we're only interested in the last eight digits. The trick is to avoid overflow by breaking the multiplication up into pieces. To multiply p by q, we write $p = 10^4 p_1 + p_0$ and $q = 10^4 q_1 + q_0$, so the product is

$$pq = (10^4 p_1 + p_0)(10^4 q_1 + q_0)$$
$$= 10^8 p_1 q_1 + 10^4 (p_1 q_0 + p_0 q_1) + p_0 q_0.$$

Now, we're only interested in eight digits for the result, so we can ignore the first term and the first four digits of the second term. This leads to the following program:

```
program random(input, output);
const m=100000000; m1=10000; b=31415821;
var i, a, N: integer;
function mult(p, q: integer): integer;
  var p1, p0, q1, q0: integer;
  begin
  p1:=p div m1; p0:=p mod m1;
  q1:=q div m1; q0:=q mod m1;
  mult:=(((p0*q1+p1*q0) mod m1)*m1+p0*q0) mod m;
  end;
function random: integer;
  begin
  a:=(mult(a, b)+1) mod m;
  random:=a;
  end;
begin
read(N, a);
for i:=1 to N do writeln(random)
end.
```

The function *mult* in this program computes $p*q$ **mod** m, with no overflow as long as m is less than half the largest integer that can be represented. The technique obviously can be applied with $m=m1*m1$ for other values of $m1$.

Here are the ten numbers produced by this program with the input $N = 10$ and $a = 1234567$:

$$
\begin{aligned}
&35884508 \\
&80001069 \\
&63512650 \\
&43635651 \\
&1034472 \\
&87181513 \\
&6917174 \\
&209855 \\
&67115956 \\
&59939877
\end{aligned}
$$

There is some obvious non-randomness in these numbers: for example, the last digits cycle through the digits 0–9. It is easy to prove from the formula that this will happen. Generally speaking, the digits on the right are not particularly random. This leads to a common and serious mistake in the use of linear congruential random number generators: the following is a bad program for producing random numbers in the range $[0, r-1]$:

```
function randombad(r: integer): integer;
  begin
  a:=(mult(b, a)+1) mod m;
  randombad:=a mod r;
  end;
```

The non-random digits on the right are the only digits that are used, so the resulting sequence has few of the desired properties. This problem is easily fixed by using the digits on the left. We want to compute a number between 0 and $r-1$ by computing $a*r$ **mod** m, but, again, overflow must be circumvented, as in the following implementation:

```
function randomint(r: integer): integer;
  begin
  a:=(mult(a, b)+1) mod m;
  randomint:=((a div m1)*r) div m1
  end;
```

Another common technique is to generate random real numbers between 0 and 1 by treating the above numbers as fractions with the decimal point to the left. This can be implemented by simply returning the real value a/m rather than the integer a. Then a user could get an integer in the range $[0, r)$ by simply multiplying this value by r and truncating to the nearest integer. Or, a random real number between 0 and 1 might be exactly what is needed.

Additive Congruential Method

Another method for generating random numbers is based on *linear feedback shift registers* which were used for early cryptographic encryption machines. The idea is to start with an register filled with some arbitrary pattern, then shift it right (say) a step at a time, filling in vacated positions from the left with a bit determined by the contents of the register. The diagram below shows a simple 4-bit register, with the new bit taken as the "exclusive or" of the two rightmost bits.

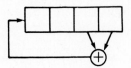

Below are listed the contents of the register for the first sixteen steps of the process:

0	1	2	3	4	5	6	7
1011	0101	1010	1101	1110	1111	0111	0011
8	9	10	11	12	13	14	15
0001	1000	0100	0010	1001	1100	0110	1011

Notice that all possible nonzero bit patterns occur, the starting value repeats after 15 steps. As with the linear congruential method, the mathematics of the properties of these registers has been studied extensively. For example, much is known about the choices of "tap" positions (the bits used for feedback) which lead to the generation of all bit patterns for registers of various sizes.

Another interesting fact is that the calculation can be done a word at a time, rather than a bit at a time, according to the same recursion formula. In our example, if we take the bitwise "exclusive or" of two successive words, we get the word which appears three places later in the list. This leads us to a random number generator suitable for easy implementation on a general-purpose computer. Using a feedback register with bits b and c tapped corresponds to using the recursion: $a[k]=(a[k-b]+a[k-c])$ **mod** m. To keep the correspondence with the shift register model, the "+" in this recursion should be a bitwise "exclusive or." However, it has been shown that good random numbers are likely to be produced even if normal integer addition is used. This is termed the *additive congruential* method.

To implement this method, we need to keep a table of size c which always has the c most recently generated numbers. The computation proceeds by replacing one of the numbers in the table by the sum of two of the other numbers in the table. Initially, the table should be filled with numbers that are not too small and not too large. (One easy way to get these numbers is to use a simple linear congruential generator!) Knuth recommends the choices $b=31$, $c=55$ will work well for most applications, which leads to the implementation below.

```
procedure randinit(s: integer);
  begin
  a[0]:=s; j:=0;
  repeat j:=j+1; a[j]:=(mult(b, a[j−1])+1) mod m until j=54;
  end;
function randomint(r: integer): integer;
  begin
  j:=(j+1) mod 55;
  a[j]:=(a[(j+23) mod 55]+a[(j+54) mod 55]) mod m;
  randomint:=((a[j] div m1)*r) div m1
  end;
```

The program maintains the 55 most recently generated numbers, with the last generated pointed to by j. Thus, the global variable a has been replaced by a full table plus a pointer (j) into it. This large amount of "global state" is a disadvantage of this generator in some applications, but it is also an advantage because it leads to an extremely long cycle even if the modulus m is small.

The function randomint returns a random integer between 0 and $r−1$. Of course, it can easily be changed, just as above, to a function which returns a random real number between 0 and 1 ($a[j]/m$).

Testing Randomness

One can easily detect numbers that are not random, but certifying that a sequence of numbers *is* random is a difficult task indeed. As mentioned above, no sequence produced by a computer can be random, but we want a sequence that exhibits many of the properties of random numbers. Unfortunately, it is often not possible to articulate exactly which properties of random numbers are important for a particular application.

On the other hand, it is always a good idea to perform some kind of test on a random number generator to be sure that no degenerate situations have turned up. Random number generators can be very, very good, but when they are bad they are horrid.

Many tests have been developed for determining whether a sequence shares various properties with a truly random sequence. Most of these tests have a substantial basis in mathematics, and it would definitely be beyond the scope of this book to examine them in detail. However, one statistical test, the χ^2 (chi-square) test, is fundamental in nature, quite easy to implement, and useful in several applications, so we'll examine it more carefully.

The idea of the χ^2 test is to check whether or not the numbers produced are spread out reasonably. If we generate N positive numbers less than r, then we'd expect to get about N/r numbers of each value. (But the frequencies of occurrence of all the values should not be exactly the same: that wouldn't be random!) It turns out that calculating whether or not a sequence of numbers is distributed as well as a random sequence is very simple, as in the following program:

```
function chisquare(N, r, s: integer): real;
  var i, t: integer;
      f: array [0..rmax] of integer;
  begin
  randinit(s);
  for i:=0 to rmax do f[i]:=0;
  for i:=1 to N do
    begin
    t:=randomint(r);
    f[t]:=f[t]+1;
    end;
  t:=0; for i:=0 to r−1 do t:=t+f[i]*f[i];
  chisquare:=((r*t/N) − N);
  end;
```

We simply calculate the sum of the squares of the frequencies of occurrence of each value, scaled by the expected frequency then subtract off the size of the sequence. This number is called the "χ^2 statistic," which may be expressed mathematically as

$$\chi^2 = \frac{\sum_{0 \leq i < r}(f_i - N/r)^2}{N/r}.$$

If the χ^2 statistic is close to r, then the numbers are random; if it is too far away, then they are not. The notions of "close" and "far away" can be more precisely defined: tables exist which tell exactly how to relate the statistic to properties of random sequences. For the simple test that we're performing, the statistic should be within $2\sqrt{r}$ of r. This is valid if N is bigger than about $10r$, and to be sure, the test should be tried a few times, since it could be wrong about one out of ten times.

This test is so simple to implement that it probably should be included with every random number generator, just to ensure that nothing unexpected can cause serious problems. All the "good generators" that we have discussed pass this test; the "bad ones" do not. Using the above generators to generate a thousand numbers less than 100, we get a χ^2 statistic of 100.8 for the linear congruential method and 105.4 for the additive congruential method, both certainly well within 20 of 100. But for the "bad" generator which uses the right-hand bits from the linear congruential generator the statistic is 0 (why?) and for a linear congruential method with a bad multiplier (101011) the statistic is 77.8, which is significantly out of range.

Implementation Notes

There are a number of facilities commonly added to make a random number generator useful for a variety of applications. Usually, it is desirable to set up the generator as a function that is initialized and then called repeatedly, returning a different random number each time. Another possibility is to call the random number generator once, having it fill up an array with all the random numbers that will be needed for a particular computation. In either case, it is desirable that the generator produce the same sequence on successive calls (for initial debugging or comparison of programs on the same inputs) and produce an arbitrary sequence (for later debugging). These facilities all involve manipulating the "state" retained by the random number generator between calls. This can be very inconvenient in some programming environments. The additive generator has the disadvantage that it has a relatively large state (the array of recently produced words), but it has the advantage of having such a long cycle that it is probably not necessary for each user to initialize it.

A conservative way to protect against eccentricities in a random number generator is to combine two generators. (The use of a linear congruential generator to initialize the table for an additive congruential generator is an elementary example of this.) An easy way to implement a combination generator is to have the first generator fill a table and the second choose random table positions to fetch numbers to output (and store new numbers from the first generator).

When debugging a program that uses a random number generator, it is usually a good idea to use a trivial or degenerate generator at first, such as one which always returns 0 or one which returns numbers in order.

As a rule, random number generators are fragile and need to be treated with respect. It's difficult to be sure that a particular generator is good without investing an enormous amount of effort in doing the various statistical tests that have been devised. The moral is: do your best to use a good generator, based on the mathematical analysis and the experience of others; just to be sure, examine the numbers to make sure that they "look" random; if anything goes wrong, blame the random number generator!

Exercises

1. Write a program to generate random four-letter words (collections of letters). Estimate how many words your program will generate before a word is repeated.

2. How would you simulate generating random numbers by throwing two dice and taking their sum, with the added complication that the dice are nonstandard (say, painted with the numbers 1,2,3,5,8, and 13)?

3. What is wrong with the following linear feedback shift register?

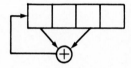

4. Why wouldn't the "or" or "and" function (instead of the "exclusive or" function) work for linear feedback shift registers?

5. Write a program to produce a random two dimensional image. (Example: generate random bits, write a "*" when 1 is generated, " " when 0 is generated. Another example: use random numbers as coordinates in a two dimensional cartesian system, write a "*" at addressed points.)

6. Use an additive congruential random number generator to generate 1000 positive integers less than 1000. Design a test to determine whether or not they're random and apply the test.

7. Use a linear congruential generator with parameters of your own choosing to generate 1000 positive integers less than 1000. Design a test to determine whether or not they're random and apply the test.

8. Why would it be unwise to use, for example, $b=3$ and $c=6$ in the additive congruential generator?

9. What is the value of the χ^2 statistic for a degenerate generator which always returns the same number?

10. Describe how you would generate random numbers with m bigger than the computer word size.

4. Polynomials

The methods for doing arithmetic operations given in Chapter 2 are simple and straightforward solutions to familiar problems. As such, they provide an excellent basis for applying algorithmic thinking to produce more sophisticated methods which are substantially more efficient. As we'll see, it is one thing to write down a formula which implies a particular mathematical calculation; it is quite another thing to write a computer program which performs the calculation efficiently.

Operations on mathematical objects are far too diverse to be catalogued here; we'll concentrate on a variety of algorithms for manipulating polynomials. The principal method that we'll study in this section is a polynomial multiplication scheme which is of no particular practical importance but which illustrates a basic design paradigm called *divide-and-conquer* which is pervasive in algorithm design. We'll see in this section how it applies to matrix multiplication as well as polynomial multiplication; in later sections we'll see it applied to most of the problems that we encounter in this book.

Evaluation

A first problem which arises naturally is to compute the value of a given polynomial at a given point. For example, to evaluate

$$p(x) = x^4 + 3x^3 - 6x^2 + 2x + 1$$

for any given x, one could compute x^4, then compute and add $3x^3$, etc. This method requires recomputation of the powers of x; an alternate method, which requires extra storage, would save the powers of x as they are computed.

A simple method which avoids recomputation and uses no extra space is known as *Horner's rule*: by alternating the multiplication and addition operations appropriately, a degree-N polynomial can be evaluated using only

$N-1$ multiplications and N additions. The parenthesization

$$p(x) = x(x(x(x+3)-6)+2)+1$$

makes the order of computation obvious:

```
y:=p[N];
for i:=N−1 downto 0 do y:=x*y+p[i];
```

This program (and the others in this section) assume the array representation for polynomials that we discussed in Chapter 2.

A more complicated problem is to evaluate a given polynomial at many different points. Different algorithms are appropriate depending on how many evaluations are to be done and whether or not they are to be done simultaneously. If a very large number of evaluations is to be done, it may be worthwhile to do some "precomputing" which can slightly reduce the cost for later evaluations. Note that using Horner's method would require about N^2 multiplications to evaluate a degree-N polynomial at N different points. Much more sophisticated methods have been designed which can solve the problem in $N(\log N)^2$ steps, and in Chapter 36 we'll see a method that uses only $N \log N$ multiplications for a specific set of N points of interest.

If the given polynomial has only one term, then the polynomial evaluation problem reduces to the *exponentiation* problem: compute x^N. Horner's rule in this case degenerates to the trivial algorithm which requires $N-1$ multiplications. For an easy example of how we can do much better, consider the following sequence for computing x^{32}:

$$x, x^2, x^4, x^8, x^{16}, x^{32}.$$

Each term is obtained by squaring the previous term, so only five multiplications are required (not 31).

The "successive squaring" method can easily be extended to general N if computed values are saved. For example, x^{55} can be computed from the above values with four more multiplications:

$$x^{55} = x^{32}x^{16}x^4x^2x^1.$$

In general, the binary representation of N can be used to choose which computed values to use. (In the example, since $55 = (110111)_2$, all but x^8 are used.) The successive squares can be computed and the bits of N tested within the same loop. Two methods are available to implement this using only one "accumulator," like Horner's method. One algorithm involves scanning the binary representation of N from left to right, starting with 1 in the accumulator. At each step, square the accumulator and also multiply by x when there is a 1 in the binary representation of N. The following sequence of values is computed by this method for $N = 55$:

$$1, 1, x, x^2, x^3, x^6, x^{12}, x^{13}, x^{26}, x^{27}, x^{54}, x^{55}.$$

Another well-known algorithm works similarly, but scans N from right to left. This problem is a standard introductory programming exercise, but it is hardly of practical interest.

Interpolation

The "inverse" problem to the problem of evaluating a polynomial of degree N at N points simultaneously is the problem of *polynomial interpolation*: given a set of N points $x_1, x_2, \ldots, x_N$ and associated values $y_1, y_2, \ldots, y_N$, find the unique polynomial of degree $N - 1$ which has

$$p(x_1) = y_1, \; p(x_2) = y_2, \; \ldots, p(x_N) = y_N.$$

The interpolation problem is to find the polynomial, given a set of points and values. The evaluation problem is to find the values, given the polynomial and the points. (The problem of finding the points, given the polynomial and the values, is *root-finding*.)

The classic solution to the interpolation problem is given by Lagrange's interpolation formula, which is often used as a proof that a polynomial of degree $N - 1$ is completely determined by N points:

$$p(x) = \sum_{1 \leq j \leq N} y_j \prod_{\substack{1 \leq i \leq N \\ i \neq j}} \frac{x - x_i}{x_j - x_i}.$$

This formula seems formidable at first but is actually quite simple. For example, the polynomial of degree 2 which has $p(1) = 3$, $p(2) = 7$, and $p(3) = 13$ is given by

$$p(x) = 3\frac{x - 2}{1 - 2}\frac{x - 3}{1 - 3} + 7\frac{x - 1}{2 - 1}\frac{x - 3}{2 - 3} + 13\frac{x - 1}{3 - 1}\frac{x - 2}{3 - 2}$$

which simplifies to

$$x^2 + x + 1.$$

For x from x_1, x_2, $\ldots$, x_N, the formula is constructed so that $p(x_k) = y_k$ for $1 \leq k \leq N$, since the product evaluates to 0 unless $j = k$, when it evaluates *see notes*

to 1. In the example, the last two terms are 0 when $x = 1$, the first and last terms are 0 when $x = 2$, and the first two terms are 0 when $x = 3$.

To convert a polynomial from the form described by Lagrange's formula to our standard coefficient representation is not at all straightforward. At least N^2 operations seem to be required, since there are N terms in the sum, each consisting of a product with N factors. Actually, it takes some cleverness to achieve a quadratic algorithm, since the factors are not just numbers, but polynomials of degree N. On the other hand, each term is very similar to the previous one. The reader might be interested to discover how to take advantage of this to achieve a quadratic algorithm. This exercise leaves one with an appreciation for the non-trivial nature of writing an efficient program to perform the calculation implied by a mathematical formula.

As with polynomial evaluation, there are more sophisticated methods which can solve the problem in $N(\log N)^2$ steps, and in Chapter 36 we'll see a method that uses only $N \log N$ multiplications for a specific set of N points of interest.

Multiplication

Our first sophisticated arithmetic algorithm is for the problem of *polynomial multiplication*: given two polynomials $p(x)$ and $q(x)$, compute their product $p(x)q(x)$. As noted in Chapter 2, polynomials of degree $N - 1$ could have N terms (including the constant) and the product has degree $2N - 2$ and as many as $2N - 1$ terms. For example,

$$(1 + x + 3x^2 - 4x^3)(1 + 2x - 5x^2 - 3x^3) = (1 + 3x - 6x^3 - 26x^4 + 11x^5 + 12x^6).$$

The naive algorithm for this problem that we implemented in Chapter 2 requires N^2 multiplications for polynomials of degree $N - 1$: each of the N terms of $p(x)$ must be multiplied by each of the N terms of $q(x)$.

To improve on the naive algorithm, we'll use a powerful technique for algorithm design called *divide-and-conquer*: split the problem into smaller parts, solve them (recursively), then put the results back together in some way. Many of our best algorithms are designed according to this principle. In this section we'll see how divide-and-conquer applies in particular to the polynomial multiplication problem. In the following section we'll look at some analysis which gives a good estimate of how much is saved.

One way to split a polynomial in two is to divide the coefficients in half: given a polynomial of degree $N-1$ (with N coefficients) we can split it into two polynomials with $N/2$ coefficients (assume that N is even): by using the $N/2$ low-order coefficients for one polynomial and the $N/2$ high-order coefficients for the other. For $p(x) = p_0 + p_1 x + \cdots + p_{N-1} x^{N-1}$, define

$$p_l(x) = p_0 + p_1 x + \cdots + p_{N/2-1} x^{N/2-1}$$
$$p_h(x) = p_{N/2} + p_{N/2+1} x + \cdots + p_{N-1} x^{N/2-1}.$$

Then, splitting $q(x)$ in the same way, we have:

$$p(x) = p_l(x) + x^{N/2} p_h(x),$$
$$q(x) = q_l(x) + x^{N/2} q_h(x).$$

Now, in terms of the smaller polynomials, the product is given by:

$$p(x)q(x) = p_l(x)q_l(x) + (p_l(x)q_h(x) + q_l(x)p_h(x))x^{N/2} + p_h(x)q_h(x)x^N.$$

(We used this same split in the previous chapter to avoid overflow.) What's interesting is that only three multiplications are necessary to compute these products, because if we compute $r_l(x) = p_l(x)q_l(x)$, $r_h(x) = p_h(x)q_h(x)$, and $r_m(x) = (p_l(x) + p_h(x))(q_l(x) + q_h(x))$, we can get the product $p(x)q(x)$ by computing

$$p(x)q(x) = r_l(x) + (r_m(x) - r_l(x) - r_h(x))x^{N/2} + r_h(x)x^N.$$

see notes

Polynomial addition requires a linear algorithm, and the straightforward polynomial multiplication algorithm of Chapter 2 is quadratic, so it's worthwhile to do a few (easy) additions to save one (difficult) multiplication. Below we'll look more closely at the savings achieved by this method.

For the example given above, with $p(x) = 1 + x + 3x^2 - 4x^3$ and $q(x) = 1 + 2x - 5x^2 - 3x^3$, we have

$$r_l(x) = (1 + x)(1 + 2x) = 1 + 3x + 2x^2,$$
$$r_h(x) = (3 - 4x)(-5 - 3x) = -15 + 11x + 12x^2,$$
$$r_m(x) = (4 - 3x)(-4 - x) = -16 + 8x + 3x^2.$$

Thus, $r_m(x) - r_l(x) - r_h(x) = -2 - 6x - 11x^2$, and the product is computed as

$$p(x)q(x) = (1 + 3x + 2x^2) + (-2 - 6x - 11x^2)x^2 + (-15 + 11x + 12x^2)x^4$$
$$= 1 + 3x - 6x^3 - 26x^4 + 11x^5 + 12x^6.$$

This divide-and-conquer approach solves a polynomial multiplication problem of size N by solving three subproblems of size $N/2$, using some polynomial addition to set up the subproblems and to combine their solutions. Thus, this procedure is easily described as a recursive program:

```
function mult(p, q: array [0..N−1] of real;
                    N: integer) : array [0..2*N−2] of real;
var pl, ql, ph, qh, t1, t2: array [0..(N div 2)−1] of real;
    rl, rm, rh: array [0..N−1] of real;
    i, N2: integer;
begin
if N=1 then mult[0]:=p[0]*q[0]
else
  begin
  N2:=N div 2;
  for i:=0 to N2−1 do
    begin pl[i]:=p[i]; ql[i]:=q[i] end;
  for i:=N2 to N−1 do
    begin ph[i−N2]:=p[i]; qh[i−N2]:=q[i] end;
  for i:=0 to N2−1 do t1[i]:=pl[i]+ph[i];
  for i:=0 to N2−1 do t2[i]:=ql[i]+qh[i];
  rm:=mult(t1, t2, N2);
  rl:=mult(pl, ql, N2);
  rh:=mult(ph, qh, N2);
  for i:=0 to N−2 do mult[i]:=rl[i]
  mult[N−1]:=0;
  for i:=0 to N−2 do mult[N+i]:=rh[i]
  for i:=0 to N−2 do
    mult[N2+i]:=mult[N2+i]+rm[i]−(rl[i]+rh[i]);
  end;
end.
```

Although the above code is a succinct description of this method, it is (unfortunately) not a legal Pascal program because functions can't dynamically declare arrays. This problem could be handled in Pascal by representing the polynomials as linked lists, as we did in Chapter 2. This program assumes that N is a power of two, though the details for general N can be worked out easily. The main complications are to make sure that the recursion terminates properly and that the polynomials are divided properly when N is odd.

The same method can be used for multiplying integers, though care must be taken to treat "carries" properly during the subtractions after the recursive calls.

As with polynomial evaluation and interpolation, there are sophisticated methods for polynomial multiplication, and in Chapter 36 we'll see a method that works in time proportional to $N \log N$.

Divide-and-conquer Recurrences

Why is the divide-and-conquer method given above an improvement? In this section, we'll look at a few simple recurrence formulas that can be used to measure the savings achieved by a divide-and-conquer algorithm.

From the recursive program, it is clear that the number of integer multiplications required to multiply two polynomials of size N is the same as the number of multiplications to multiply three pairs of polynomials of size $N/2$. (Note that, for example, no multiplications are required to compute $r_h(x)x^N$, just data movement.) If $M(N)$ is the number of multiplications required to multiply two polynomials of size N, we have

$$M(N) = 3M(N/2)$$

for $N > 1$ with $M(1) = 1$. Thus $M(2) = 3$, $M(4) = 9$, $M(8) = 27$, etc. In general, if we take $N = 2^n$, then we can repeatedly apply the recurrence to itself to find the solution:

$$M(2^n) = 3M(2^{n-1}) = 3^2 M(2^{n-2}) = 3^3 M(2^{n-3}) = \cdots = 3^n M(1) = 3^n.$$

If $N = 2^n$, then $3^n = 2^{(\lg 3)n} = 2^{n \lg 3} = N^{\lg 3}$. Although this solution is exact only for $N = 2^n$, it works out in general that

$$M(N) \approx N^{\lg 3} \approx N^{1.58},$$

which is a substantial savings over the N^2 naive method. Note that if we were to have used all four multiplications in the simple divide-and-conquer method, the recurrence would be $M(N) = 4M(N/2)$ with the solution $M(2^n) = 4^n = N^2$.

The method described in the previous section nicely illustrates the divide-and-conquer technique, but it is seldom used in practice because a much better divide-and-conquer method is known, which we'll study in Chapter 36. This method gets by with dividing the original into only two subproblems, with a little extra processing. The recurrence describing the number of multiplications required is

$$M(N) = 2M(N/2) + N.$$

Though we don't want to dwell on the mathematics of solving such recurrences, formulas of this particular form arise so frequently that it will be worthwhile to examine the development of an approximate solution. First, as above, we write $N = 2^n$:

$$M(2^n) = 2M(2^{n-1}) + 2^n.$$

The trick to making it simple to apply this same recursive formula to itself is to divide both sides by 2^n:

$$\frac{M(2^n)}{2^n} = \frac{M(2^{n-1})}{2^{n-1}} + 1.$$

Now, applying this same formula to itself n times ends up simply giving n copies of the "1," from which it follows immediately that $M(2^n) = n2^n$. Again, it turns out that this holds true (roughly) for all N, and we have the solution

$$M(N) \approx N \lg N.$$

We'll see several algorithms from different applications areas whose performance characteristics are described by recurrences of this type. Fortunately, many of the recurrences that come up are so similar to those above that the same techniques can be used.

For another example, consider the situation when an algorithm divides the problem to be solved in half, then is able to ignore one half and (recursively) solve the other. The running time of such an algorithm might be described by the recurrence

$$M(N) = M(N/2) + 1.$$

This is easier to solve than the one in the previous paragraph. We immediately have $M(2^n) = n$ and, again, it turns out that $M(N) \approx \lg N$.

Of course, it's not always possible to get by with such trivial manipulations. For a slightly more difficult example, consider an algorithm of the type described in the previous paragraph which must somehow examine each element before or after the recursive step. The running time of such an algorithm is described by the recurrence

$$M(N) = M(N/2) + N.$$

Substituting $N = 2^n$ and applying the same recurrence to itself n times now gives

$$M(2^n) = 2^n + 2^{n-1} + 2^{n-2} + \cdots + 1.$$

This must be evaluated to get the result $M(2^n) = 2^{n+1} - 1$ which translates to $M(N) \approx 2N$ for general N.

To summarize, many of the most interesting algorithms that we will encounter are based on the divide-and-conquer technique of combining the solutions of recursively solved smaller subproblems. The running time of such algorithms can usually be described by recurrence relationships which are a direct mathematical translation of the structure of the algorithm. Though such relationships can be challenging to solve precisely, they are often easy to solve for some particular values of N to get solutions which give reasonable estimates for all values of N. Our purpose in this discussion is to gain some intuitive feeling for how divide-and-conquer algorithms achieve efficiency, not to do detailed analysis of the algorithms. Indeed, the particular recurrences that we've just solved are sufficient to describe the performance of most of the algorithms that we'll be studying, and we'll simply be referring back to them.

Matrix Multiplication

The most famous application of the divide-and-conquer technique to an arithmetic problem is Strassen's method for matrix multiplication. We won't go into the details here, but we can sketch the method, since it is very similar to the polynomial multiplication method that we have just studied.

The straightforward method for multiplying two N-by-N matrices requires N^3 scalar multiplications, since each of the N^2 elements in the product matrix is obtained by N multiplications.

Strassen's method is to divide the size of the problem in half; this corresponds to dividing each of the matrices into quarters, each $N/2$ by $N/2$. The remaining problem is equivalent to multiplying 2-by-2 matrices. Just as we were able to reduce the number of multiplications required from four to three by combining terms in the polynomial multiplication problem, Strassen was able to find a way to combine terms to reduce the number of multiplications required for the 2-by-2 matrix multiplication problem from 8 to 7. The rearrangement and the terms required are quite complicated.

The number of multiplications required for matrix multiplication using Strassen's method is therefore defined by the divide-and-conquer recurrence

$$M(N) = 7M(N/2)$$

which has the solution

$$M(N) \approx N^{\lg 7} \approx N^{2.81}.$$

This result was quite surprising when it first appeared, since it had previously been thought that N^3 multiplications were absolutely necessary for matrix multiplication. The problem has been studied very intensively in recent years, and slightly better methods than Strassen's have been found. The "best" algorithm for matrix multiplication has still not been found, and this is one of the most famous outstanding problems of computer science.

It is important to note that we have been counting multiplications only. Before choosing an algorithm for a practical application, the costs of the extra additions and subtractions for combining terms and the costs of the recursive calls must be considered. These costs may depend heavily on the particular implementation or computer used. But certainly, this overhead makes Strassen's method less efficient than the standard method for small matrices. Even for large matrices, in terms of the number of data items input, Strassen's method really represents an improvement only from $N^{1.5}$ to $N^{1.41}$. This improvement is hard to notice except for very large N. For example, N would have to be more than a million for Strassen's method to use four times as few multiplications as the standard method, even though the overhead per multiplication is likely to be four times as large. Thus the algorithm is a theoretical, not practical, contribution.

This illustrates a general tradeoff which appears in all applications (though the effect is not always so dramatic): simple algorithms work best for small problems, but sophisticated algorithms can reap tremendous savings for large problems.

Exercises

1. Give a method for evaluating a polynomial with known roots r_1, r_2, ..., r_N, and compare your method with Horner's method.

2. Write a program to evaluate polynomials using Horner's method, where a linked list representation is used for the polynomials. Be sure that your program works efficiently for sparse polynomials.

3. Write an N^2 program to do Lagrangian interpolation.

4. Suppose that we know that a polynomial to be interpolated is sparse (has few non-zero coefficients). Describe how you would modify Lagrangian interpolation to run in time proportional to N times the number of non-zero coefficients.

5. Write out all of the polynomial multiplications performed when the divide-and-conquer polynomial multiplication method described in the text is used to square $1 + x + x^2 + x^3 + x^4 + x^5 + x^6 + x^7 + x^8$.

6. The polynomial multiplication routine *mult* could be made more efficient for sparse polynomials by returning 0 if all coefficients of either input are 0. About how many multiplications (to within a constant factor) would such a program use to square $1 + x^N$?

7. Can x^{32} be computed with less than five multiplications? If so, say which ones; if not, say why not.

8. Can x^{55} be computed with less than nine multiplications? If so, say which ones; if not, say why not.

9. Describe exactly how you would modify *mult* to multiply a polynomial of degree N by another of degree M, with $N > M$.

10. Give the representation that you would use for programs to add and multiply multivariate polynomials such as $xy^2z + 31wx^3y^3z^{50} + w$. Give the single most important reason for choosing this representation.

5. Gaussian Elimination

Certainly one of the most fundamental scientific computations is the solution of systems of simultaneous equations. The basic algorithm for solving systems of equations, *Gaussian elimination*, is relatively simple and has changed little in the 150 years since it was invented. This algorithm has come to be well understood, especially in the past twenty years, so that it can be used with some confidence that it will efficiently produce accurate results.

This is an example of an algorithm that will surely be available in most computer installations; indeed, it is a primitive in several computer languages, notably APL and Basic. However, the basic algorithm is easy to understand and implement, and special situations do arise where it might be desirable to implement a modified version of the algorithm rather than work with a standard subroutine. Also, the method deserves to be learned as one of the most important numeric methods in use today.

As with the other mathematical material that we have studied so far, our treatment of the method will highlight only the basic principles and will be self-contained. Familiarity with linear algebra is not required to understand the basic method. We'll develop a simple Pascal implementation that might be easier to use than a library subroutine for simple applications. However, we'll also see examples of problems which could arise. Certainly for a large or important application, the use of an expertly tuned implementation is called for, as well as some familiarity with the underlying mathematics.

A Simple Example

Suppose that we have three variables x, y and z and the following three equations:

$$x + 3y - 4z = 8,$$
$$x + y - 2z = 2,$$
$$-x - 2y + 5z = -1.$$

Our goal is to compute the values of the variables which simultaneously satisfy the equations. Depending on the particular equations there may not always be a solution to this problem (for example, if two of the equations are contradictory, such as $x + y = 1$, $x + y = 2$) or there may be many solutions (for example, if two equations are the same, or there are more variables than equations). We'll assume that the number of equations and variables is the same, and we'll look at an algorithm that will find a unique solution if one exists.

To make it easier to extend the formulas to cover more than just three points, we'll begin by renaming the variables, using subscripts:

$$x_1 + 3x_2 - 4x_3 = 8,$$
$$x_1 + x_2 - 2x_3 = 2,$$
$$-x_1 - 2x_2 + 5x_3 = -1.$$

To avoid writing down variables repeatedly, it is convenient to use matrix notation to express the simultaneous equations. The above equations are exactly equivalent to the matrix equation

$$\begin{pmatrix} 1 & 3 & -4 \\ 1 & 1 & -2 \\ -1 & -2 & 5 \end{pmatrix} \begin{pmatrix} x_1 \\ x_2 \\ x_3 \end{pmatrix} = \begin{pmatrix} 8 \\ 2 \\ -1 \end{pmatrix}.$$

There are several operations which can be performed on such equations which will not alter the solution:

Interchange equations: Clearly, the order in which the equations are written down doesn't affect the solution. In the matrix representation, this operation corresponds to interchanging rows in the matrix (and the vector on the right hand side).

Rename variables: This corresponds to interchanging columns in the matrix representation. (If columns i and j are switched, then variables x_i and x_j must also be considered switched.)

Multiply equations by a constant: Again, in the matrix representation, this corresponds to multiplying a row in the matrix (and the corresponding element in the vector on the right-hand side) by a constant.

Add two equations and replace one of them by the sum. (It takes a little thought to convince oneself that this will not affect the solution.)

For example, we can get a system of equations equivalent to the one above by replacing the second equation by the difference between the first two:

$$\begin{pmatrix} 1 & 3 & -4 \\ 0 & 2 & -2 \\ -1 & -2 & 5 \end{pmatrix} \begin{pmatrix} x_1 \\ x_2 \\ x_3 \end{pmatrix} = \begin{pmatrix} 8 \\ 6 \\ -1 \end{pmatrix}.$$

Notice that this eliminates x_1 from the second equation. In a similar manner, we can eliminate x_1 from the third equation by replacing the third equation by the sum of the first and third:

$$\begin{pmatrix} 1 & 3 & -4 \\ 0 & 2 & -2 \\ 0 & 1 & 1 \end{pmatrix} \begin{pmatrix} x_1 \\ x_2 \\ x_3 \end{pmatrix} = \begin{pmatrix} 8 \\ 6 \\ 7 \end{pmatrix}.$$

Now the variable x_1 is eliminated from all but the first equation. By systematically proceeding in this way, we can transform the original system of equations into a system with the same solution that is much easier to solve. For the example, this requires only one more step which combines two of the operations above: replacing the third equation by the difference between the second and twice the third. This makes all of the elements below the main diagonal 0: systems of equations of this form are particularly easy to solve. The simultaneous equations which result in our example are:

$$x_1 + 3x_2 - 4x_3 = 8,$$
$$2x_2 - 2x_3 = 6,$$
$$-4x_3 = -8.$$

Now the third equation can be solved immediately: $x_3 = 2$. If we substitute this value into the second equation, we can compute the value of x_2:

$$2x_2 - 4 = 6,$$
$$x_2 = 5.$$

Similarly, substituting these two values in the first equation allows the value of x_1 to be computed:

$$x_1 + 15 - 8 = 8,$$
$$x_1 = 1,$$

which completes the solution of the equations.

This example illustrates the two basic phases of Gaussian elimination. The first is the *forward elimination* phase, where the original system is transformed, by systematically eliminating variables from equations, into a system with all zeros below the diagonal. This process is sometimes called *triangulation*. The second phase is the *backward substitution phase*, where the values of the variables are computed using the triangulated matrix produced by the first phase.

Outline of the Method

In general, we want to solve a system of N equations in N unknowns:

$$a_{11}x_1 + a_{12}x_2 + \cdots + a_{1N}x_N = b_1,$$
$$a_{21}x_1 + a_{22}x_2 + \cdots + a_{2N}x_N = b_2,$$
$$\vdots$$
$$a_{N1}x_1 + a_{N2}x_2 + \cdots + a_{NN}x_N = b_N.$$

In matrix form, these equations are written as a single matrix equation:

$$\begin{pmatrix} a_{11} & a_{12} & \ldots & a_{1N} \\ a_{21} & a_{22} & \ldots & a_{2N} \\ & \vdots & & \\ a_{N1} & a_{N2} & \ldots & a_{NN} \end{pmatrix} \begin{pmatrix} x_1 \\ x_2 \\ \vdots \\ x_N \end{pmatrix} = \begin{pmatrix} b_1 \\ b_2 \\ \vdots \\ b_N \end{pmatrix}$$

or simply $Ax = b$, where A represents the matrix, x represents the variables, and b represents the right-hand sides of the equations. Since the rows of A are manipulated along with the elements of b, it is convenient to regard b as the $(N+1)$st column of A and use an N-by-$(N+1)$ array to hold both.

Now the forward elimination phase can be summarized as follows: first eliminate the first variable in all but the first equation by adding the appropriate multiple of the first equation to each of the other equations, then eliminate the second variable in all but the first two equations by adding the appropriate multiple of the second equation to each of the third through the Nth equations, then eliminate the third variable in all but the first three equations, etc. To eliminate the ith variable in the jth equation (for j between $i+1$ and N) we multiply the ith equation by a_{ji}/a_{ii} and subtract it from the jth equation. This process is perhaps more succinctly described by the following program, which reads in N followed by an N-by-$(N+1)$ matrix, performs the forward elimination, and writes out the triangulated result. In the input and in the output the ith line contains the ith row of the matrix followed by b_i.

```
program gauss(input, output);
const maxN=50;
var a: array [1..maxN, 1..maxN] of real;
    i, j, k, N: integer;
begin
readln(N);
for j:=1 to N do
   begin for k:=1 to N+1 do read(a[j, k]);  readln end;
for i:=1 to N do
   for j:=i+1 to N do
      for k:=N+1 downto i do
         a[j, k]:=a[j, k]−a[i, k]*a[j, i]/a[i, i];
for j:=1 to N do
   begin for k:=1 to N+1 do write(a[j, k]);  writeln end;
end.
```

(As we found with polynomials, if we want to have a program that takes N as input, it is necessary in Pascal to first decide how large a value of N will be "legal," and declare the array suitably.) Note that the code consists of three nested loops, so that the total running time is essentially proportional to N^3. The third loop goes backwards so as to avoid destroying $a[j, i]$ before it is needed to adjust the values of other elements in the same row.

The program in the above paragraph is too simple to be quite right: $a[i, i]$ might be zero, so division by zero could occur. This is easily fixed, because we can exchange any row (from $i+1$ to N) with the ith row to make $a[i, i]$ non-zero in the outer loop. If no such row can be found, then the matrix is *singular*: there is no unique solution.

In fact, it is advisable to do slightly more than just find a row with a non-zero entry in the ith column. It's best to use the row (from $i+1$ to N) whose entry in the ith column is the largest in absolute value. The reason for this is that severe computational errors can arise if the $a[i, i]$ value which is used to scale a row is very small. If $a[i, i]$ is very small, then the scaling factor $a[j, i]/a[i, i]$ which is used to eliminate the ith variable from the jth equation (for j from $i+1$ to N) will be very large. In fact, it could get so large as to dwarf the actual coefficients $a[j, k]$, to the point where the $a[j, k]$ value gets distorted by "round-off error."

Put simply, numbers which differ greatly in magnitude can't be accurately added or subtracted in the floating point number system commonly used to represent real numbers, but using a small $a[i, i]$ value greatly increases the likelihood that such operations will have to be performed. Using the largest value in the ith column from rows $i+1$ to N will ensure that the scaling factor is always less than 1, and will prevent the occurrence of this type of error. One might contemplate looking beyond the ith column to find a large element, but it has been shown that accurate answers can be obtained without resorting to this extra complication.

The following code for the forward elimination phase of Gaussian elimination is a straightforward implementation of this process. For each i from 1 to N, we scan down the ith column to find the largest element (in rows past the ith). The row containing this element is exchanged with the ith , then the ith variable is eliminated in the equations $i+1$ to N exactly as before:

```
procedure eliminate;
  var i, j, k, max: integer;
      t: real;
  begin
  for i:=1 to N do
    begin
    max:=i;
    for j:=i+1 to N do
      if abs(a[j, i])>abs(a[max, i]) then max:=j;
    for k:=i to N+1 do
      begin t:=a[i, k]; a[i, k]:=a[max, k]; a[max, k]:=t end;
    for j:=i+1 to N do
      for k:=N+1 downto i do
        a[j, k]:=a[j, k]−a[i, k]*a[j, i]/a[i, i];
    end
  end;
```

(A call to eliminate should replace the three nested **for** loops in the program gauss given above.) There are some algorithms where it is required that the pivot $a[i, i]$ be used to eliminate the ith variable from every equation but the ith (not just the $(i+1)$st through the Nth). This process is called *full pivoting*; for forward elimination we only do part of this work hence the process is called *partial pivoting* .

After the forward elimination phase has completed, the array a has all zeros below the diagonal, and the backward substitution phase can be executed. The code for this is even more straightforward:

```
procedure substitute;
  var j, k: integer;
      t: real;
  begin
  for j:=N downto 1 do
    begin
    t:=0.0;
    for k:=j+1 to N do t:=t+a[j, k]*x[k];
    x[j]:=(a[j, N+1]−t)/a[j, j]
    end
  end;
```

A call to eliminate followed by a call to substitute computes the solution in the N-element array x. Division by 0 could still occur for singular matrices.

Obviously a "library" routine would check for this explicitly.

An alternate way to proceed after forward elimination has created all zeros below the diagonal is to use precisely the same method to produce all zeros above the diagonal: first make the last column zero except for $a[N, N]$ by adding the appropriate multiple of $a[N, N]$, then do the same for the next-to-last column, etc. That is, we do "partial pivoting" again, but on the other "part" of each column, working backwards through the columns. After this process, called *Gauss-Jordan* reduction, is complete, only diagonal elements are non-zero, which yields a trivial solution.

Computational errors are a prime source of concern in Gaussian elimination. As mentioned above, we should be wary of situations when the magnitudes of the coefficients vastly differ. Using the largest available element in the column for partial pivoting ensures that large coefficients won't be arbitrarily created in the pivoting process, but it is not always possible to avoid severe errors. For example, very small coefficients turn up when two different equations have coefficients which are quite close to one another. It is actually possible to determine in advance whether such problems will cause inaccurate answers in the solution. Each matrix has an associated numerical quantity called the *condition number* which can be used to estimate the accuracy of the computed answer. A good library subroutine for Gaussian elimination will compute the condition number of the matrix as well as the solution, so that the accuracy of the solution can be known. Full treatment of the issues involved would be beyond the scope of this book.

Gaussian elimination with partial pivoting using the largest available pivot is "guaranteed" to produce results with very small computational errors. There are quite carefully worked out mathematical results which show that the calculated answer is quite accurate, except for ill-conditioned matrices (which might be more indicative of problems in the system of equations than in the method of solution). The algorithm has been the subject of fairly detailed theoretical studies, and can be recommended as a computational procedure of very wide applicability.

Variations and Extensions

The method just described is most appropriate for N-by-N matrices with most of the N^2 elements non-zero. As we've seen for other problems, special techniques are appropriate for *sparse* matrices where most of the elements are 0. This situation corresponds to systems of equations in which each equation has only a few terms.

If the non-zero elements have no particular structure, then the linked list representation discussed in Chapter 2 is appropriate, with one node for each non-zero matrix element, linked together by both row and column. The standard method can be implemented for this representation, with the usual extra complications due to the need to create and destroy non-zero elements. This technique is not likely to be worthwhile if one can afford the memory to hold the whole matrix, since it is much more complicated than the standard method. Also, sparse matrices become substantially less sparse during the Gaussian elimination process.

Some matrices not only have just a few non-zero elements but also have a simple structure, so that linked lists are not necessary. The most common example of this is a "band" matrix, where the non-zero elements all fall very close to the diagonal. In such cases, the inner loops of the Gaussian elimination algorithms need only be iterated a few times, so that the total running time (and storage requirement) is proportional to N, not N^3.

An interesting special case of a band matrix is a "tridiagonal" matrix, where only elements directly on, directly above, or directly below the diagonal are non-zero. For example, below is the general form of a tridiagonal matrix for $N = 5$:

$$\begin{pmatrix} a_{11} & a_{12} & 0 & 0 & 0 \\ a_{21} & a_{22} & a_{23} & 0 & 0 \\ 0 & a_{32} & a_{33} & a_{34} & 0 \\ 0 & 0 & a_{43} & a_{44} & a_{45} \\ 0 & 0 & 0 & a_{54} & a_{55} \end{pmatrix}$$

For such matrices, forward elimination and backward substitution each reduce to a single **for** loop:

```
for i:=1 to N−1 do
   begin
   a[i+1,N+1]:=a[i+1,N+1]−a[i,N+1]*a[i+1,i]/a[i,i];
   a[i+1,i+1]:=a[i+1,i+1]−a[i,i+1]*a[i+1,i]/a[i,i]
   end;
for j:=N downto 1 do
   x[j]:=(a[j,N+1]−a[j,j+1]*x[j+1])/a[j,j];
```

For forward elimination, only the case $j=i+1$ and $k=i+1$ needs to be included, since $a[i,k]=0$ for $k>i+1$. (The case $k=i$ can be skipped since it sets to 0 an array element which is never examined again —this same change could be made to straight Gaussian elimination.) Of course, a two-dimensional array of size N^2 wouldn't be used for a tridiagonal matrix. The storage required for the above program can be reduced to be linear in N by maintaining four arrays instead of the a matrix: one for each of the three nonzero diagonals and one for the $(N+1)$st column. Note that this program doesn't necessarily pivot on the largest available element, so there is no insurance against division by zero or the accumulation of computational errors. For some types of tridiagonal matrices which arise commonly, it can be proven that this is not a reason for concern.

Gauss-Jordan reduction can be implemented with full pivoting to replace a matrix by its *inverse* in one sweep through it. The inverse of a matrix A, written A^{-1}, has the property that a system of equations $Ax = b$ could be solved just by performing the matrix multiplication $x = A^{-1}b$. Still, N^3 operations are required to compute x given b. However, there is a way to preprocess a matrix and "decompose" it into component parts which make it possible to solve the corresponding system of equations with any given right-hand side in time proportional to N^2, a savings of a factor of N over using Gaussian elimination each time. Roughly, this involves remembering the operations that are performed on the $(N+1)$st column during the forward elimination phase, so that the result of forward elimination on a new $(N+1)$st column can be computed efficiently and then back-substitution performed as usual.

Solving systems of linear equations has been shown to be computationally equivalent to multiplying matrices, so there exist algorithms (for example, Strassen's matrix multiplication algorithm) which can solve systems of N equations in N variables in time proportional to $N^{2.81\cdots}$. As with matrix multiplication, it would not be worthwhile to use such a method unless very large systems of equations were to be processed routinely. As before, the actual running time of Gaussian elimination in terms of the number of inputs is $N^{3/2}$, which is difficult to improve upon in practice.

Exercises

1. Give the matrix produced by the forward elimination phase of Gaussian elimination (*gauss*, with *eliminate*) when used to solve the equations $x + y + z = 6$, $2x + y + 3z = 12$, and $3x + y + 3z = 14$.

2. Give a system of three equations in three unknowns for which *gauss* as is (without *eliminate*) fails, even though there is a solution.

3. What is the storage requirement for Gaussian elimination on an N-by-N matrix with only $3N$ nonzero elements?

4. Describe what happens when *eliminate* is used on a matrix with a row of all 0's.

5. Describe what happens when *eliminate* then *substitute* are used on a matrix with a column of all 0's.

6. Which uses more arithmetic operations: Gauss-Jordan reduction or back substitution?

7. If we interchange columns in a matrix, what is the effect on the corresponding simultaneous equations?

8. How would you test for contradictory or identical equations when using *eliminate*.

9. Of what use would Gaussian elimination be if we were presented with a system of M equations in N unknowns, with $M < N$? What if $M > N$?

10. Give an example showing the need for pivoting on the largest available element, using a mythical primitive computer where numbers can be represented with only two significant digits (all numbers must be of the form $x.y \times 10^z$ for single digit integers x, y, and z).

6. Curve Fitting

The term *curve fitting* (or *data fitting*) is used to describe the general problem of finding a function which matches a set of observed values at a set of given points. Specifically, given the points

$$x_1, x_2, \ldots, x_N$$

and the corresponding values

$$y_1, y_2, \ldots, y_N,$$

the goal is to find a function (perhaps of a specified type) such that

$$f(x_1) = y_1, f(x_2) = y_2, \ldots, f(x_N) = y_N$$

and such that $f(x)$ assumes "reasonable" values at other data points. It could be that the x's and y's are related by some unknown function, and our goal is to find that function, but, in general, the definition of what is "reasonable" depends upon the application. We'll see that it is often easy to identify "unreasonable" functions.

Curve fitting has obvious application in the analysis of experimental data, and it has many other uses. For example, it can be used in computer graphics to produce curves that "look nice" without the overhead of storing a large number of points to be plotted. A related application is the use of curve fitting to provide a fast algorithm for computing the value of a known function at an arbitrary point: keep a short table of exact values, curve fit to find other values.

Two principal methods are used to approach this problem. The first is *interpolation*: a smooth function is to be found which exactly matches the given values at the given points. The second method, *least squares data fitting*, is used when the given values may not be exact, and a function is sought which matches them as well as possible.

Polynomial Interpolation

We've already seen one method for solving the data-fitting problem: if f is known to be a polynomial of degree $N-1$, then we have the polynomial inter-polation problem of Chapter 4. Even if we have no particular knowledge about f, we could solve the data-fitting problem by letting $f(x)$ be the interpolating polynomial of degree $N-1$ for the given points and values. This could be computed using methods outlined elsewhere in this book, but there are many reasons not to use polynomial interpolation for data fitting. For one thing, a fair amount of computation is involved (advanced $N(\log N)^2$ methods are available, but elementary techniques are quadratic). Computing a polynomial of degree 100 (for example) seems overkill for interpolating a curve through 100 points.

The main problem with polynomial interpolation is that high-degree polynomials are relatively complicated functions which may have unexpected properties not well suited to the function being fitted. A result from classical mathematics (the Weierstrass approximation theorem) tells us that it is pos-sible to approximate any reasonable function with a polynomial (of sufficiently high degree). Unfortunately, polynomials of very high degree tend to fluctuate wildly. It turns out that, even though most functions are closely approximated almost everywhere on a closed interval by an interpolation polynomial, there are always some places where the approximation is terrible. Furthermore, this theory assumes that the data values are exact values from some unknown function when it is often the case that the given data values are only ap-proximate. If the y's were approximate values from some unknown low-degree polynomial, we would hope that the coefficients for the high-degree terms in the interpolating polynomial would be 0. It doesn't usually work out this way; instead the interpolating polynomial tries to use the high-degree terms to help achieve an exact fit. These effects make interpolating polynomials inappropriate for many curve-fitting applications.

Spline Interpolation

Still, low-degree polynomials are simple curves which are easy to work with analytically, and they are widely used for curve fitting. The trick is to abandon the idea of trying to make *one* polynomial go through all the points and instead use different polynomials to connect adjacent points, piecing them together smoothly. An elegant special case of this, which also involves relatively straightforward computation, is called *spline interpolation*.

A "spline" is a mechanical device used by draftsmen to draw aesthetically pleasing curves: the draftsman fixes a set of points (*knots*) on his drawing, then bends a flexible strip of plastic or wood (the spline) around them and traces it to produce the curve. Spline interpolation is the mathematical equivalent of this process and results in the same curve.

It can be shown from elementary mechanics that the shape assumed by the spline between two adjacent knots is a third-degree (cubic) polynomial. Translated to our data-fitting problem, this means that we should consider the curve to be $N-1$ different cubic polynomials

$$s_i(x) = a_i x^3 + b_i x^2 + c_i x + d_i, \quad i = 1, 2, \ldots, N-1,$$

with $s_i(x)$ defined to be the cubic polynomial to be used in the interval between x_i and x_{i+1}, as shown in the following diagram:

The spline can be represented in the obvious way as four one-dimensional arrays (or a 4-by-$(N-1)$ two-dimensional array). Creating a spline consists of computing the necessary a, b, c, d coefficients from the given x points and y values. The physical constraints on the spline correspond to simultaneous equations which can be solved to yield the coefficients.

For example, we obviously must have $s_i(x_i) = y_i$ and $s_i(x_{i+1}) = y_{i+1}$ for $i = 1, 2, \ldots, N-1$ because the spline must touch the knots. Not only does the spline touch the knots, but also it curves smoothly around them with no sharp bends or kinks. Mathematically, this means that the first derivatives of the spline polynomials must be equal at the knots ($s'_{i-1}(x_i) = s'_i(x_i)$ for $i = 2, 3, \ldots,$ $N-1$). In fact, it turns out that the second derivatives of the polynomials must be equal at the knots. These conditions give a total of $4N-6$ equations in the $4(N-1)$ unknown coefficients. Two more conditions need to be specified to describe the situation at the endpoints of the spline. Several options are available; we'll use the so-called "natural" spline which derives from $s''_1(x_1) =$ 0 and $s''_{N-1}(x_N) = 0$. These conditions give a full system of $4N-4$ equations in $4N-4$ unknowns, which could be solved using Gaussian elimination to calculate all the coefficients that describe the spline.

The same spline can be computed somewhat more efficiently because there are actually only $N-2$ "unknowns": most of the spline conditions are redundant. For example, suppose that p_i is the value of the second derivative of the spline at x_i, so that $s''_{i-1}(x_i) = s''_i(x_i) = p_i$ for $i = 2, \ldots, N-1$, with $p_1 = p_N = 0$. If the values of $p_1, \ldots, p_N$ are known, then all of the a, b, c, d coefficients can be computed for the spline segments, since we have four equations in four unknowns for each spline segment: for $i = 1,2,\ldots,N-1$, we must have

$$s_i(x_i) = y_i$$
$$s_i(x_{i+1}) = y_{i+1}$$
$$s_i''(x_i) = p_i$$
$$s_i''(x_{i+1}) = p_{i+1}.$$

Thus, to fully determine the spline, we need only compute the values of $p_2,\ldots,p_{N-1}$. But this discussion hasn't even considered the conditions that the first derivatives must match. These $N-2$ conditions provide exactly the $N-2$ equations needed to solve for the $N-2$ unknowns, the p_i second derivative values.

To express the a, b, c, and d coefficients in terms of the p second derivative values, then substitute those expressions into the four equations listed above for each spline segment, leads to some unnecessarily complicated expressions. Instead it is convenient to express the equations for the spline segments in a certain canonical form that involves fewer unknown coefficients. If we change variables to $t = (x - x_i)/(x_{i+1} - x_i)$ then the spline can be expressed in the following way:

$$s_i(t) = ty_{i+1} + (1-t)y_i + (x_{i+1} - x_i)^2[(t^3 - t)p_{i+1} - ((1-t)^3 - (1-t))p_i]$$

Now each spline is defined on the interval $[0,1]$. This equation is less formidable than it looks because we're mainly interested in the endpoints 0 and 1, and either t or $(1-t)$ is 0 at these points. It's trivial to check that the spline interpolates and is continuous because $s_{i-1}(1) = s_i(0) = y_i$ for $i = 2,\ldots,N-1$, and it's only slightly more difficult to verify that the second derivative is continuous because $s_i''(1) = s_{i+1}''(0) = p_{i+1}$. These are cubic polynomials which satisfy the requisite conditions at the endpoints, so they are equivalent to the spline segments described above. If we were to substitute for t and find the coefficient of x^3, etc., then we would get the same expressions for the a's, b's, c's, and d's in terms of the x's, y's, and p's as if we were to use the method described in the previous paragraph. But there's no reason to do so, because we've checked that these spline segments satisfy the end conditions, and we can evaluate each at any point in its interval by computing t and using the above formula (once we know the p's).

To solve for the p's we need to set the first derivatives of the spline segments equal at the endpoints. The first derivative (with respect to x) of the above equation is

$$s_i'(t) = z_i + (x_{i+1} - x_i)[(3t^2 - 1)p_{i+1} + (3(1-t)^2 + 1)p_i]$$

where $z = (y_{i+1} - y_i)/(x_{i+1} - x_i)$. Now, setting $s'_{i-1}(1) = s'_i(0)$ for $i = 2, \ldots, N-1$ gives our system of $N - 2$ equations to solve:

$$(x_i - x_{i-1})p_{i-1} + 2(x_{i+1} - x_{i-1})p_i + (x_{i+1} - x_i)p_{i+1} = z_i - z_{i-1}.$$

This system of equations is a simple "tridiagonal" form which is easily solved with a degenerate version of Gaussian elimination as we saw in Chapter 5. If we let $u_i = x_{i+1} - x_i$, $d_i = 2(x_{i+1} - x_{i-1})$, and $w_i = z_i - z_{i-1}$, we have, for example, the following simultaneous equations for $N = 7$:

$$\begin{pmatrix} d_2 & u_2 & 0 & 0 & 0 \\ u_2 & d_3 & u_3 & 0 & 0 \\ 0 & u_3 & d_4 & u_4 & 0 \\ 0 & 0 & u_4 & d_5 & u_5 \\ 0 & 0 & 0 & u_5 & d_6 \end{pmatrix} \begin{pmatrix} p_2 \\ p_3 \\ p_4 \\ p_5 \\ p_6 \end{pmatrix} = \begin{pmatrix} w_2 \\ w_3 \\ w_4 \\ w_5 \\ w_6 \end{pmatrix}.$$

In fact, this is a symmetric tridiagonal system, with the diagonal below the main diagonal equal to the diagonal above the main diagonal. It turns out that pivoting on the largest available element is not necessary to get an accurate solution for this system of equations.

The method described in the above paragraph for computing a cubic spline translates very easily into Pascal:

```
procedure makespline;
    var i: integer;
    begin
    readln(N);
    for i:=1 to N do readln(x[i], y[i]);
    for i:=2 to N−1 do d[i]:=2*(x[i+1]−x[i−1]);
    for i:=1 to N−1 do u[i]:=x[i+1]−x[i];
    for i:=2 to N−1 do
        w[i]:=(y[i+1]−y[i])/u[i]−(y[i]−y[i−1])/u[i−1];
    p[1]:=0.0;  p[N]:=0.0;
    for i:=2 to N−2 do
        begin
        w[i+1]:=w[i+1]−w[i]*u[i]/d[i];
        d[i+1]:=d[i+1]−u[i]*u[i]/d[i]
        end;
    for i:=N−1 downto 2 do
        p[i]:=(w[i]−u[i]*p[i+1])/d[i];
    end;
```

The arrays d and u are the representation of the tridiagonal matrix that is solved using the program in Chapter 5. We use $d[i]$ where $a[i,i]$ is used in that program, $u[i]$ where $a[i+1,i]$ or $a[i,i+1]$ is used, and $z[i]$ where $a[i,N+1]$ is used.

For an example of the construction of a cubic spline, consider fitting a spline to the five data points

$$(1.0,2.0), \quad (2.0,1.5), \quad (4.0,1.25), \quad (5.0,1.2), \quad (8.0,1.125), \quad (10.0,1.1).$$

(These come from the function $1 + 1/x$.) The spline parameters are found by solving the system of equations

$$\begin{pmatrix} 6 & 2 & 0 & 0 \\ 2 & 6 & 1 & 0 \\ 0 & 1 & 8 & 3 \\ 0 & 0 & 3 & 10 \end{pmatrix} \begin{pmatrix} p_2 \\ p_3 \\ p_4 \\ p_5 \end{pmatrix} = \begin{pmatrix} .3750 \\ .0750 \\ .0250 \\ .0125 \end{pmatrix}$$

with the result $p_2 = 0.06590$, $p_3 = -0.01021$, $p_4 = 0.00443$, $p_5 = -0.00008$.

To evaluate the spline for any value of x in the range $[x_1, x_N]$, we simply find the interval $[x_i, x_{i+1}]$ containing x, then compute t and use the formula above for $s_i(x)$ (which, in turn, uses the computed values for p_i and p_{i+1}).

```
function eval(v: real): real;
  var t: real; i: integer;
  function f(x: real): real;
     begin f:=x*x*x−x end;
  begin
  i:=0; repeat i:=i+1 until v<=x[i+1];
  t:=(v−x[i])/u[i];
  eval:=t*y[i+1]+(1−t)*y[i]
       +u[i]*u[i]*(f(t)*p[i+1]−f(1−t)*p[i])
  end;
```

This program does not check for the error condition when v is not between $x[1]$ and $x[N]$. If there are a large number of spline segments (that is, if N is large), then there are more efficient "searching" methods for finding the interval containing v, which we'll study in Chapter 14.

There are many variations on the idea of curvefitting by piecing together polynomials in a "smooth" way: the computation of splines is a quite well-developed field of study. Other types of splines involve other types of smoothness criteria as well as changes such as relaxing the condition that the spline must exactly touch each data point. Computationally, they involve exactly the same steps of determining the coefficients for each of the spline pieces by solving the system of linear equations derived from imposing constraints on how they are joined.

Method of Least Squares

A very common experimental situation is that, while the data values that we have are not exact, we do have some idea of the form of the function which is to fit the data. The function might depend on some parameters

$$f(x) = f(c_1, c_2, \ldots, c_M, x)$$

and the curve fitting procedure is to find the choice of parameters that "best" matches the observed values at the given points. If the function were a polynomial (with the parameters being the coefficients) and the values were exact, then this would be interpolation. But now we are considering more general functions and inaccurate data. To simplify the discussion, we'll concentrate on fitting to functions which are expressed as a linear combination of simpler functions, with the unknown parameters being the coefficients:

$$f(x) = c_1 f_1(x) + c_2 f_2(x) + \cdots + c_M f_M(x).$$

This includes most of the functions that we'll be interested in. After studying this case, we'll consider more general functions.

A common way of measuring how well a function fits is the *least-squares criterion*: the error is calculated by adding up the squares of the errors at each of the observation points:

$$E = \sum_{1 \leq j \leq N} (f(x_j) - y_j)^2.$$

This is a very natural measure: the squaring is done to stop cancellations among errors with different signs. Obviously, it is most desirable to find the choice of parameters that minimizes E. It turns out that this choice can be computed efficiently: this is the so-called *method of least squares*.

The method follows quite directly from the definition. To simplify the derivation, we'll do the case $M = 2, N = 3$, but the general method will follow directly. Suppose that we have three points x_1, x_2, x_3 and corresponding values y_1, y_2, y_3 which are to be fitted to a function of the form $f(x) = c_1 f_1(x) + c_2 f_2(x)$. Our job is to find the choice of the coefficients c_1, c_2 which minimizes the least-squares error

$$\begin{aligned} E =& (c_1 f_1(x_1) + c_2 f_2(x_1) - y_1)^2 \\ &+ (c_1 f_1(x_2) + c_2 f_2(x_2) - y_2)^2 \\ &+ (c_1 f_1(x_3) + c_2 f_2(x_3) - y_3)^2. \end{aligned}$$

To find the choices of c_1 and c_2 which minimize this error, we simply need to set the derivatives dE/dc_1 and dE/dc_2 to zero. For c_1 we have:

$$\begin{aligned}
\frac{dE}{dc_1} =\;& 2(c_1 f_1(x_1) + c_2 f_2(x_1) - y_1)f_1(x_1) \\
& + 2(c_1 f_1(x_2) + c_2 f_2(x_2) - y_2)f_1(x_2) \\
& + 2(c_1 f_1(x_3) + c_2 f_2(x_3) - y_3)f_1(x_3).
\end{aligned}$$

Setting the derivative equal to zero leaves an equation which the variables c_1 and c_2 must satisfy ($f_1(x_1)$, etc. are all "constants" with known values):

$$c_1[f_1(x_1)f_1(x_1) + f_1(x_2)f_1(x_2) + f_1(x_3)f_1(x_3)]$$

$$+ c_2[f_2(x_1)f_1(x_1) + f_2(x_2)f_1(x_2) + f_2(x_3)f_1(x_3)]$$

$$= y_1 f_1(x_1) + y_2 f_1(x_2) + y_3 f_1(x_3).$$

We get a similar equation when we set the derivative dE/dc_2 to zero. These rather formidable-looking equations can be greatly simplified using vector notation and the "dot product" operation that we encountered briefly in Chapter 2. If we define the vectors $\mathbf{x} = (x_1, x_2, x_3)$ and $\mathbf{y} = (y_1, y_2, y_3)$ and then the dot product of $\mathbf{x}$ and $\mathbf{y}$ is the real number defined by

$$\mathbf{x} \cdot \mathbf{y} = x_1 y_1 + x_2 y_2 + x_3 y_3.$$

Now, if we define the vectors $\mathbf{f}_1 = (f_1(x_1), f_1(x_2), f_1(x_3))$ and $\mathbf{f}_2 = (f_2(x_1), f_2(x_2), f_2(x_3))$ then our equations for the coefficients c_1 and c_2 can be very simply expressed:

$$c_1 \mathbf{f}_1 \cdot \mathbf{f}_1 + c_2 \mathbf{f}_1 \cdot \mathbf{f}_2 = \mathbf{y} \cdot \mathbf{f}_1$$

$$c_1 \mathbf{f}_2 \cdot \mathbf{f}_1 + c_2 \mathbf{f}_2 \cdot \mathbf{f}_2 = \mathbf{y} \cdot \mathbf{f}_2.$$

These can be solved with Gaussian elimination to find the desired coefficients.

For example, suppose that we know that the data points

$$(1.0, 2.05), \quad (2.0, 1.53), \quad (4.0, 1.26), \quad (5.0, 1.21), \quad (8.0, 1.13), \quad (10.0, 1.1).$$

should be fit by a function of the form $c_1 + c_2/x$. (These data points are slightly perturbed from the exact values for $1 + 1/x$). In this case, we have $f_1 = (1.0, 1.0, 1.0, 1.0, 1.0, 1.0)$ and $f_2 = (1.0, 0.5, 0.25, 0.2, 0.125, 0.1)$ so we have to solve the system of equations

$$\begin{pmatrix} 6.000 & 2.175 \\ 2.175 & 1.378 \end{pmatrix} \begin{pmatrix} c_1 \\ c_2 \end{pmatrix} = \begin{pmatrix} 8.280 \\ 3.623 \end{pmatrix}$$

with the result $c_1 = 0.998$ and $c_2 = 1.054$ (both close to 1, as expected).

The method outlined above easily generalizes to find more than two coefficients. To find the constants $c_1, c_2, \ldots, c_M$ in

$$f(x) = c_1 f_1(x) + c_2 f_2(x) + \cdots + c_M f_M(x)$$

which minimize the least squares error for the point and observation vectors

$$\mathbf{x} = (x_1, x_2, \ldots, x_N)$$
$$\mathbf{y} = (y_1, y_2, \ldots, y_N),$$

first compute the function component vectors

$$\mathbf{f}_1 = (f_1(x_1), f_1(x_2), \ldots, f_1(x_N)),$$
$$\mathbf{f}_2 = (f_2(x_1), f_2(x_2), \ldots, f_2(x_N)),$$
$$\vdots$$
$$\mathbf{f}_M = (f_M(x_1), f_M(x_2), \ldots, f_M(x_N)).$$

Then make up an M-by-M linear system of equations $Ac = b$ with

$$a_{ij} = \mathbf{f}_i \cdot \mathbf{f}_j,$$
$$b_j = \mathbf{f}_j \cdot \mathbf{y}.$$

The solution to this system of simultaneous equations yields the required coefficients.

This method is easily implemented by maintaining a two dimensional array for the $\mathbf{f}$ vectors, considering $\mathbf{y}$ as the $(M+1)$st vector. Then an array $a[1..M, 1..M+1]$ can be filled as follows:

```
for i:=1 to M do
  for j:=1 to M+1 do
    begin
    t:= 0.0;
    for k:=1 to N do t:=t+f[i, k]*f[j, k];
    a[i, j]:=t;
    end;
```

and then solved using the Gaussian elimination procedure from Chapter 5.

The method of least squares can be extended to handle nonlinear functions (for example a function such as $f(x) = c_1 e^{-c_2 x} \sin c_3 x$), and it is often used for this type of application. The idea is fundamentally the same; the problem is that the derivatives may not be easy to compute. What is used is an *iterative* method: use some estimate for the coefficients, then use these within the method of least squares to compute the derivatives, thus producing a better estimate for the coefficients. This basic method, which is widely used today, was outlined by Gauss in the 1820s.

Exercises

1. Approximate the function $\lg x$ with a degree 4 interpolating polynomial at the points 1,2,3,4, and 5. Estimate the quality of the fit by computing the sum of the squares of the errors at 1.5, 2.5, 3.5, and 4.5.

2. Solve the previous problem for the function $\sin x$. Plot the function and the approximation, if that's possible on your computer system.

3. Solve the previous problems using a cubic spline instead of an interpolating polynomial.

4. Approximate the function $\lg x$ with a cubic spline with knots at 2^N for N between 1 and 10. Experiment with different placements of knots in the same range to try to obtain a better fit.

5. What would happen in least squares data fitting if one of the functions was the function $f_i(x) = 0$ for some i?

6. What would happen in least squares data-fitting if all the observed values were 0?

7. What values of a, b, c minimize the least-squares error in using the function $f(x) = ax \log x + bx + c$ to approximate the observations $f(1) = 0, f(4) = 13, f(8) = 41$?

8. Excluding the Gaussian elimination phase, how many multiplications are involved in using the method of least squares to find M coefficients based on N observations?

9. Under what circumstances would the matrix which arises in least-squares curve fitting be singular?

10. Does the least-squares method work if two different observations are included for the same point?

7. Integration

Computing the integral is a fundamental analytic operation often performed on functions being processed on computers. One of two completely different approaches can be used, depending on the way the function is represented. If an explicit representation of the function is available, then it may be possible to do *symbolic integration* to compute a similar representation for the integral. At the other extreme, the function may be defined by a table, so that function values are known for only a few points. The most common situation is between these: the function to be integrated is represented in such a way that its value at any particular point can be computed. In this case, the goal is to compute a reasonable approximation to the integral of the function, without performing an excessive number of function evaluations. This computation is often called *quadrature* by numerical analysts.

Symbolic Integration

If full information is available about a function, then it may be worthwhile to consider using a method which involves manipulating some representation of the function rather than working with numeric values. The goal is to transform a representation of the function into a representation of the integral, in much the same way that indefinite integration is done by hand.

A simple example of this is the integration of polynomials. In Chapters 2 and 4 we examined methods for "symbolically" computing sums and products of polynomials, with programs that worked on a particular representation for the polynomials and produced the representation for the answers from the representation for the inputs. The operation of integration (and differentiation) of polynomials can also be done in this way. If a polynomial

$$p(x) = p_0 + p_1 x + p_2 x^2 + \cdots + p_{N-1} x^{N-1}$$

79

is represented simply by keeping the values of the coefficients in an array *p* then the integral can be easily computed as follows:

---

for *i*:=*N* **downto** *1* **do** *p*[*i*]:=*p*[*i*−*1*]/*i*;
p[*0*]:=*0*;

This is a direct implementation of the well-known symbolic integration rule $\int_0^x t^{i-1}\,dt = x^i/i$ for $i > 0$.

Obviously a wider class of functions than just polynomials can be handled by adding more symbolic rules. The addition of composite rules such as *integration by parts*,

$$\int u\,dv = uv - \int v\,du,$$

can greatly expand the set of functions which can be handled. (Integration by parts requires a differentiation capability. Symbolic differentiation is somewhat easier than symbolic integration, since a reasonable set of elementary rules plus the composite *chain rule* will suffice for most common functions.)

The large number of rules available to be applied to a particular function makes symbolic integration a difficult task. Indeed, it has only recently been shown that there is an *algorithm* for this task: a procedure which either returns the integral of any given function or says that the answer cannot be expressed in terms of elementary functions. A description of this algorithm in its full generality would be beyond the scope of this book. However, when the functions being processed are from a small restricted class, symbolic integration can be a powerful tool.

Of course, symbolic techniques have the fundamental limitation that there are a great many integrals (many of which occur in practice) which can't be evaluated symbolically. Next, we'll examine some techniques which have been developed to compute approximations to the values of real integrals.

Simple Quadrature Methods

Perhaps the most obvious way to approximate the value of an integral is the *rectangle method*: evaluating an integral is the same as computing the area under a curve, and we can estimate the area under a curve by summing the areas of small rectangles which nearly fit under the curve, as diagrammed below.

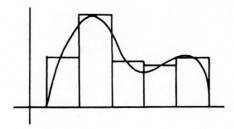

To be precise, suppose that we are to compute $\int_a^b f(x)\,dx$, and that the interval $[a, b]$ over which the integral is to be computed is divided into N parts, delimited by the points $x_1, x_2,\ldots,x_{N+1}$. Then we have N rectangles, with the width of the ith rectangle $(1 \le i \le N)$ given by $x_{i+1} - x_i$. For the height of the ith rectangle, we could use $f(x_i)$ or $f(x_{i+1})$, but it would seem that the result would be more accurate if the value of f at the midpoint of the interval $(f((x_i + x_{i+1})/2))$ is used, as in the above diagram. This leads to the quadrature formula

$$r = \sum_{1 \le i \le N} (x_{i+1} - x_i) f\left(\frac{x_i + x_{i+1}}{2}\right)$$

which estimates the value of the integral of $f(x)$ over the interval from $a = x_1$ to $b = x_{N+1}$. In the common case where all the intervals are to be the same size, say $x_{i+1} - x_i = w$, we have $x_{i+1} + x_i = (2i + 1)w$, so the approximation r to the integral is easily computed.

```
function intrect(a, b: real; N: integer): real;
   var i: integer; w, r: real;
   begin
   r:=0; w:=(b−a)/N;
   for i:=1 to N do r:=r+w*f(a−w/2+i*w);
   intrect :=r;
   end;
```

Of course, as N gets larger, the answer becomes more accurate, For example, the following table shows the estimate produced by this function for $\int_1^2 dx/x$ (which we know to be $\ln 2 = 0.6931471805599\ldots$) when invoked with the call $intrect(1.0, 2.0,N)$ for $N = 10, 100, 1000$:

$$
\begin{array}{ll}
10 & 0.6928353604100 \\
100 & 0.6931440556283 \\
1000 & 0.6931471493100
\end{array}
$$

When $N = 1000$, our answer is accurate to about seven decimal places. More sophisticated quadrature methods can achieve better accuracy with much less work.

It is not difficult to derive an analytic expression for the error made in the rectangle method by expanding $f(x)$ in a Taylor series about the midpoint of each interval, integrating, then summing over all intervals. We won't go through the details of this calculation: our purpose is not to derive detailed error bounds, but rather to show error estimates for simple methods and how these estimates suggest more accurate methods. This can be appreciated even by a reader not familiar with Taylor series. It turns out that

$$
\int_a^b f(x)\,dx = r + w^3 e_3 + w^5 e_5 + \cdots
$$

where w is the interval width $((b - a)/N)$ and e_3 depends on the value of the third derivative of f at the interval midpoints, etc. (Normally, this is a good approximation because most "reasonable" functions have small high-order derivatives, though this is not always true.) For example, if we choose to make $w = .01$ (which would correspond to $N = 200$ in the example above), this formula says the integral computed by the procedure above should be accurate to about six places.

Another way to approximate the integral is to divide the area under the curve into trapezoids, as diagrammed below.

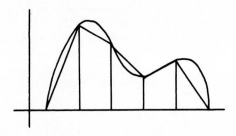

This *trapezoid method* leads to the quadrature formula

$$
t = \sum_{1 \le i \le N} (x_{i+1} - x_i)\frac{f(x_i) + f(x_{i+1})}{2}.
$$

(Recall that the area of a trapezoid is one-half the product of the height and the sum of the lengths of the two bases.) The error for this method can be derived in a similar way as for the rectangle method. It turns out that

$$\int_a^b f(x)\,dx = t - 2w^3 e_3 - 4w^5 e_5 + \cdots.$$

Thus the rectangle method is twice as accurate as the trapezoid method. This is borne out by our example. The following procedure implements the trapezoid method in the common case where all the intervals are the same width:

```
function inttrap(a, b: real; N: integer): real;
  var i: integer; w, t: real;
  begin
  t:=0; w:=(b−a)/N;
  for i:=1 to N do t:=t+w*(f(a+(i−1)*w)+f(a+i*w))/2;
  inttrap:=t;
  end;
```

This procedure produces the following estimates for $\int_1^2 dx/x$:

$$
\begin{array}{rl}
10 & 0.6937714031754 \\
100 & 0.6931534304818 \\
1000 & 0.6931472430599
\end{array}
$$

It may seem surprising at first that the rectangle method is more accurate than the trapezoid method: the rectangles tend to fall partly under the curve, partly over (so that the error can cancel out within an interval), while the trapezoids tend to fall either completely under or completely over the curve.

Another perfectly reasonable method is *spline quadrature*: spline interpolation is performed using methods we have discussed and then the integral is computed by piecewise application of the trivial symbolic polynomial integration technique described above. Below, we'll see how this relates to the other methods.

Compound Methods

Examination of the formulas given above for the error of the rectangle and trapezoid methods leads to a simple method with much greater accuracy, called *Simpson's method*. The idea is to eliminate the leading term in the error by combining the two methods. Multiplying the formula for the rectangle method by 2, adding the formula for the trapezoid method then dividing by 3 gives the equation

$$\int_a^b f(x)\,dx = \frac{1}{3}(2r + t - 2w^5 e_5 + \cdots).$$

The w^3 term has disappeared, so this formula tells us that we can get a method that is accurate to within w^5 by combining the quadrature formulas in the same way:

$$s = \sum_{1 \le i \le N} \frac{x_{i+1} - x_i}{6} \left[f(x_i) + 4f(\frac{x_i + x_{i+1}}{2}) + f(x_{i+1}) \right].$$

If an interval size of .01 is used for Simpson's rule, then the integral can be computed to about ten-place accuracy. Again, this is borne out in our example. The implementation of Simpson's method is only slightly more complicated than the others (again, we consider the case where the intervals are the same width):

```
function intsimp(a, b: real; N: integer): real;
  var i: integer; w, s: real;
  begin
  s:=0; w:=(b−a)/N;
  for i:=1 to N do
    s:=s+w*(f(a+(i−1)*w)+4*f(a−w/2+i*w)+f(a+i*w))/6;
  intsimp:=s;
  end;
```

This program requires three "function evaluations" (rather than two) in the inner loop, but it produces far more accurate results than do the previous two methods.

$$
\begin{array}{rl}
10 & 0.6931473746651 \\
100 & 0.6931471805795 \\
1000 & 0.6931471805599
\end{array}
$$

More complicated quadrature methods have been devised which gain accuracy by combining simpler methods with similar errors. The most well-known is *Romberg integration*, which uses two different sets of subintervals for its two "methods."

It turns out that Simpson's method is exactly equivalent to interpolating the data to a piecewise quadratic function, then integrating. It is interesting to note that the four methods we have discussed all can be cast as piecewise interpolation methods: the rectangle rule interpolates to a constant (degree-0 polynomial); the trapezoid rule to a line (degree-1 polynomial); Simpson's rule to a quadratic polynomial; and spline quadrature to a cubic polynomial.

Adaptive Quadrature

A major flaw in the methods that we have discussed so far is that the errors involved depend not only upon the subinterval size used, but also upon the value of the high-order derivatives of the function being integrated. This implies that the methods will not work well at all for certain functions (those with large high-order derivatives). But few functions have large high-order derivatives everywhere. It is reasonable to use small intervals where the derivatives are large and large intervals where the derivatives are small. A method which does this in a systematic way is called an *adaptive quadrature* routine.

The general approach in adaptive quadrature is to use two different quadrature methods for each subinterval, compare the results, and subdivide the interval further if the difference is too great. Of course some care should be exercised, since if two equally bad methods are used, they might agree quite closely on a bad result. One way to avoid this is to ensure that one method always overestimates the result and that the other always underestimates the result. Another way to avoid this is to ensure that one method is more accurate than the other. A method of this type is described next.

There is significant overhead involved in recursively subdividing the interval, so it pays to use a good method for estimating the integrals, as in the following implementation:

```
function adapt(a, b: real): real;
  begin
  if abs(intsimp(a, b, 10)−intsimp(a, b, 5))<tolerance
    then adapt:=intsimp(a, b, 10)
    else  adapt:=adapt(a, (a+b)/2) + adapt((a+b)/2, b);
  end;
```

Both estimates for the integral are derived from Simpson's method, one using twice as many subdivisions as the other. Essentially, this amounts to checking the accuracy of Simpson's method over the interval in question and then subdividing if it is not good enough.

Unlike our other methods, where we decide how much work we want to do and then take whatever accuracy results, in adaptive quadrature we do however much work is necessary to achieve a degree of accuracy that we decide upon ahead of time. This means that *tolerance* must be chosen carefully, so that the routine doesn't loop indefinitely to achieve an impossibly high tolerance. The number of steps required depends very much on the nature of the function being integrated. A function which fluctuates wildly will require a large number of steps, but such a function would lead to a very inaccurate answer for the "fixed interval" methods. A smooth function such as our example can be handled with a reasonable number of steps. The following table gives, for various values of t, the value produced and the number of recursive calls required by the above routine to compute $\int_1^2 dx/x$:

$$
\begin{array}{lll}
0.00001000000 & 0.6931473746651 & 1 \\
0.00000010000 & 0.6931471829695 & 5 \\
0.00000000100 & 0.6931471806413 & 13 \\
0.00000000001 & 0.6931471805623 & 33 \\
\end{array}
$$

The above program can be improved in several ways. First, there's certainly no need to call *intsimp*(a, b, 10) twice. In fact, the function values for this call can be shared by *intsimp*(a, b, 5). Second, the tolerance bound can be related to the accuracy of the answer more closely if the *tolerance* is scaled by the ratio of the size of the current interval to the size of the full interval. Also, a better routine can obviously be developed by using an even better quadrature rule than Simpson's (but it is a basic law of recursion that another *adaptive* routine wouldn't be a good idea). A sophisticated adaptive quadrature routine can provide very accurate results for problems which can't be handled any other way, but careful attention must be paid to the types of functions to be processed.

We will be seeing several algorithms that have the same recursive structure as the adaptive quadrature method given above. The general technique of adapting simple methods to work hard only on difficult parts of complex problems can be a powerful one in algorithm design.

Exercises

1. Write a program to symbolically integrate (and differentiate) polynomials in x and $\ln x$. Use a recursive implementation based on integration by parts.

2. Which quadrature method is likely to produce the best answer for integrating the following functions: $f(x) = 5x$, $f(x) = (3 - x)(4 + x)$, $f(x) = \sin(x)$?

3. Give the result of using each of the four elementary quadrature methods (rectangle, trapezoid, Simpson's, spline) to integrate $y = 1/x$ in the interval $[.1,10]$.

4. Answer the previous question for the function $y = \sin x$.

5. Discuss what happens if adaptive quadrature is used to integrate the function $y = 1/x$ in the interval $[-1,2]$.

6. Answer the previous question for the elementary quadrature methods.

7. Give the points of evaluation when adaptive quadrature is used to integrate the function $y = 1/x$ in the interval $[.1,10]$ with a tolerance of .1.

8. Compare the accuracy of an adaptive quadrature based on Simpson's method to an adaptive quadrature based on the rectangle method for the integral given in the previous problem.

9. Answer the previous question for the function $y = \sin x$.

10. Give a specific example of a function for which adaptive quadrature would be likely to give a drastically more accurate result than the other methods.

SOURCES for Mathematical Algorithms

Much of the material in this section falls within the domain of numerical analysis, and several excellent textbooks are available. One which pays particular attention to computational issues is the 1977 book by Forsythe, Malcomb and Moler. In particular, much of the material given here in Chapters 5, 6, and 7 is based on the presentation given in that book.

The second major reference for this section is the second volume of D. E. Knuth's comprehensive treatment of "The Art of Computer Programming." Knuth uses the term "seminumerical" to describe algorithms which lie at the interface between numerical and symbolic computation, such as random number generation and polynomial arithmetic. Among many other topics, Knuths volume 2 covers in great depth the material given here in Chapters 1, 3, and 4. The 1975 book by Borodin and Munroe is an additional reference for Strassen's matrix multiplication method and related topics. Many of the algorithms that we've considered (and many others, principally symbolic methods as mentioned in Chapter 7) are embodied in a computer system called MACSYMA, which is regularly used for serious mathematical work.

Certainly, a reader seeking more information on mathematical algorithms should expect to find the topics treated at a much more advanced mathematical level in the references than the material we've considered here.

Chapter 2 is concerned with elementary data structures, as well as polynomials. Beyond the references mentioned in the previous part, a reader interested in learning more about this subject might study how elementary data structures are handled in modern programming languages such as Ada, which have facilities for building abstract data structures.

A. Borodin and I. Munroe, *The Computational Complexity of Algebraic and Numerical Problems*, American Elsevier, New York, 1975.

G. E. Forsythe, M. A. Malcomb, and C. B. Moler, *Computer Methods for Mathematical Computations*, Prentice-Hall, Englewood Cliffs, NJ, 1977.

D. E. Knuth, *The Art of Computer Programming. Volume 2: Seminumerical Algorithms*, Addison-Wesley, Reading, MA (second edition), 1981.

MIT Mathlab Group, *MACSYMA Reference Manual*, Laboratory for Computer Science, Massachusetts Institute of Technology, 1977.

P. Wegner, *Programming with ada: an introduction by means of graduated examples*, Prentice-Hall, Englewood Cliffs, NJ, 1980.

SORTING

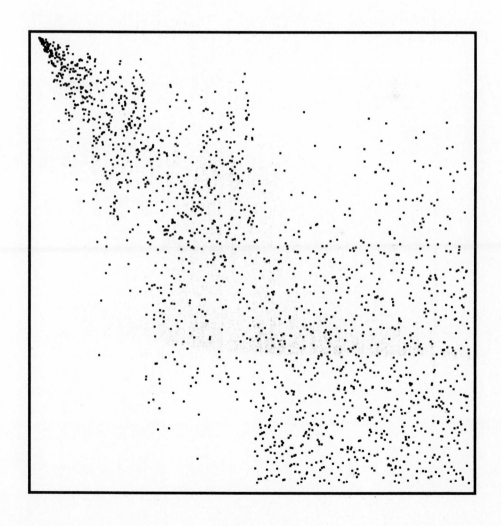

8. Elementary Sorting Methods

As our first excursion into the area of sorting algorithms, we'll study some "elementary" methods which are appropriate for small files or files with some special structure. There are several reasons for studying these simple sorting algorithms in some detail. First, they provide a relatively painless way to learn terminology and basic mechanisms for sorting algorithms so that we get an adequate background for studying the more sophisticated algorithms. Second, there are a great many applications of sorting where it's better to use these simple methods than the more powerful general-purpose methods. Finally, some of the simple methods extend to better general-purpose methods or can be used to improve the efficiency of more powerful methods. The most prominent example of this is seen in recursive sorts which "divide and conquer" big files into many small ones. Obviously, it is advantageous to know the best way to deal with small files in such situations.

As mentioned above, there are several sorting applications in which a relatively simple algorithm may be the method of choice. Sorting programs are often used only once (or only a few times). If the number of items to be sorted is not too large (say, less than five hundred elements), it may well be more efficient just to run a simple method than to implement and debug a complicated method. Elementary methods are always suitable for small files (say, less than fifty elements); it is unlikely that a sophisticated algorithm would be justified for a small file, unless a very large number of such files are to be sorted. Other types of files that are relatively easy to sort are ones that are already almost sorted (or already sorted!) or ones that contain large numbers of equal keys. Simple methods can do much better on such well-structured files than general-purpose methods.

As a rule, the elementary methods that we'll be discussing take about N^2 steps to sort N randomly arranged items. If N is small enough, this may not be a problem, and if the items are not randomly arranged, some of the methods might run much faster than more sophisticated ones. However, it must be emphasized that these methods (with one notable exception) should *not* be used for large, randomly arranged files.

Rules of the Game

Before considering some specific algorithms, it will be useful to discuss some general terminology and basic assumptions for sorting algorithms. We'll be considering methods of sorting *files* of *records* containing *keys*. The keys, which are only part of the records (often a small part), are used to control the sort. The objective of the sorting method is to rearrange the records so that their keys are in order according to some well-defined ordering rule (usually numerical or alphabetical order).

If the file to be sorted will fit into memory (or, in our context, if it will fit into a Pascal **array**), then the sorting method is called *internal*. Sorting files from tape or disk is called *external* sorting. The main difference between the two is that any record can easily be accessed in an internal sort, while an external sort must access records sequentially, or at least in large blocks. We'll look at a few external sorts in Chapter 13, but most of the algorithms that we'll consider are internal sorts.

As usual, the main performance parameter that we'll be interested in is the running time of our sorting algorithms. As mentioned above, the elementary methods that we'll examine in this chapter require time proportional to N^2 to sort N items, while more advanced methods can sort N items in time proportional to $N \log N$. It can be shown that no sorting algorithm can use less than $N \log N$ comparisons between keys, but we'll see that there are methods that use digital properties of keys to get a total running time proportional to N.

The amount of extra memory used by a sorting algorithm is the second important factor we'll be considering. Basically, the methods divide into three types: those that sort in place and use no extra memory except perhaps for a small stack or table; those that use a linked-list representation and so use N extra words of memory for list pointers; and those that need enough extra memory to hold another copy of the array to be sorted.

A characteristic of sorting methods which is sometimes important in practice is *stability*: a sorting method is called *stable* if it preserves the relative order of equal keys in the file. For example, if an alphabetized class list is sorted by grade, then a stable method will produce a list in which students with the same grade are still in alphabetical order, but a non-stable method is likely to produce a list with no evidence of the original alphabetic order. Most of the simple methods are stable, but most of the well-known sophisticated algorithms are not. If stability is vital, it can be forced by appending a small index to each key before sorting or by lengthening the sort key in some other way. It is easy to take stability for granted: people often react to the unpleasant effects of instability with disbelief. Actually there are few methods which achieve stability without using significant extra time or space.

The following program, for sorting three records, is intended to illustrate the general conventions that we'll be using. (In particular, the main program is a peculiar way to exercise a program that is known to work only for $N = 3$: the point is that most of the sorting programs we'll consider could be substituted for *sort3* in this "driver" program.)

```
program threesort(input, output);
const maxN=100;
var a: array[1..maxN] of integer;
    N, i: integer;
procedure sort3;
  var t: integer;
  begin
  if a[1]>a[2] then
    begin t:=a[1]; a[1]:=a[2]; a[2]:=t end;
  if a[1]>a[3] then
    begin t:=a[1]; a[1]:=a[3]; a[3]:=t end;
  if a[2]>a[3] then
    begin t:=a[2]; a[2]:=a[3]; a[3]:=t end;
  end;
begin
readln(N);
for i:=1 to N do read(a[i]);
if N=3 then sort3;
for i:=1 to N do write(a[i]);
writeln
end.
```

The three assignment statements following each **if** actually implement an "exchange" operation. We'll write out the code for such exchanges rather than use a procedure call because they're fundamental to many sorting programs and often fall in the inner loop.

In order to concentrate on algorithmic issues, we'll work with algorithms that simply sort arrays of integers into numerical order. It is generally straightforward to adapt such algorithms for use in a practical application involving large keys or records. Basically, sorting programs access records in one of two ways: either keys are accessed for comparison, or entire records are accessed to be moved. Most of the algorithms that we will study can be recast in terms of performing these two operations on arbitrary records. If the records to be sorted are large, it is normally wise to do an "indirect sort": here the records themselves are not necessarily rearranged, but rather an array of pointers (or indices) is rearranged so that the first pointer points to the smallest record, etc. The keys can be kept either with the records (if they are large) or with the pointers (if they are small).

By using programs which simply operate on a global array, we're ignoring "packaging problems" that can be troublesome in some programming environments. Should the array be passed to the sorting routine as a parameter? Can the same sorting routine be used to sort arrays of integers and arrays of reals (and arrays of arbitrarily complex records)? Even with our simple assumptions, we must (as usual) circumvent the lack of dynamic array sizes in Pascal by predeclaring a maximum. Such concerns will be easier to deal with in programming environments of the future than in those of the past and present. For example, some modern languages have quite well-developed facilities for packaging together programs into large systems. On the other hand, such mechanisms are not truly required for many applications: small programs which work directly on global arrays have many uses; and some operating systems make it quite easy to put together simple programs like the one above, which serve as "filters" between their input and their output. Obviously, these comments apply to many of the other algorithms that we'll be examining, though their effects are perhaps most acutely felt for sorting algorithms.

Some of the programs use a few other global variables. Declarations which are not obvious will be included with the program code. Also, we'll sometimes assume that the array bounds go to 0 or $N+1$, to hold special keys used by some of the algorithms. We'll frequently use letters from the alphabet rather than numbers for examples: these are handled in the obvious way using Pascal's *ord* and *chr* "transfer functions" between integers and characters.

The *sort3* program above uses an even more constrained access to the file: it is three instructions of the form "compare two records and exchange them if necessary to put the one with the smaller key first." Programs which use only this type of instruction are interesting because they are well suited for hardware implementation. We'll study this issue in more detail in Chapter 35.

Selection Sort

One of the simplest sorting algorithms works as follows: first find the smallest element in the array and exchange it with the element in the first position, then find the second smallest element and exchange it with the element in the second position, continuing in this way until the entire array is sorted. This method is called *selection sort* because it works by repeatedly "selecting" the smallest remaining element. The following program sorts $a[1..N]$ into numerical order:

```
procedure selection;
  var i, j, min, t: integer;
  begin
  for i:=1 to N do
    begin
    min:=i;
    for j:=i+1 to N do
      if a[j]<a[min] then min:=j;
    t:=a[min]; a[min]:=a[i]; a[i]:=t
    end;
  end;
```

This is among the simplest of sorting methods, and it will work very well for small files. Its running time is proportional to N^2: the number of comparisons between array elements is about $N^2/2$ since the outer loop (on i) is executed N times and the inner loop (on j) is executed about $N/2$ times on the average. It turns out that the statement $min:=j$ is executed only on the order of $N \log N$ times, so it is not part of the inner loop.

Despite its simplicity, selection sort has a quite important application: it is the method of choice for sorting files with very large records and small keys. If the records are M words long (but the keys are only a few words long), then the exchange takes time proportional to M, so the total running time is proportional to N^2 (for the comparisons) plus NM (for the exchanges). If M is proportional to N then the running time is linear in the amount of data input, which is difficult to beat even with an advanced method. Of course if it is not absolutely required that the records be actually rearranged, then an "indirect sort" can be used to avoid the NM term entirely, so a method which uses less comparisons would be justified. Still selection sort is quite attractive for sorting (say) a thousand 1000-word records on one-word keys.

Insertion Sort

An algorithm almost as simple as selection sort but perhaps more flexible is *insertion sort*. This is the method often used by people to sort bridge hands: consider the elements one at a time, inserting each in its proper place among those already considered (keeping them sorted). The element being considered is inserted merely by moving larger elements one position to the right, then inserting the element into the vacated position. The code for this algorithm is straightforward:

```
procedure insertion;
  var v: integer;
  begin
  for i:=2 to N do
    begin
    v:=a[i]; j:=i;
    while a[j−1]>v do
        begin a[j]:=a[j−1]; j:=j−1 end;
    a[j]:=v
    end;
  end;
```

As is, this code doesn't work, because the **while** will run past the left end of the array if t is the smallest element in the array. One way to fix this is to put a "sentinel" key in a[0], making it at least as small as the smallest element in the array. Using sentinels in situations like this is common in sorting programs to avoid including a test (in this case j>1) which almost always succeeds within the inner loop. If for some reason it is inconvenient to use a sentinel and the array really must have the bounds [1..N], then standard Pascal does not allow a clean alternative, since it does not have a "conditional" **and** instruction: the test **while** (j>1) **and** (a[j−1]>v) won't work because even when j=1, the second part of the **and** will be evaluated and will cause an out-of-bounds array access. A **goto** out of the loop seems to be required. (Some programmers prefer to goto some lengths to avoid **goto** instructions, for example by performing an action within the loop to ensure that the loop terminates. In this case, such a solution seems hardly justified, since it makes the program no clearer, and it adds extra overhead everytime through the loop to guard against a rare event.)

On the average, the inner loop of insertion sort is executed about $N^2/2$ times: The "average" insertion goes about halfway into a subfile of size $N/2$. This is inherent in the method. The point of insertion can be found more efficiently using the searching techniques in Chapter 14, but $N^2/2$ moves (to make room for each element being inserted) are still required; or the number of moves can be lowered by using a linked list instead of an array, but then the methods of Chapter 14 don't apply and $N^2/2$ comparisons are required (to find each insertion point).

Shellsort

Insertion sort is slow because it exchanges only adjacent elements. For example, if the smallest element happens to be at the end of the array, it takes N steps to get it where it belongs. *Shellsort* is a simple extension of insertion sort which gets around this problem by allowing exchanges of elements that are far apart.

If we replace every occurrence of "1" by "h" (and "2" by "$h+1$") in insertion sort, the resulting program rearranges a file to give it the property that taking every hth element (starting anywhere) yields a sorted file. Such a file is said to be *h-sorted*. Put another way, an h-sorted file is h independent sorted files, interleaved together. By h-sorting for some large values of h, we can move elements in the array long distances and thus make it easier to h-sort for smaller values of h. Using such a procedure for any sequence of values of h which ends in 1 will produce a sorted file: this is Shellsort.

The following example shows how a sample file of fifteen elements is sorted using the increments 13, 4, 1:

	1	2	3	4	5	6	7	8	9	10	11	12	13	14	15
	A	S	O	R	T	I	N	G	E	X	A	M	P	L	E
13															
	A	E	O	R	T	I	N	G	E	X	A	M	P	L	S
4															
	A	E	A	G	E	I	N	M	P	L	O	R	T	X	S
1															
	A	A	E	E	G	I	L	M	N	O	P	R	S	T	X

In the first pass, the A in position 1 is compared to the L in position 14, then the S in position 2 is compared (and exchanged) with the E in position 15. In the second pass, the A T E P in positions 1, 5, 9, and 13 are rearranged to put A E P T in those positions, and similarly for positions 2, 6, 10, and 14, etc. The last pass is just insertion sort, but no element has to move very far.

The above description of how Shellsort gains efficiency is necessarily imprecise because no one has been able to analyze the algorithm. Some sequences of values of h work better than others, but no explanation for this has been discovered. A sequence which has been shown empirically to do well is ...,1093,364,121,40,13,4,1, as in the following program:

```
procedure shellsort;
   label 0;
   var i, j, h, v: integer;
   begin
   h:=1; repeat h:=3*h+1 until h>N;
   repeat
      h:=h div 3;
      for i:=h+1 to N do
      begin
         v:=a[i]; j:=i;
         while a[j−h]>v do
            begin
            a[j]:=a[j−h]; j:=j−h;
            if j<=h then goto 0
            end;
         0: a[j]:=v
      end;
   until h=1;
   end;
```

Note that sentinels are not used because there would have to be *h* of them, for the largest value of *h* used.

The increment sequence in this program is easy to use and leads to an efficient sort. There are many other increment sequences which lead to a more efficient sort (the reader might be amused to try to discover one), but it is difficult to beat the above program by more than 20% even for relatively large N. (The possibility that much better increment sequences exist is still, however, quite real.) On the other hand, there are some bad increment sequences. Shellsort is sometimes implemented by starting at $h=N$ (instead of initializing so as to ensure the same sequence is always used as above). This virtually ensures that a bad sequence will turn up for some N.

Comparing Shellsort with other methods analytically is difficult because the functional form of the running time for Shellsort is not even known (and depends on the increment sequence). For the above program, two conjectures are $N(\log N)^2$ and $N^{1.25}$. The running time is not particularly sensitive to the initial ordering of the file, especially in contrast to, say, insertion sort, which is linear for a file already in order but quadratic for a file in reverse order.

Shellsort is the method of choice for many sorting applications because it has acceptable running time even for moderately large files (say, five thousand elements) and requires only a very small amount of code, which is easy to get working. We'll see methods that are more efficient in the next few chapters, but they're perhaps only twice as fast (if that much) except for large N, and they're significantly more complicated. In short, if you have a sorting problem, *use the above program*, then determine whether the extra effort required to replace it with a sophisticated method will be worthwhile. (On the other hand, the Quicksort algorithm of the next chapter is not *that* much more difficult to implement...)

Digression: Bubble Sort

An elementary sorting method that is often taught in introductory classes is *bubble sort*: keep passing through the file, exchanging adjacent elements, if necessary; when no exchanges are required on some pass, the file is sorted. An implementation of this method is given below.

```
procedure bubblesort;
  var j, t: integer;
  begin
  repeat
    t:=a[1];
    for j:=2 to N do
      if a[j-1]>a[j] then
        begin t:=a[j-1]; a[j-1]:=a[j]; a[j]:=t end
    until t=a[1];
  end;
```

It takes a moment's reflection to convince oneself first that this works at all, second that the running time is quadratic. It is not clear why this method is so often taught, since insertion sort seems simpler and more efficient by almost any measure. The inner loop of bubble sort has about twice as many instructions as either insertion sort or selection sort.

Distribution Counting

A very special situation for which there is a simple sorting algorithm is the following: "sort a file of N records whose keys are distinct integers between 1 and N." The algorithm for this problem is

```
for i:=1 to N do t[a[i]]:=a[i];
for i:=1 to N do a[i]:=t[i];
```

This algorithm uses a temporary array t. It is possible (but much more complicated) to solve this problem without an auxiliary array.

A more realistic problem solved by an algorithm in the same spirit is: "sort a file of N records whose keys are integers between 0 and $M-1$." If M is not too large, an algorithm called *distribution counting* can be used to solve this problem. The idea is to count the number of keys with each value, then use the counts to move the records into position on a second pass through the file, as in the following code:

```
for j:=0 to M-1 do count[j]:=0;
for i:=1 to N do
   count[a[i]]:=count[a[i]]+1;
for j:=1 to M-1 do
   count[j]:=count[j-1]+count[j];
for i:=N downto 1 do
   begin
   t[count[a[i]]]:=a[i];
   count[a[i]]:=count[a[i]]-1
   end;
for i:=1 to N do a[i]:=t[i];
```

To see how this code works, consider the following sample file of integers:

1	2	3	4	5	6	7	8	9	10	11	12	13	14	15	16
3	1	1	0	3	7	5	5	2	4	2	1	0	2	6	4

The first **for** loop initializes the counts to 0; the second produces the counts

0	1	2	3	4	5	6	7
2	3	3	2	2	2	1	1

This says that there are two 0's, three 1's, etc. The third **for** loop adds these numbers to produce

0	1	2	3	4	5	6	7
2	5	8	10	12	14	15	16

That is, there are two numbers less than 1, five numbers less than 2, etc.

Now, these can be used as addresses to sort the array:

1	2	3	4	5	6	7	8	9	10	11	12	13	14	15	16
0	0	1	1	1	2	2	2	3	3	4	4	5	5	6	7

For example, when the 4 at the end of the file is encountered, it's put into location 12, since *count*[4] says that there are 12 keys less than or equal to 4. Then *count*[4] is decremented, since there's now one less key less than or equal to 4. The inner loop goes from N down to 1 so that the sort will be stable. (The reader may wish to check this.)

This method will work very well for the type of files postulated. Furthermore, it can be extended to produce a much more powerful method that we'll examine in Chapter 10.

Non-Random Files

We usually think of sorting files that are in some arbitrarily scrambled order. However, it is quite often the case that we have a lot of information about a file to be sorted. For example, one often wants to add a few elements to a sorted file and thus produce a larger sorted file. One way to do so is to simply append the new elements to the end of the file, then call a sorting algorithm. General-purpose sorts are commonly misused for such applications; actually, elementary methods can take advantage of the order present in the file.

For example, consider the operation of insertion sort on a file which is already sorted. Each element is immediately determined to be in its proper in the file, and the total running time is linear. The same is true for bubble sort, but selection sort is still quadratic. (The leading term in the running time of selection sort does not depend on the order in the file to be sorted.)

Even if a file is not completely sorted, insertion sort can be quite useful because the running time of insertion sort depends quite heavily on the order present in the file. The running time depends on the number of *inversions*: for each element count up the number of elements to its left which are greater. This is the distance the elements have to move when inserted into the file during insertion sort. A file which has some order in it will have fewer inversions in it than one which is arbitrarily scrambled.

The example cited above of a file formed by tacking a few new elements onto a large sorted file is clearly a case where the number of the inversions is low: a file which has only a constant number of elements out of place will have only a linear number of inversions. Another example is a file where each element is only a constant distance from its final position. Files like this can be created in the intial stages of some advanced sorting methods: at a certain point it is worthwhile to switch over to insertion sort.

In short, insertion sort is the method of choice for "almost sorted" files with few inversions: for such files, it will outperform even the sophisticated methods in the next few chapters.

Exercises

1. Give a sequence of "compare-exchange" operations for sorting four records.

2. Which of the three elementary methods runs fastest for a file which is already sorted?

3. Which of the three elementary methods runs fastest for a file in reverse order?

4. Test the hypothesis that selection sort is the fastest of the three elementary methods, then insertion sort, then bubble sort.

5. Give a good reason why it might be inconvenient to use a sentinel key for insertion sort (aside from the one that comes up in the implementation of Shellsort).

6. How many comparisons are used by Shellsort to 7-sort, then 3-sort the keys EASYQUESTION?

7. Give an example to show why $8, 4, 2, 1$ would not be a good way to finish off a Shellsort increment sequence.

8. Is selection sort stable? How about insertion sort and bubble sort?

9. Give a specialized version of distribution counting for sorting files where elements have only one of two values (x or y).

10. Experiment with different increment sequences for Shellsort: find one that runs faster than the one given for a random file of 1000 elements.

9. Quicksort

In this chapter, we'll study the sorting algorithm which is probably more widely used than any other, *Quicksort*. The basic algorithm was invented in 1960 by C. A. R. Hoare, and it has been studied by many people since that time. Quicksort is popular because it's not difficult to implement, it's a good "general-purpose" sort (works well in a variety of situations), and it consumes less resources than any other sorting method in many situations.

The desirable features of the Quicksort algorithm are that it is in-place (uses only a small auxiliary stack), requires only about $N \log N$ operations on the average to sort N items, and has an extremely short inner loop. The drawbacks of the algorithm are that it is recursive (implementation is complicated if recursion is not available), has a worst case where it takes about N^2 operations, and is fragile: a simple mistake in the implementation might go unnoticed and could cause it to perform badly for some files.

The performance of Quicksort is very well understood. It has been subjected to a thorough mathematical analysis and very precise statements can be made about performance issues. The analysis has been verified by extensive empirical experience, and the algorithm has been refined to the point where it is the method of choice in a broad variety of practical sorting applications. This makes it worthwhile to look somewhat more carefully at ways of efficiently implementing Quicksort than we have for other algorithms. Similar implementation techniques are appropriate for other algorithms; with Quicksort we can use them with confidence because the performance is so well understood.

It is tempting to try to develop ways to improve Quicksort: a faster sorting algorithm is computer science's "better mousetrap." Almost from the moment Hoare first published the algorithm, "improved" versions have been appearing in the literature. Many ideas have been tried and analyzed, but it is easy to be deceived, because the algorithm is so well balanced that the effects of improvements in one part of the program can be more than offset by the effects of bad performance in another part of the program. We'll examine in some detail three modifications which do improve Quicksort substantially.

A carefully tuned version of Quicksort is likely to run significantly faster than any other sorting method on most computers. However, it must be cautioned that tuning any algorithm can make it more fragile, leading to undesirable and unexpected effects for some inputs. Once a version has been developed which seems free of such effects, this is likely to be the program to use for a library sort utility or for a serious sorting application. But if one is not willing to invest the effort to be sure that a Quicksort implementation is not flawed, Shellsort is a much safer choice and will perform adequately for significantly less implementation effort.

The Basic Algorithm

Quicksort is a "divide-and-conquer" method for sorting. It works by *partitioning* a file into two parts, then sorting the parts independently. As we will see, the exact position of the partition depends on the file, so the algorithm has the following recursive structure:

```
procedure quicksort(l, r: integer);
  var i;
  begin
  if r>l then
    begin
    i:=partition(l, r)
    quicksort(l, i−1);
    quicksort(i+1, r);
    end
  end;
```

The parameters l and r delimit the subfile within the original file that is to be sorted: the call $quicksort(1, N)$ sorts the whole file.

The crux of the method is the *partition* procedure, which must rearrange the array to make the following three conditions hold:

(i) the element $a[i]$ is in its final place in the array for some i,
(ii) all the elements in $a[l],\ldots,a[i-1]$ are less than or equal to $a[i]$,
(iii) all the elements in $a[i+1], \ldots,a[r]$ are greater than or equal to $a[i]$.

This can be simply and easily implemented through the following general strategy. First, arbitrarily choose $a[r]$ to be the element that will go into its final position. Next, scan from the left end of the array until finding an element greater than $a[r]$ and scan from the right end of the array until finding an element less than $a[r]$. The two elements which stopped the scans are obviously out of place in the final partitioned array, so exchange them. (Actually, it turns out, for reasons described below, to be best to also stop the scans for elements equal to $a[r]$, even though this might seem to involve some unnecessary exhanges.) Continuing in this way ensures that all array elements to the left of the left pointer are less than $a[r]$, and array elements to the right of the right pointer are greater than $a[r]$. When the scan pointers cross, the partitioning process is nearly complete: all that remains is to exchange $a[r]$ with the leftmost element of the right subfile.

The following table shows how our sample file of keys is partitioned using this method:

1	2	3	4	5	6	7	8	9	10	11	12	13	14	15
A	S	O	R	T	I	N	G	E	X	A	M	P	L	E
A	A									S	M	P	L	E
A	A	E						O	X	S	M	P	L	E
A	A	E	E	T	I	N	G	O	X	S	M	P	L	R

The rightmost element, E, is chosen as the partitioning element. First the scan from the left stops at the S, then the scan from the right stops at the A, then these two are exchanged, as shown on the second line of the table. Next the scan from the left stops at the O, then the scan from the right stops at the E, then these two are exchanged, as shown on the third line of the table. Next the pointers cross. The scan from the left stops at the R, and the scan from the right stops at the E. The proper move at this point is to exchange the E at the right with the R, leaving the partitioned file shown on the last line of the table. The sort is finished by sorting the two subfiles on either side of the partitioning element (recursively).

The following program gives a full implementation of the method.

```
procedure quicksort(l, r: integer);
  var v, t, i, j: integer;
  begin
  if r>l then
    begin
    v:=a[r]; i:=l−1; j:=r;
    repeat
      repeat i:=i+1 until a[i]>=v;
      repeat j:=j−1 until a[j]<=v;
      t:=a[i]; a[i]:=a[j]; a[j]:=t;
    until j<=i;
    a[j]:=a[i]; a[i]:=a[r]; a[r]:=t;
    quicksort(l, i−1);
    quicksort(i+1, r)
    end
  end;
```

In this implementation, the variable v holds the current value of the "partitioning element" $a[r]$, and i and j are the left and right scan pointers, respectively. An extra exchange of $a[i]$ with $a[j]$ is done with $j<i$ just after the pointers cross but before the crossing is detected and the outer **repeat** loop exited. (This could be avoided with a **goto**.) The three assignment statements following that loop implement the exchanges $a[i]$ with $a[j]$ (to undo the extra exchange) and $a[i]$ with $a[r]$ (to put the partitioning element into position).

As in insertion sort, a sentinel key is needed to stop the scan in the case that the partitioning element is the smallest element in the file. In this implementation, no sentinel is needed to stop the scan when the partitioning element is the largest element in the file, because the partitioning element itself is at the right end of the file to stop the scan. We'll shortly see an easy way to avoid having either sentinel key.

The "inner loop" of Quicksort consists simply of incrementing a pointer and comparing an array element against a fixed value. This is really what makes Quicksort quick: it's hard to imagine a simpler inner loop.

Now the two subfiles are sorted recursively, finishing the sort. The following table traces through these recursive calls. Each line depicts the result of partitioning the displayed subfile, using the boxed partitioning element.

1	2	3	4	5	6	7	8	9	10	11	12	13	14	15
A	S	O	R	T	I	N	G	E	X	A	M	P	L	E
A	A	E	E̲	T	I	N	G	O	X	S	M	P	L	R
A	A	E̲												
A̲	A													
	A̲													
				L	I	N	G	O	P	M	R̲	X	T	S
				L	I	G	M̲	O	P	N				
			G̲	I	L									
				I	L̲									
			I̲											
						N̲	P	O						
							O̲	P						
								P̲						
												S̲	T	X
													T	X̲
												T̲		
A	A	E	E	G	I	L	M	N	O	P	R	S	T	X

Note that every element is (eventually) put into place by being used as a partitioning element.

The most disturbing feature of the program above is that it runs very inefficiently on simple files. For example, if it is called with a file that is already sorted, the partitions will be degenerate, and the program will call itself N times, only knocking off one element for each call. This means not only that the time required will be about $N^2/2$, but also that the space required to handle the recursion will be about N (see below), which is unacceptable. Fortunately, there are relatively easy ways to ensure that this worst case doesn't occur in actual applications of the program.

When equal keys are present in the file, two subtleties become apparent. First, there is the question of whether to have both pointers stop on keys equal to the partitioning element, or to have one pointer stop and the other scan over them, or to have both pointers scan over them. This question has actually been studied in some detail mathematically, with the result that it's best to have both pointers stop. This tends to balance the partitions in the presence of many equal keys. Second, there is the question of properly handling the pointer crossing in the presence of equal keys. Actually, the program above can be slightly improved by terminating the scans when $j<i$, then using $quicksort(l, j)$ for the first recursive call. This is an improvement because when $j=i$ we can put two elements into position with the partitioning, by letting the loop iterate one more time. (This case occurs, for example, if R were E in the example above.) It is probably worth making this change because the program given leaves a record with a key equal to the partitioning key in $a[r]$, which makes the first partition in the call $quicksort(i+1, r)$ degenerate because its rightmost key is its smallest. The implementation of partitioning given above is a bit easier to understand, so we'll leave it as is in the discussions below, with the understanding that this change should be made when large numbers of equal keys are present.

The best thing that could happen would be for each partitioning stage to divide the file exactly in half. This would make the number of comparisons used by Quicksort satisfy the divide-and-conquer recurrence

$$C(N) = 2C(N/2) + N.$$

(The $2C(N/2)$ covers the cost of doing the two subfiles; the N is the cost of examining each element, using one partitioning pointer or the other.) From Chapter 4, we know this recurrence has the solution

$$C(N) \approx N \lg N.$$

Though things don't always go this well, it is true that the partition falls in the middle *on the average*. Taking the precise probability of each partition position into account makes the recurrence more complicated, and more difficult to solve, but the final result is similar. It turns out that

$$C(N) \approx 2N \ln N,$$

which implies that the total running time of Quicksort is proportional to $N \log N$ (on the average).

Thus, the implementation above will perform very well for many applications and is a very reasonable general-purpose sort. However, if the sort is to be used a great many times, or if it is to be used to sort a very large file, then it might be worthwhile to implement several of the improvements discussed below which can ensure that the worst case won't occur, reduce the average running time by 20–30%, and easily eliminate the need for a sentinel key.

Removing Recursion

In Chapter 1 we saw that the recursive call could be removed from Euclid's algorithm to yield a non-recursive program controlled by a simple loop. This can be done for other programs with one recursive call, but the situation is more complicated when two or more recursive calls are involved, as in Quicksort. Before dealing with one recursive call, enough information must be saved to allow processing of later recursive calls.

The Pascal programming environment uses a *pushdown stack* to manage this. Each time a procedure call is made, the values of all the variables are *pushed* onto the stack (saved). Each time a procedure returns, the stack is *popped*: the information that was most recently put on it is removed.

A stack may be represented as a linked list, in which case a *push* is implemented by linking a new node onto the front of the list and a *pop* by removing the first node on the list, or as an array, in which case a pointer is maintained which points to the top of the stack, so that a *push* is implemented by storing the information and incrementing the pointer, and a *pop* by decrementing the pointer and retrieving the information.

There is a companion data structure called a *queue*, where items are returned in the order they were added. In a linked list implementation of a queue new items are added at the end, not the beginning. The array implementation of queues is slightly more complicated. Later in this book we'll see other examples of data structures which support the twin operations of inserting new items and deleting items according to a prescribed rule (most notably in Chapters 11 and 20).

When we use recursive calls, the values of *all* variables are saved on an implicit stack by the programming environment; when we want an improved program, we use an explicit stack and save only necessary information. It is usually possible to determine which variables must be saved by examining the program carefully; another approach is to rework the algorithm based on using an explicit stack rather than explicit recursion.

This second approach is particularly appropriate for Quicksort and many similar algorithms. We think of the stack as containing "work to be done," in the form of subfiles to be sorted. Any time we need a subfile to process, we pop the stack. When we partition, we create two subfiles to be processed, which can be pushed on the stack. This leads to the following non-recursive implementation of Quicksort:

```
procedure quicksort;
   var t, i, l, r: integer;
      stack: array [0..50] of integer; p: integer;
   begin
   l:=1; r:=N; p:=2;
   repeat
      if r>l then
         begin
         i:=partition(l, r);
         if (i−l)>(r−i)
            then begin stack[p]:=l; stack[p+1]:=i−1; l:=i+1 end
            else begin stack[p]:=i+1; stack[p+1]:=r; r:=i−1 end;
         p:=p+2;
         end
      else
         begin p:=p−2; l:=stack[p]; r:=stack[p+1] end;
   until p=0
   end;
```

This program differs from the description above in two important ways. First, rather than simply putting two subfiles on the stack in some arbitrary order, their sizes are checked and the larger of the two is put on the stack first. Second, the smaller of the two subfiles is not put on the stack at all; the values of the parameters are simply reset, just as we did for Euclid's algorithm. This technique, called "end-recursion removal" can be applied to any procedure whose last action is a recursive call. For Quicksort, the combination of end-recursion removal and a policy of processing the smaller of the two subfiles first turns out to ensure that the stack need only contain room for about $\lg N$ entries, since each entry on the stack after the top one must represent a subfile less than half the size of the previous entry.

This is in sharp contrast to the size of the stack in the worst case in the recursive implementation, which could be as large as N (for example, in the case that the file is already sorted). This is a subtle but real difficulty with a recursive implementation of Quicksort: there's always an underlying stack, and a degenerate case on a large file could cause the program to terminate abnormally because of lack of memory. This behavior is obviously undesirable for a library sorting routine. Below we'll see ways to make degenerate cases extremely unlikely, but there's no way to avoid this problem completely in a recursive implementation (even switching the order in which subfiles are processed doesn't help, without end-recursion removal).

Of course the non-recursive method processes the same subfiles as the recursive method for our example; it just does them in a different order, as shown in the following table:

1	2	3	4	5	6	7	8	9	10	11	12	13	14	15
A	S	O	R	T	I	N	G	E	X	A	M	P	L	E
A	A	E	[E]	I	N	G	O	X	S	M	P	L	R	
A	A	[E]												
[A]	A													
	[A]													
				L	I	N	G	O	P	M	[R]	X	T	S
											[S]	T	X	
												T	[X]	
												[T]		
				L	I	G	[M]	O	P	N				
				[G]	I	L								
					I	[L]								
					[I]									
								[N]	P	O				
									[O]	P				
									[P]					
A	A	E	E	G	I	L	M	N	O	P	R	S	T	X

The simple use of an explicit stack above leads to a far more efficient program than the direct recursive implementation, but there is still overhead that could be removed. The problem is that, if both subfiles have only one element, entries with $r=l$ are put on the stack only to be immediately taken off and discarded. It is straightforward to change the program to simply not put any such files on the stack. This change is more important when the next improvement is included, which involves ignoring small subfiles in the same way.

Small Subfiles

The second improvement stems from the observation that a recursive program is guaranteed to call itself for many small subfiles, so it should be changed to use a better method when small subfiles are encountered. One obvious way to do this is to change the test at the beginning of the recursive routine from "**if** *r>l* **then**" to a call on insertion sort (modified to accept parameters defining the subfile to be sorted), that is "**if** *r−l* <= *M* **then** *insertion*(*l, r*)." Here *M* is some parameter whose exact value depends upon the implementation. The value chosen for *M* need not be the best possible: the algorithm works about the same for *M* in the range from about 5 to about 25. The reduction in the running time is on the order of 20% for most applications.

A slightly easier method, which is also slightly more efficient, is to just change the test at the beginning to "**if** *r−l* > *M* **then**": that is, simply ignore small subfiles during partitioning. In the non-recursive implementation, this would be done by not putting any files of less than *M* on the stack. After partitioning, what is left is a file that is almost sorted. As mentioned in the previous chapter, insertion sort is the method of choice for such files. That is, insertion sort will work about as well for such a file as for the collection of little files that it would get if it were being used directly. This method should be used with caution, because the insertion sort is likely always to sort even if the Quicksort has a bug which causes it not to work at all. The excessive cost may be the only sign that something went wrong.

Median-of-Three Partitioning

The third improvement is to use a better partitioning element. There are several possibilities here. The safest thing to do to avoid the worst case would be to use a random element from the array for a partitioning element. Then the worst case will happen with negligibly small probability. This is a simple example of a "probabilistic algorithm," which uses randomness to achieve good performance almost always, regardless of the arrangement of the input. This can be a useful tool in algorithm design, especially if some bias in the input is suspected. However, for Quicksort it is probably overkill to put a full random-number generator in just for this purpose: an arbitrary number will do just as well.

A more useful improvement is to take three elements from the file, then use the median of the three for the partitioning element. If the three elements chosen are from the left, middle, and right of the array, then the use of sentinels can be avoided as follows: sort the three elements (using the three-exchange method in the last chapter), then exchange the one in the middle with *a*[*r−1*], then run the partitioning algorithm on *a*[*l+1*, ..., *r−2*]. This improvement is called the *median-of-three* partitioning method.

The median-of-three method helps Quicksort in three ways. First, it makes the worst case much more unlikely to occur in any actual sort. In order for the sort to take N^2 time, two out of the three elements examined must be among the largest or among the smallest elements in the file, and this must happen consistently through most of the partitions. Second, it eliminates the need for a sentinel key for partitioning, since this function is served by the three elements examined before partitioning. Third, it actually reduces the total running time of the algorithm by about 5%.

The combination of a nonrecursive implementation of the median-of-three method with a cutoff for small subfiles can improve the running time of Quicksort from the naive recursive implementation by 25% to 30%. Further algorithmic improvements are possible (for example the median of five or more elements could be used), but the amount of time saved will be marginal. More significant time savings can be realized (with less effort) by coding the inner loops (or the whole program) in assembly or machine language. Neither path is recommended except for experts with serious sorting applications.

Exercises

1. Implement a recursive Quicksort with a cutoff to insertion sort for subfiles with less than M elements and empirically determine the value of M for which it runs fastest on a random file of 1000 elements.

2. Solve the previous problem for a nonrecursive implementation.

3. Solve the previous problem also incorporating the median-of-three improvement.

4. About how long will Quicksort take to sort a file of N equal elements?

5. What is the maximum number of times that the largest element could be moved during the execution of Quicksort?

6. Show how the file ABABABA is partitioned, using the two methods suggested in the text.

7. How many comparisons does Quicksort use to sort the keys EASYQUE STION?

8. How many "sentinel" keys are needed if insertion sort is called directly from within Quicksort?

9. Would it be reasonable to use a queue instead of a stack for a non-recursive implementation of Quicksort? Why or why not?

10. Use a least squares curvefitter to find values of a and b that give the best formula of the form $aN \ln N + bN$ for describing the total number of instructions executed when Quicksort is run on a random file.

10. Radix Sorting

The "keys" used to define the order of the records in files for many sorting applications can be very complicated. (For example, consider the ordering function used in the telephone book or a library catalogue.) Because of this, it is reasonable to define sorting methods in terms of the basic operations of "comparing" two keys and "exchanging" two records. Most of the methods we have studied can be described in terms of these two fundamental operations. For many applications, however, it is possible to take advantage of the fact that the keys can be thought of as numbers from some restricted range. Sorting methods which take advantage of the digital properties of these numbers are called *radix sorts*. These methods do not just compare keys: they process and compare pieces of keys.

Radix sorting algorithms treat the keys as numbers represented in a base-M number system, for different values of M (the *radix*) and work with individual digits of the numbers. For example, consider an imaginary problem where a clerk must sort a pile of cards with three-digit numbers printed on them. One reasonable way for him to proceed is to make ten piles: one for the numbers less than 100, one for the numbers between 100 and 199, etc., place the cards in the piles, then deal with the piles individually, either by using the same method on the next digit or by using some simpler method if there are only a few cards. This is a simple example of a radix sort with $M = 10$. We'll examine this and some other methods in detail in this chapter. Of course, with most computers it's more convenient to work with $M = 2$ (or some power of 2) rather than $M = 10$.

Anything that's represented inside a digital computer can be treated as a binary number, so many sorting applications can be recast to make feasible the use of radix sorts operating on keys which are binary numbers. Unfortunately, Pascal and many other languages intentionally make it difficult to write a program that depends on the binary representation of numbers.

(The reason is that Pascal is intended to be a language for expressing programs in a machine-independent manner, and different computers may use different representations for the same numbers.) This philosophy eliminates many types of "bit-flicking" techniques in situations better handled by fundamental Pascal constructs such as records and sets, but radix sorting seems to be a casualty of this progressive philosophy. Fortunately, it's not too difficult to use arithmetic operations to simulate the operations needed, and so we'll be able to write (inefficient) Pascal programs to describe the algorithms that can be easily translated to efficient programs in programming languages that support bit operations on binary numbers.

Given a (key represented as a) binary number, the fundamental operation needed for radix sorts is extracting a contiguous set of bits from the number. Suppose we are to process keys which we know to be integers between 0 and 1000. We may assume that these are represented by ten-bit binary numbers. In machine language, bits are extracted from binary numbers by using bitwise "and" operations and shifts. For example, the leading two bits of a ten-bit number are extracted by shifting right eight bit positions, then doing a bitwise "and" with the mask 0000000011. In Pascal, these operations can be simulated with **div** and **mod**. For example, the leading two bits of a ten-bit number x are given by $(x \ \textbf{div} \ 256) \bmod 4$. In general, "shift x right k bit positions" can be simulated by computing $x \ \textbf{div} \ 2^k$, and "zero all but the j rightmost bits of x" can be simulated by computing $x \bmod 2^j$. In our description of the radix sort algorithms, we'll assume the existence of a **function** $bits(x, k, j$: *integer*): *integer* which combines these operations to return the j bits which appear k bits from the right in x by computing $(x \ \textbf{div} \ 2^k) \bmod 2^j$. For example, the rightmost bit of x is returned by the call $bits(x, 0, 1)$. This function can be made efficient by precomputing (or defining as constants) the powers of 2. Note that a program which uses only this function will do radix sorting whatever the representation of the numbers, though we can hope for much improved efficiency if the representation is binary and the compiler is clever enough to notice that the computation can actually be done with machine language "shift" and "and" instructions. Many Pascal implementations have extensions to the language which allow these operations to be specified somewhat more directly.

Armed with this basic tool, we'll consider two different types of radix sorts which differ in the order in which they examine the bits of the keys. We assume that the keys are not short, so that it is worthwhile to go to the effort of extracting their bits. If the keys are short, then the distribution counting method in Chapter 8 can be used. Recall that this method can sort N keys known to be integers between 0 and $M - 1$ in linear time, using one auxiliary table of size M for counts and another of size N for rearranging records. Thus, if we can afford a table of size 2^b, then b-bit keys can easily be sorted in linear time. Radix sorting comes into play if the keys are sufficiently long (say $b = 32$) that this is not possible.

The first basic method for radix sorting that we'll consider examines the bits in the keys from left to right. It is based on the fact that the outcome of "comparisons" between two keys depends only on the value of the bits at the first position at which they differ (reading from left to right). Thus, all keys with leading bit 0 appear before all keys with leading bit 1 in the sorted file; among the keys with leading bit 1, all keys with second bit 0 appear before all keys with second bit 1, and so forth. The left-to-right radix sort, which is called *radix exchange sort*, sorts by systematically dividing up the keys in this way.

The second basic method that we'll consider, called *straight radix sort*, examines the bits in the keys from right to left. It is based on an interesting principle that reduces a sort on b-bit keys to b sorts on 1-bit keys. We'll see how this can be combined with distribution counting to produce a sort that runs in linear time under quite generous assumptions.

The running times of both basic radix sorts for sorting N records with b-bit keys is essentially Nb. On the one hand, one can think of this running time as being essentially the same as $N \log N$, since if the numbers are all different, b must be at least $\log N$. On the other hand, both methods usually use many fewer than Nb operations: the left-to-right method because it can stop once differences between keys have been found; and the right-to-left method, because it can process many bits at once.

Radix Exchange Sort

Suppose we can rearrange the records of a file so that all those whose keys begin with a 0 bit come before all those whose keys begin with a 1 bit. This immediately defines a recursive sorting method: if the two subfiles are sorted independently, then the whole file is sorted. The rearrangement (of the file) is done very much like the partitioning in Quicksort: scan from the left to find a key which starts with a 1 bit, scan from the right to find a key which starts with a 0 bit, exchange, and continue the process until the scanning pointers cross. This leads to a recursive sorting procedure that is very similar to Quicksort:

```
procedure radixexchange(l, r, b: integer);
  var t, i, j: integer;
  begin
  if (r>l) and (b>=0) then
    begin
    i:=l;  j:=r;
    repeat
      while (bits(a[i], b, 1)=0) and (i<j) do i:=i+1;
      while (bits(a[j], b, 1)=1) and (i<j) do j:=j−1;
      t:=a[i];  a[i]:=a[j];  a[j]:=t;
    until j=i;
    if bits(a[r], b, 1)=0 then j:=j+1;
    radixexchange(l, j−1, b−1);
    radixexchange(j, r, b−1);
    end
  end;
```

For simplicity, assume that $a[1..N]$ contains positive integers less than 2^{32} (that is, they could be represented as 31-bit binary numbers). Then the call *radixexchange*$(1, N, 30)$ will sort the array. The variable b keeps track of the bit being examined, ranging from 30 (leftmost) down to 0 (rightmost). (It is normally possible to adapt the implementation of *bits* to the machine representation of negative numbers so that negative numbers are handled in a uniform way in the sort.)

This implementation is obviously quite similar to the recursive implementation of Quicksort in the previous chapter. Essentially, the partitioning in radix exchange sort is like partitioning in Quicksort except that the number 2^b is used instead of some number from the file as the partitioning element. Since 2^b may not be in the file, there can be no guarantee that an element is put into its final place during partitioning. Also, since only one bit is being examined, we can't rely on sentinels to stop the pointer scans; therefore the tests $(i<j)$ are included in the scanning loops. As with Quicksort, an extra exchange is done for the case $j=i$, but it is not necessary to undo this exchange outside the loop because the "exchange" is $a[i]$ with itself. Also as with Quicksort, some care is necessary in this algorithm to ensure that the nothing ever "falls between the cracks" when the recursive calls are made. The partitioning stops with $j=i$ and all elements to the right of $a[i]$ having 1 bits in the bth position and all elements to the left of $a[i]$ having 0 bits in the bth position. The element $a[i]$ itself will have a 1 bit *unless* all keys in the file have a 0 in position b. The implementation above has an extra test just after the partitioning loop to cover this case.

The following table shows how our sample file of keys is partitioned and sorted by this method. This table ~~is~~ can be compared with the table given in Chapter 9 for Quicksort, though the operation of the partitioning method is completely opaque without the binary representation of the keys.

1	2	3	4	5	6	7	8	9	10	11	12	13	14	15
A	S	O	R	T	I	N	G	E	X	A	M	P	L	E
A	E	O	L	M	I	N	G	E	A	X	T	P	R	S
A	E	A	E	G	I	N	M	L	O					
A	A	E	E	G										
A	A													
A	A													
		E	E	G										
		E	E											
					I	N	M	L	O					
						L	M	N	O					
						L	M							
								N	O					
										S	T	P	R	X
										S	R	P	T	
										P	R	S		
											R	S		
A	A	E	E	G	I	L	M	N	O	P	R	S	T	X

The binary representation of the keys used for this example is a simple five-bit code with the ith letter in the alphabet represented by the binary representation of the number i. This is a simplified version of real character codes, which use more bits (seven or eight) and represent more characters (upper/lower case letters, numbers, special symbols). By translating the keys in this table to this five-bit character code, compressing the table so that the subfile partitioning is shown "in parallel" rather than one per line, and then transposing rows and columns, we can see how the leading bits of the keys control partitioning:

```
A 00001    A 0|0001    A 00|001    A 000|01    A 0000|1    A 00001
S 10011    E 0|0101    E 00|101    A 000|01    A 0000|1    A 00001
O 01111    O 0|1111    A 00|001    E 001|01    E 0010|1    E 00101
R 10010    L 0|1100    E 00|101    E 001|01    E 0010|1    E 00101
T 10100    M 0|1101    G 00|111    G 001|11    G 0011|1
I 01001    I 0|1001    I 01|001    I 010|01
N 01110    N 0|1110    N 01|110    N 011|10    L 0110|0    L 01100
G 00111    G 0|0111    M 01|101    M 011|01    M 0110|1    M 01101
E 00101    E 0|0101    L 01|100    L 011|00    N 0111|0    N 01110
X 11000    A 0|0001    O 01|111    O 011|11    O 0111|1    O 01111
A 00001    X 1|1000    S 10|011    S 100|11    P 1000|0
M 01101    T 1|0100    T 10|100    R 100|10    R 1001|0    R 10010
P 10000    P 1|0000    P 10|000    P 100|00    S 1001|1    S 10011
L 01100    R 1|0010    R 10|010    T 101|00
E 00101    S 1|0011    X 11|000
```

One serious potential problem for radix sort not brought out in this example is that degenerate partitions (with all keys having the same value for the bit being used) can happen frequently. For example, this arises commonly in real files when small numbers (with many leading zeros) are being sorted. It also occurs for characters: for example suppose that 32-bit keys are made up from four characters by encoding each in a standard eight-bit code then putting them together. Then degenerate partitions are likely to happen at the beginning of each character position, since, for example, lower case letters all begin with the same bits in most character codes. Many other similar effects are obviously of concern when sorting encoded data.

From the example, it can be seen that once a key is distinguished from all the other keys by its left bits, no further bits are examined. This is a distinct advantage in some situations, a disadvantage in others. When the keys are truly random bits, each key should differ from the others after about $\lg N$ bits, which could be many fewer than the number of bits in the keys. This is because, in a random situation, we expect each partition to divide the subfile in half. For example, sorting a file with 1000 records might involve only examining about ten or eleven bits from each key (even if the keys are, say, 32-bit keys). On the other hand, notice that all the bits of equal keys are examined. Radix sorting simply does not work well on files which contain many equal keys. Radix exchange sort is actually slightly faster than Quicksort if the keys to be sorted are comprised of truly random bits, but Quicksort can adapt better to less random situations.

Straight Radix Sort

An alternative radix sorting method is to examine the bits from right to left. This is the method used by old computer-card-sorting machines: a deck of cards was run through the machine 80 times, once for each column, proceeding from right to left. The following example shows how a right-to-left bit-by-bit radix sort works on our file of sample keys.

A 00001	R 1001|0	T 101|00	X 11|000	P 1|0000	A 00001
S 10011	T 1010|0	X 110|00	P 10|000	A 0|0001	A 00001
O 01111	N 0111|0	P 100|00	A 00|001	A 0|0001	E 00101
R 10010	X 1100|0	L 011|00	I 01|001	R 1|0010	E 00101
T 10100	P 1000|0	A 000|01	A 00|001	S 1|0011	G 00111
I 01001	L 0110|0	I 010|01	R 10|010	T 1|0100	I 01001
N 01110	A 0000|1	E 001|01	S 10|011	E 0|0101	L 01100
G 00111	S 1001|1	A 000|01	T 10|100	E 0|0101	M 01101
E 00101	O 0111|1	M 011|01	L 01|100	G 0|0111	N 01110
X 11000	I 0100|1	E 001|01	E 00|101	X 1|1000	O 01111
A 00001	G 0011|1	R 100|10	M 01|101	I 0|1001	P 10000
M 01101	E 0010|1	N 011|10	E 00|101	L 0|1100	R 10010
P 10000	A 0000|1	S 100|11	N 01|110	M 0|1101	S 10011
L 01100	M 0110|1	O 011|11	O 01|111	N 0|1110	T 10100
E 00101	E 0010|1	G 001|11	G 00|111	O 0|1111	X 11000

The ith column in this table is sorted on the trailing i bits of the keys. The ith column is derived from the $(i-1)$st column by extracting all the keys with a 0 in the ith bit, then all the keys with a 1 in the ith bit.

It's not easy to be convinced that the method works; in fact it doesn't work at all unless the one-bit partitioning process is stable. Once stability has been identified as being important, a trivial proof that the method works can be found: after putting keys with ith bit 0 before those with ith bit 1 (in a stable manner) we know that any two keys appear in proper order (on the basis of the bits so far examined) in the file either because their ith bits are different, in which case partitioning puts them in the proper order, or because their ith bits are the same, in which case they're in proper order because of stability. The requirement of stability means, for example, that the partitioning method used in the radix exchange sort can't be used for this right-to-left sort.

The partitioning is like sorting a file with only two values, and the distribution counting sort that we looked at in Chapter 8 is entirely appropriate for this. If we assume that $M = 2$ in the distribution counting program and replace $a[i]$ by $bits(a[i], k, 1)$, then that program becomes a method for sorting the elements of the array a on the bit k positions from the right and putting the result in a temporary array t. But there's no reason to use $M = 2$; in fact we should make M as large as possible, realizing that we need a table of M counts. This corresponds to using m bits at a time during the sort, with $M = 2^m$. Thus, straight radix sort becomes little more than a generalization of distribution counting sort, as in the following implementation for sorting $a[1..N]$ on the b rightmost bits:

```
procedure straightradix(b: integer);
  var i, j, pass: integer;
  begin
  for pass:=0 to (b div m)−1 do
    begin
    for j:=0 to M−1 do count[j]:=0;
    for i:=1 to N do
      count[bits(a[i], pass*m, m)]:=count[bits(a[i], pass*m, m)]+1;
    for j:=1 to M−1 do
      count[j]:=count[j−1]+count[j];
    for i:=N downto 1 do
      begin
      t[count[bits(a[i], pass*m, m)]]:=a[i];
      count[bits(a[i], pass*m, m)]:=count[bits(a[i], pass*m, m)]−1;
      end;
    for i:=1 to N do a[i]:=t[i];
    end;
  end;
```

For clarity, this procedure uses two calls on *bits* to increment and decrement *count*, when one would suffice. Also, the correspondence $M = 2^m$ has been preserved in the variable names, though some versions of "pascal" can't tell the difference between m and M.

The procedure above works properly only if b is a multiple of m. Normally, this is not a particularly restrictive assumption for radix sort: it simply corresponds to dividing the keys to be sorted into an integral number of equal size pieces. When $m=b$ we have distribution counting sort; when $m=1$ we have *straight radix sort*, the right-to-left bit-by-bit radix sort described in the example above.

The implementation above moves the file from a to t during each distribution counting phase, then back to a in a simple loop. This "array copy" loop could be eliminated if desired by making two copies of the distribution counting code, one to sort from a into t, the other to sort from t into a.

A Linear Sort

The straight radix sort implementation given in the previous section makes b/m passes through the file. By making m large, we get a very efficient sorting method, as long as we have $M = 2^m$ words of memory available. A reasonable choice is to make m about one-fourth the word-size ($b/4$), so that the radix sort is four distribution counting passes. The keys are treated as base-M numbers, and each (base-M) digit of each key is examined, but there are only four digits per key. (This directly corresponds with the architectural organization of many computers: one typical organization is to have 32-bit words, each consisting of four 8-bit bytes. The *bits* procedure then winds up extracting particular bytes from words in this case, which obviously can be done very efficiently on such computers.) Now, each distribution counting pass is linear, and since there are only four of them, the entire sort is linear, certainly the best performance we could hope for in a sort.

In fact, it turns out that we can get by with only two distribution counting passes. (Even a careful reader is likely to have difficulty telling right from left by this time, so some caution is called for in trying to understand this method.) This can be achieved by taking advantage of the fact that the file will be *almost* sorted if only the leading $b/2$ bits of the b-bit keys are used. As with Quicksort, the sort can be completed efficiently by using insertion sort on the whole file afterwards. This method is obviously a trivial modification to the implementation above: to do a right-to-left sort using the leading half of the keys, we simply start the outer loop at $k=b$ **div** ($2*m$) rather than $k=1$. Then a conventional insertion sort can be used on the nearly-ordered file that results. To become convinced that a file sorted on its leading bits is quite well-ordered, the reader should examine the first few columns of the table for radix exchange sort above. For example, insertion sort run on the the file sorted on the first three bits would require only six exchanges.

Using two distribution counting passes (with m about one-fourth the word size), then using insertion sort to finish the job will yield a sorting method that is likely to run faster than any of the others that we've seen for large files whose keys are random bits. Its main disadvantage is that it requires an extra array of the same size as the array being sorted. It is possible to eliminate the extra array using linked-list techniques, but extra space proportional to N (for the links) is still required.

A linear sort is obviously desirable for many applications, but there are reasons why it is not the panacea that it might seem. First, it really does depend on the keys being random bits, randomly ordered. If this condition is not satisfied, severely degraded performance is likely. Second, it requires extra space proportional the size of the array being sorted. Third, the "inner loop" of the program actually contains quite a few instructions, so even though it's linear, it won't be as much faster than Quicksort (say) as one might expect, except for quite large files (at which point the extra array becomes a real liability). The choice between Quicksort and radix sort is a difficult one that is likely to depend not only on features of the application such as key, record, and file size, but also on features of the programming and machine environment that relate to the efficiency of access and use of individual bits. Again, such tradeoffs need to be studied by an expert and this type of study is likely to be worthwhile only for serious sorting applications.

Exercises

1. Compare the number of exchanges used by radix exchange sort with the number of exchanges used by Quicksort for the file 001,011,101,110, 000,001,010,111,110,010.

2. Why is it not as important to remove the recursion from the radix exchange sort as it was for Quicksort?

3. Modify radix exchange sort to skip leading bits which are identical on all keys. In what situations would this be worthwhile?

4. True or false: the running time of straight radix sort does not depend on the order of the keys in the input file. Explain your answer.

5. Which method is likely to be faster for a file of all equal keys: radix exchange sort or straight radix sort?

6. True or false: both radix exchange sort and straight radix sort examine all the bits of all the keys in the file. Explain your answer.

7. Aside from the extra memory requirement, what is the major disadvantage to the strategy of doing straight radix sorting on the leading bits of the keys, then cleaning up with insertion sort afterwards?

8. Exactly how much memory is required to do a 4-pass straight radix sort of N b-bit keys?

9. What type of input file will make radix exchange sort run the most slowly (for very large N)?

10. Empirically compare straight radix sort with radix exchange sort for a random file of 1000 32-bit keys.

11. Priority Queues

In many applications, records with keys must be processed in order, but not necessarily in full sorted order and not necessarily all at once. Often a set of records must be collected, then the largest processed, then perhaps more records collected, then the next largest processed, and so forth. An appropriate data structure in such an environment is one which supports the operations of inserting a new element and deleting the largest element. This can be contrasted with queues (delete the oldest) and stacks (delete the newest). Such a data structure is called a *priority queue*. In fact, the priority queue might be thought of as a generalization of the stack and the queue (and other simple data structures), since these data structures can be implemented with priority queues, using appropriate priority assignments.

Applications of priority queues include simulation systems (where the keys might correspond to "event times" which must be processed in order), job scheduling in computer systems (where the keys might correspond to "priorities" which indicate which users should be processed first), and numerical computations (where the keys might be computational errors, so the largest can be worked on first).

Later on in this book, we'll see how to use priority queues as basic building blocks for more advanced algorithms. In Chapter 22, we'll develop a file compression algorithm using routines from this chapter, and in Chapters 31 and 33, we'll see how priority queues can serve as the basis for several fundamental graph searching algorithms. These are but a few examples of the important role served by the priority queue as a basic tool in algorithm design.

It is useful to be somewhat more precise about how a priority queue will be manipulated, since there are several operations we may need to perform on priority queues in order to maintain them and use them effectively for applications such as those mentioned above. Indeed, the main reason that priority queues are so useful is their flexibility in allowing a variety of different operations to be efficiently performed on set of records with keys. We want to build and maintain a data structure containing records with numerical keys (*priorities*), supporting some of the following operations:

> *Construct* a priority queue from N given items.
>
> *Insert* a new item.
>
> *Remove* the largest item.
>
> *Replace* the largest item with a new item (unless the new item is larger).
>
> *Change* the priority of an item.
>
> *Delete* an arbitrary specified item.
>
> *Join* two priority queues into one large one.

(If records can have duplicate keys, we take "largest" to mean "any record with the largest key value.")

The *replace* operation is almost equivalent to an *insert* followed by a *remove* (the difference being that the *insert/remove* requires the priority queue to grow temporarily by one element). Note that this is quite different from doing a *remove* followed by an *insert*. This is included as a separate capability because, as we will see, some implementations of priority queues can do the *replace* operation quite efficiently. Similarly, the *change* operation could be implemented as a *delete* followed by an *insert* and the *construct* could be implemented with repeated uses of the *insert* operation, but these operations can be directly implemented more efficiently for some choices of data structure. The *join* operation requires quite advanced data structures for efficient implementation; we'll concentrate instead on a "classical" data structure, called a *heap*, which allows efficient implementations of the first five operations.

The priority queue as described above is an excellent example of an *abstract data structure*: it is very well defined in terms of the operations performed on it, independent of the way the data is organized and processed in any particular implementation. The basic premise of an abstract data structure is that nothing outside of the definitions of the data structure and the algorithms operating on it should refer to anything inside, except through function and procedure calls for the fundamental operations. The main motivation for the development of abstract data structures has been as a mechanism for organizing large programs. They provide a way to limit the size and complexity of the interface between (potentially complicated) algorithms and associated data structures and (a potentially large number of) programs which use the algorithms and data structures. This makes it easier to understand the large program, and makes it more convenient to change or improve the fundamental algorithms. For example, in the present context, there are several methods for implementing the various operations listed above that can have quite different performance characteristics. Defining priority queues in terms of operations on an abstract data structure provides the flexibility necessary to allow experimentation with various alternatives.

Different implementations of priority queues involve different performance characteristics for the various operations to be performed, leading to cost tradeoffs. Indeed, performance differences are really the only differences allowed by the abstract data structure concept. First, we'll illustrate this point by examining a few elementary data structures for implementing priority queues. Next, we'll examine a more advanced data structure, and then show how the various operations can be implemented efficiently using this data structure. Also, we'll examine an important sorting algorithm that follows naturally from these implementations.

Elementary Implementations

One way to organize a priority queue is as an *unordered list*, simply keeping the items in an array $a[1..N]$ without paying attention to the keys. Thus *construct* is a "no-op" for this organization. To *insert* simply increment N and put the new item into $a[N]$, a constant-time operation. But *replace* requires scanning through the array to find the element with the largest key, which takes linear time (all the elements in the array must be examined). Then *remove* can be implemented by exchanging $a[N]$ with the element with the largest key and decrementing N.

Another organization is to use a *sorted list*, again using an array $a[1..N]$ but keeping the items in increasing order of their keys. Now *remove* simply involves returning $a[N]$ and decrementing N (constant time), but *insert* involves moving larger elements in the array right one position, which could take linear time.

Linked lists could also be used for the unordered list or the sorted list. This wouldn't change the fundamental performance characteristics for *insert*, *remove*, or *replace*, but it would make it possible to do *delete* and *join* in constant time.

Any priority queue algorithm can be turned into a sorting algorithm by successively using *insert* to build a priority queue containing all the items to be sorted, then successively using *remove* to empty the priority queue, receiving the items in reverse order. Using a priority queue represented as an unordered list in this way corresponds to selection sort; using the sorted list corresponds to insertion sort.

As usual, it is wise to keep these simple implementations in mind because they can outperform more complicated methods in many practical situations. For example, the first method might be appropriate in an application where only a few "remove largest" operations are performed as opposed to a large number of insertions, while the second method would be appropriate if the items inserted always tended to be close to the largest element in the priority queue. Implementations of methods similar to these for the searching problem (find a record with a given key) are given in Chapter 14.

Heap Data Structure

The data structure that we'll use to support the priority queue operations involves storing the records in an array in such a way that each key is guaranteed to be larger than the keys at two other specific positions. In turn, each of those keys must be larger than two more keys, and so forth. This ordering is very easy to see if we draw the array in a two-dimensional "tree" structure with lines down from each key to the two keys known to be smaller.

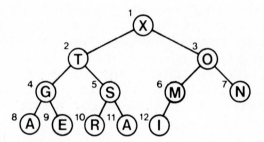

This structure is called a "complete binary tree": place one node (called the *root*), then, proceeding down the page and from left to right, connect two nodes beneath each node on the previous level until *N* nodes have been placed. The nodes below each node are called its *sons*; the node above each node is called its *father*. (We'll see other kinds of "binary trees" and "trees" in Chapter 14 and later chapters of this book.) Now, we want the keys in the tree to satisfy the *heap condition*: the key in each node should be larger than (or equal to) the keys in its sons (if it has any). Note that this implies in particular that the largest key is in the root.

We can represent complete binary trees sequentially within an array by simply putting the root at position 1, its sons at positions 2 and 3, the nodes at the next level in positions 4, 5, 6 and 7, etc., as numbered in the diagram above. For example, the array representation for the tree above is the following:

1	2	3	4	5	6	7	8	9	10	11	12
X	T	O	G	S	M	N	A	E	R	A	I

This natural representation is useful because it is very easy to get from a node to its father and sons. The father of the node in position j is in position j **div** 2, and, conversely, the two sons of the node in position j are in position $2j$ and $2j + 1$. This makes traversal of such a tree even easier than if the tree were implemented with a standard linked representation (with each element containing a pointer to its father and sons). The rigid structure of complete binary trees represented as arrays does limit their utility as data structures, but there is just enough flexibility to allow the implementation of efficient priority queue algorithms. A *heap* is a complete binary tree, represented as an array, in which every node satisfies the heap condition. In particular, the largest key is always in the first position in the array.

All of the algorithms operate along some *path* from the root to the bottom of the heap (just moving from father to son or from son to father). It is easy to see that, in a heap of N nodes, all paths have about $\lg N$ nodes on them. (There are about $N/2$ nodes on the bottom, $N/4$ nodes with sons on the bottom, $N/8$ nodes with grandsons on the bottom, etc. Each "generation" has about half as many nodes as the next, which implies that there can be at most $\lg N$ generations.) Thus all of the priority queue operations (except *join*) can be done in logarithmic time using heaps.

Algorithms on Heaps

The priority queue algorithms on heaps all work by first making a simple structural modification which could violate the heap condition, then traveling through the heap modifying it to ensure that the heap condition is satisfied everywhere. Some of the algorithms travel through the heap from bottom to top, others from top to bottom. In all of the algorithms, we'll assume that the records are one-word integer keys stored in an array a of some maximum size, with the current size of the heap kept in an integer N. Note that N is as much a part of the definition of the heap as the keys and records themselves.

To be able to build a heap, it is necessary first to implement the *insert* operation. Since this operation will increase the size of the heap by one, N must be incremented. Then the record to be inserted is put into $a[N]$, but this may violate the heap property. If the heap property is violated (the new node is greater than its father), then the violation can be fixed by exchanging the new node with its father. This may, in turn, cause a violation, and thus can be fixed in the same way. For example, if P is to be inserted in the heap above, it is first stored in $a[N]$ as the right son of M. Then, since it is greater than M, it is exchanged with M, and since it is greater than O, it is exchanged with O, and the process terminates since it is less that X. The following heap results:

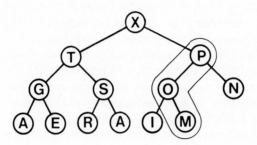

The code for this method is straightforward. In the following implementation, *insert* adds a new item to $a[N]$, then calls *upheap*(N) to fix the heap condition violation at N:

```
procedure upheap(k: integer);
   var v: integer;
   begin
   v:=a[k];  a[0]:=maxint;
   while a[k div 2]<=v do
      begin a[k]:=a[k div 2];  k:=k div 2 end;
   a[k]:=v;
   end;
procedure insert(v: integer);
   begin
   N:=N+1;  a[N]:=v;
   upheap(N)
   end;
```

As with insertion sort, it is not necessary to do a full exchange within the loop, because v is always involved in the exchanges. A sentinel key must be put in $a[0]$ to stop the loop for the case that v is greater than all the keys in the heap.

The *replace* operation involves replacing the key at the root with a new key, then moving down the heap from top to bottom to restore the heap condition. For example, if the X in the heap above is to be replaced with C, the first step is to store C at the root. This violates the heap condition, but the violation can be fixed by exchanging C with T, the larger of the two sons of the root. This creates a violation at the next level, which can be fixed again by exchanging C with the larger of its two sons (in this case S). The process continues until the heap condition is no longer violated at the node occupied by C. In the example, C makes it all the way to the bottom of the heap, leaving:

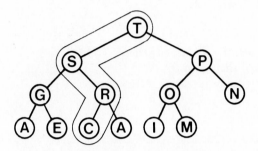

The "*remove* the largest" operation involves almost the same process. Since the heap will be one element smaller after the operation, it is necessary to decrement N, leaving no place for the element that was stored in the last position. But the largest element is to be removed, so the *remove* operation amounts to a *replace*, using the element that was in $a[N]$. For example, the following heap results from removing the T from the heap above:

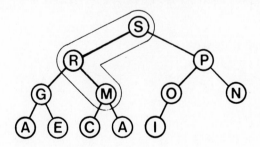

The implementation of these procedures is centered around the operation of fixing up a heap which satisfies the heap condition everywhere except possibly at the root. The same operation can be used to fix up the heap after the value in any position is lowered. It may be implemented as follows:

```
procedure downheap(k: integer);
  label 0;
  var i, j, v: integer;
  begin
  v:=a[k];
  while k<= N div 2 do
    begin
    j:=k+k;
    if j<N then if a[j]<a[j+1] then j:=j+1;
    if v>=a[j] then goto 0;
    a[k]:=a[j]; k:=j;
    end;
0:a[k]:=v
  end;
```

This procedure moves down the heap (starting from position k), exchanging the node at position j with the larger of its two sons if necessary, stopping when j is larger than both sons or the bottom is reached. As above, a full exchange is not needed because v is always involved in the exchanges. The inner loop in this program is an example of a loop which really has two distinct exits: one for the case that the bottom of the heap is hit (as in the first example above), and another for the case that the heap condition is satisfied somewhere in the interior of the heap.

Now the implementation of the *remove* operation is simple:

```
function remove: integer;
  begin
  remove:=a[1];
  a[1]:=a[N]; N:=N−1;
  downheap(1);
  end;
```

The return value is set from $a[1]$, then the element from $a[N]$ is put into $a[1]$ and the size of the heap decremented, leaving only a call to *downheap* to fix up the heap condition everywhere.

The implementation of the *replace* operation is only slightly more complicated:

```
function replace(v: integer):integer;
begin
a[0]:=v;
downheap(0);
replace:=a[0];
end;
```

This code uses $a[0]$ in an artificial way: its sons are 0 (itself) and 1, so if v is larger than the largest element in the heap, the heap is not touched; otherwise v is put into the heap and $a[1]$ returned.

The *delete* operation for an arbitrary element from the heap and the *change* operation can also be implemented by using a simple combination of the methods above. For example, if the priority of the element at position k is raised, then $upheap(k)$ can be called, and if it is lowered then $downheap(k)$ does the job. On the other hand, the *join* operation is far more difficult and seems to require a much more sophisticated data structure.

All of the basic operations *insert, remove, replace,* (*downheap* and *upheap*), *delete,* and *change* involve moving along a path between the root and the bottom of the heap, which includes no more than about $\log N$ elements for a heap of size N. Thus the running times of the above programs are logarithmic.

Heapsort

An elegant and efficient sorting method can be defined from the basic operations on heaps outlined above. This method, called *Heapsort,* uses no extra memory and is guaranteed to sort M elements in about $M \log M$ steps no matter what the input. Unfortunately, its inner loop is quite a bit longer than the inner loop of Quicksort, and it is about twice as slow as Quicksort on the average.

The idea is simply to build a heap containing the elements to be sorted and then to remove them all in order. In this section, N will continue to be the size of the heap, so we will use M for the number of elements to be sorted. One way to sort is to implement the *construct* operation by doing M *insert* operations, as in the first two lines of the following code, then do M *remove* operations, putting the element removed into the place just vacated by the shrinking heap:

```
N:=0;
for k:=1 to M do insert(a[k]);
for k:=M downto 1 do a[k]:=remove;
```

This code breaks all the rules of abstract data structures by assuming a particular representation for the priority queue (during each loop, the priority queue resides in $a[1], \ldots, a[k-1]$), but it is reasonable to do this here because we are implementing a sort, not a priority queue. The priority queue procedures are being used only for descriptive purposes: in an actual implementation of the sort, we would simply use the code from the procedures to avoid doing so many unnecessary procedure calls.

It is actually a little better to build the heap by going backwards through it, making little heaps from the bottom up. Note that every position in the array is the root of a small heap, and *downheap* will work equally well for such small heaps as for the big heap. Also, we've noted that *remove* can be implemented by exchanging the first and last elements, decrementing *N*, and calling *downheap(1)*. This leads to the following implementation of Heapsort:

```
procedure heapsort;
  var k, t: integer;
  begin
  N:=M;
  for k:=M div 2 downto 1 do downheap(k);
  repeat
    t:=a[1]; a[1]:=a[N]; a[N]:=t;
    N:=N-1; downheap(1)
  until N<=1;
  end;
```

The first two lines of this code constitute an implementation of *construct(M: integer)* to build a heap of *M* elements. (The keys in $a[(M \text{ div } 2)+1..M]$ each form heaps of one element, so they trivially satisfy the heap condition and don't need to be checked.) It is interesting to note that, though the loops in this program seem to do very different things, they can be built around the same fundamental procedure.

The following table shows the contents of each heap operated on by *downheap* for our sorting example, just after *downheap* has made the heap condition hold everywhere.

| | | j | | | | | | | | | | | | |
1	2	3	4	5	6	7	8	9	10	11	12	13	14	15
A	S	O	R	T	I	N	G	E	X	A	M	P	L	E
						N							L	E
				P							M	I		
			X						T	A				
			R				G	E						
		P			O	N					M	I	L	E
	X		R	T			G	E	S	A				
X	T	P	R	S	O	N	G	E	A	A	M	I	L	E
T	S	P	R	E	O	N	G	E	A	A	M	I	L	
S	R	P	L	E	O	N	G	E	A	A	M	I		
R	L	P	I	E	O	N	G	E	A	A	M			
P	L	O	I	E	M	N	G	E	A	A				
O	L	N	I	E	M	A	G	E	A					
N	L	M	I	E	A	A	G	E						
M	L	E	I	E	A	A	G							
L	I	E	G	E	A	A								
I	G	E	A	E	A									
G	E	E	A	A										
E	A	E	A											
E	A	A												
A	A													
A														
A	A	E	E	G	I	L	M	N	O	P	R	S	T	X

As mentioned above, the primary reason that Heapsort is of practical interest is that the number of steps required to sort M elements is *guaranteed* to be proportional to $M \log M$, no matter what the input. Unlike the other methods that we've seen, there is no "worst-case" input that will make Heapsort run slower. The proof of this is simple: we make about $3M/2$ calls to *downheap* (about $M/2$ to construct the heap and M for the sort), each of which examines less than $\log M$ heap elements, since the heap never has more than M elements.

Actually, the above proof uses an overestimate. In fact, it can be proven that the construction process takes *linear* time since so many small heaps are processed. This is not of particular importance to Heapsort, since this time is still dominated by the $M \log M$ time for sorting, but it is important for other priority queue applications, where a linear time *construct* can lead to a linear time algorithm. Note that constructing a heap with M successive *inserts* requires $M \log M$ steps in the worst case (though it turns out to be linear on the average).

Indirect Heaps

For many applications of priority queues, we don't want the records moved around at all. Instead, we want the priority queue routine to tell us *which* of the records is the largest, etc., instead of returning values. This is akin to the "indirect sort" or the "pointer sort" concept described at the beginning of Chapter 8. Modification of the above programs to work in this way is straightforward, though sometimes confusing. It will be worthwhile to examine this in more detail here because it is so convenient to use heaps in this way.

Specifically, instead of rearranging the keys in the array *a* the priority queue routines will work with an array *heap* of indices into the array *a*, such that *a*[*heap*[*k*]] is the key of the *k*th element of the heap, for *k* between 1 and *N*. Moreover, we want to maintain another array *inv* which keeps the heap position of the *k*th array element. Thus the *inv* entry for the largest element in the array is *1*, etc. For example, if we wished to change the value of *a*[*k*] we could find its heap position in *inv*[*k*], for use by *upheap* or *downheap*. The following table gives the values in these arrays for our sample heap:

k:	1	2	3	4	5	6	7	8	9	10	11	12	13	14	15
a[k]:	A	S	O	R	T	I	N	G	E	X	A	M	P	L	E
heap[k]:	10	5	13	4	2	3	7	8	9	1	11	12	6	14	15
a[heap[k]]:	X	T	P	R	S	O	N	G	E	A	A	M	I	L	E
a[k]:	A	S	O	R	T	I	N	G	E	X	A	M	P	L	E
inv[k]:	10	5	6	4	2	13	7	8	9	1	11	12	3	14	15

Note that *heap*[*inv*[*k*]]=*inv*[*heap*[*k*]]=*k* for all *k* from 1 to *N*.

We start with *heap*[*k*]=*inv*[*k*]=*k* for *k* from 1 to *N*, which indicates that no rearrangement has been done. The code for heap construction looks much the same as before:

```
procedure pqconstruct;
  var k: integer;
  begin
  N:=M;
  for k:=1 to N do
    begin heap[k]:=k; inv[k]:=k end;
  for k:=M div 2 downto 1 do pqdownheap(k);
  end;
```

We'll prefix implementations of priority queue routines based on indirect heaps with "*pq*" for indentification when they are used in later chapters.

Now, to modify *downheap* to work indirectly, we need only examine the places where it references a. Where it did a *comparison* before, it must now access a indirectly through *heap*. Where it did a *move* before, it must now make the move in *heap*, not a, and it must modify *inv* accordingly. This leads to the following implementation:

```
procedure pqdownheap(k: integer);
  label 0;
  var j, v: integer;
  begin
  v:=heap[k];
  while k<= N div 2 do
    begin
    j:=k+k;
    if j<N then if a[heap[j]]<a[heap[j+1]] then j:=j+1;
    if a[v]>=a[heap[j]] then goto 0;
    heap[k]:=heap[j]; inv[heap[j]]:=k; k:=j;
    end;
0:heap[k]:=v; inv[v]:=k
  end;
```

The other procedures given above can be modified in a similar fashion to implement "*pqinsert*," "*pqchange*," etc.

A similar indirect implementation can be developed based on maintaining *heap* as an array of pointers to separately allocated records. In this case, a little more work is required to implement the function of *inv* (find the heap position, given the record).

Advanced Implementations

If the *join* operation must be done efficiently, then the implementations that we have done so far are insufficient and more advanced techniques are needed. Although we don't have space here to go into the details of such methods, we can discuss some of the considerations that go into their design.

By "efficiently," we mean that a *join* should be done in about the same time as the other operations. This immediately rules out the linkless representation for heaps that we have been using, since two large heaps can be joined only by moving all the elements in at least one of them to a large array. It is easy to translate the algorithms we have been examining to use linked representations; in fact, sometimes there are other reasons for doing so (for example, it might be inconvenient to have a large contiguous array). In a direct linked representation, links would have to be kept in each node pointing to the father and both sons.

It turns out that the heap condition itself seems to be too strong to allow efficient implementation of the *join* operation. The advanced data structures designed to solve this problem all weaken either the heap or the balance condition in order to gain the flexibility needed for the *join*. These structures allow all the operations be completed in logarithmic time.

Exercises

1. Draw the heap that results when the following operations are performed on an intitially empty heap: *insert(10)*, *insert(5)*, *insert(2)*, *replace(4)*, *insert(6)*, *insert(8)*, *remove*, *insert(7)*, *insert(3)*.

2. Is a file in reverse sorted order a heap?

3. Give the heap constructed by successive application of *insert* on the keys EASYQUESTION.

4. Which positions could be occupied by the 3rd largest key in a heap of size 32? Which positions could not be occupied by the 3rd smallest key in a heap of size 32?

5. Why not use a sentinel to avoid the $j<N$ test in *downheap*?

6. Show how to obtain the functions of stacks and normal queues as special cases of priority queues.

7. What is the minimum number of keys that must be moved during a *remove* the largest operation in a heap? Draw a heap of size 15 for which the minimum is achieved.

8. Write a program to delete the element at postion d in a heap.

9. Empirically compare the two methods of heap construction described in the text, by building heaps with 1000 random keys.

10. Give the contents of *inv* after *pqconstruct* is used on the keys E A S Y Q U E S T I O N.

12. Selection and Merging

Sorting programs are often used for applications in which a full sort is not necessary. Two important operations which are similar to sorting but can be done much more efficiently are *selection*, finding the kth smallest element (or finding the k smallest elements) in a file, and *merging*, combining two sorted files to make one larger sorted file. Selection and merging are intimately related to sorting, as we'll see, and they have wide applicability in their own right.

An example of selection is the process of finding the median of a set of numbers, say student test scores. An example of a situation where merging might be useful is to find such a statistic for a large class where the scores are divided up into a number of individually sorted sections.

Selection and merging are complementary operations in the sense that selection splits a file into two independent files and merging joins two independent files to make one file. The relationship between these operations also becomes evident if one tries to apply the "divide-and-conquer" paradigm to create a sorting method. The file can either be rearranged so that when two parts are sorted the whole file is sorted, or broken into two parts to be sorted and then combined to make the sorted whole file. We've already seen what happens in the first instance: that's Quicksort, which consists basically of a selection procedure followed by two recursive calls. Below, we'll look at mergesort, Quicksort's complement in that it consists basically of two recursive calls followed by a merging procedure.

Both selection and merging are easier than sorting in the sense that their running time is essentially linear: the programs take time proportional to N when operating on N items. But available methods are not perfect in either case: the only known ways to merge in place (without using extra space) are too complex to be reduced to practical programs, as are the only known selection methods which are guaranteed to be linear even in the worst case.

Selection

Selection has many applications in the processing of experimental and other data. The most prominent use is the special case mentioned above of finding the *median* element of a file: that item which is greater than half the items in the file and smaller than half the items in the file. The use of the median and other *order statistics* to divide a file up into smaller percentile groups is very common. Often only a small part of a large file is to be saved for further processing; in such cases, a program which can select, say, the top ten percent of the elements of the file might be more appropriate than a full sort.

An algorithm for selection must find the kth smallest item out of a file of N items. Since an algorithm cannot guarantee that a particular item is the kth smallest without having examined and identified the $k-1$ items which are smaller and the $N - k$ elements which are larger, most selection algorithms can return all of the k smallest elements of a file without a great deal of extra calculation.

We've already seen two algorithms which are suitable for direct adaptation to selection methods. If k is very small, then *selection sort* will work very well, requiring time proportional to Nk: first find the smallest element, then find the second smallest by finding the smallest among the remaining items, etc. For slightly larger k, priority queues provide a selection mechanism: first *insert* k items, then *replace* the largest $N-k$ times using the remaining items, leaving the k smallest items in the priority queue. If heaps are used to implement the priority queue, everything can be done in place, with an approximate running time proportional to $N \log k$.

An interesting method which will run much faster on the average can be formulated from the partitioning procedure used in Quicksort. Recall that Quicksort's partitioning method rearranges an array $a[1..N]$ and returns an integer i such that $a[1],\ldots,a[i-1]$ are less than or equal to $a[i]$ and $a[i+1],\ldots,$ $a[N]$ are greater than or equal to $a[i]$. If we're looking for the kth smallest element in the file, and we're fortunate enough to have $k=i$, then we're done. Otherwise, if $k<i$ then we need to look for the kth smallest element in the left subfile, and if $k>i$ then we need to look for the $(k-i)$th smallest element in the right subfile. This leads immediately to the recursive formulation:

```
procedure select(l, r, k: integer);
  var i;
  begin
  if r>l then
    begin
    i:=partition(l, r);
    if i>l+k−1 then select(l, i−1, k);
    if i<l+k−1 then select(i+1, r, k−i);
    end
  end;
```

This procedure rearranges the array so that $a[l], \ldots, a[k−1]$ are less than or equal to $a[k]$ and $a[k+1], \ldots, a[r]$ are greater than or equal to $a[k]$. For example, the call $select(1, N, (N+1) \textbf{ div } 2)$ partitions the array on its median value. For the keys in our sorting example, this program uses only three recursive calls to find the median, as shown in the following table:

1	2	3	4	5	6	7	8	9	10	11	12	13	14	15
A	S	O	R	T	I	N	G	E	X	A	M	P	L	E
A	A	E	[E]	T	I	N	G	O	X	S	M	P	L	R
			L	I	N	G	O	P	M	[R]	X	T	S	
			L	I	G	[M]	O	P	N					
						[M]								

The file is rearranged so that the median is in place with all smaller elements to the left and all larger elements to the right (and equal elements on either side), but it is not fully sorted.

Since the *select* procedure always ends with only one call on itself, it is not really recursive in the sense that no stack is needed to remove the recursion: when the time comes for the recursive call, we can simply reset the parameters and go back to the beginning, since there is nothing more to do.

```
procedure select(k: integer);
  var v, t, i, j, l, r: integer;
  begin
  l:=1;  r:=N;
  while r>l do
    begin
    v:=a[r];  i:=l−1;  j:=r;
    repeat
      repeat i:=i+1 until a[i]>=v;
      repeat j:=j−1 until a[j]<=v;
      t:=a[i];  a[i]:=a[j];  a[j]:=t;
    until j<=i;
    a[j]:=a[i];  a[i]:=a[r];  a[r]:=t;
    if i>=k then r:=i−1;
    if i<=k then l:=i+1;
    end;
  end;
```

We use the identical partitioning procedure to Quicksort: as with Quicksort, this could be changed slightly if many equal keys are expected. Note that in this non-recursive program, we've eliminated the simple calculations involving k.

This method has about the same worst case as Quicksort: using it to find the smallest element in an already sorted file would result in a quadratic running time. It is probably worthwhile to use an arbitrary or a random partitioning element (but not the median-of-three: for example, if the smallest element is sought, we probably don't want the file split near the middle). The average running time is proportional to about $N + k \log(N/k)$, which is linear for any allowed value of k.

It is possible to modify this Quicksort-based selection procedure so that its running time is *guaranteed* to be linear. These modifications, while important from a theoretical standpoint, are extremely complex and not at all practical.

Merging

It is common in many data processing environments to maintain a large (sorted) data file to which new entries are regularly added. Typically, a number of new entries are "batched," appended to the (much larger) main file, and the whole thing resorted. This situation is tailor-made for merging: a much better strategy is to sort the (small) batch of new entries, then merge it with the large main file. Merging has many other similar applications which make it worthwhile to study. Also, we'll examine a sorting method based on merging.

In this chapter we'll concentrate on programs for *two-way merging*: programs which combine two sorted input files to make one sorted output file. In the next chapter, we'll look in more detail at *multiway merging*, when more than two files are involved. (The most important application of multiway merging is external sorting, the subject of that chapter.)

To begin, suppose that we have two sorted arrays $a[1..M]$ and $b[1..N]$ of integers which we wish to merge into a third array $c[1..M+N]$. The following is a direct implementation of the obvious method of successively choosing for c the smallest remaining element from a and b:

```
i:=1; j:=1;
a[M+1]:=maxint; b[N+1]:=maxint;
for k:=1 to M+N do
  if a[i]<b[j]
    then begin c[k]:=a[i]; i:=i+1 end
    else begin c[k]:=b[j]; j:=j+1 end;
```

The implementation is simplified by making room in the a and b arrays for sentinel keys with values larger than all the other keys. When the $a(b)$ array is exhausted, the loop simply moves the rest of the $b(a)$ array into the c array. The time taken by this method is obviously proportional to $M+N$.

The above implementation uses extra space proportional to the size of the merge. It would be desirable to have an in-place method which uses $c[1..M]$ for one input and $c[M+1..M+N]$ for the other. While such methods exist, they are so complicated that an $(N + M)\log(N + M)$ inplace sort would be more efficient for practical values of N and M.

Since extra space appears to be required for a practical implementation, we might as well consider a linked-list implementation. In fact, this method is very well suited to linked lists. A full implementation which illustrates all the conventions we'll use is given below; note that the code for the actual merge is just about as simple as the code above:

```
program listmerge(input, output);
type  link=↑node;
      node=record k: integer;  next: link end;
var N, M: integer;  z: link;
function merge(a, b: link): link;
  var c: link;
  begin
  c:=z;
  repeat
    if a↑.k<=b↑.k
      then begin c↑.next:=a;  c:=a;  a:=a↑.next end
      else  begin c↑.next:=b;  c:=b;  b:=b↑.next end
    until c↑.k=maxint;
  merge:=z↑.next;  z↑.next:=z
  end;
begin
readln(N, M);
new(z);  z↑.k:=maxint;  z↑.next:=z;
writelist(merge(readlist(N), readlist(M)))
end.
```

This program merges the list pointed to by a with the list pointed to by b, with the help of an auxiliary pointer c. The lists are initially built with the *readlist* routine from Chapter 2. All lists are defined to end with the dummy node z, which normally points to itself, and also serves as a sentinel. During the merge, z points to the beginning of the newly merged list (in a manner similar to the implementation of *readlist*), and c points to the end of the newly merged list (the node whose link field must be changed to add a new element to the list). After the merged list is built, the pointer to its first node is retrieved from z and z is reset to point to itself.

The key comparison in *merge* includes equality so that the merge will be stable, if the b list is considered to follow the a list. We'll see below how this stability in the merge implies stability in the sorting programs which use this merge.

Once we have a merging procedure, it's not difficult to use it as the basis for a recursive sorting procedure. To sort a given file, divide it in half, sort the two halves (recursively), then merge the two halves together. This involves enough data movement that a linked list representation is probably the most convenient. The following program is a direct recursive implementation of a function which takes a pointer to an unsorted list as input and returns as its value a pointer to the sorted version of the list. The program does this by rearranging the nodes of the list: no temporary nodes or lists need be allocated.

(It is convenient to pass the list length as a parameter to the recursive program: alternatively, it could be stored with the list or the program could scan the list to find its length.)

```
function sort(c: link; N: integer): link;
  var a, b: link;
      i: integer;
  begin
  if c↑.next=z then sort:=c else
    begin
    a:=c;
    for i:= 2 to N div 2 do c:=c↑.next;
    b:=c↑.next; c↑.next:=z;
    sort:=merge(sort(a, N div 2), sort(b, N−(N div 2)));
    end;
  end;
```

This program sorts by splitting the list pointed to by c into two halves, pointed to by a and b, sorting the two halves recursively, then using *merge* to produce the final result. Again, this program adheres to the convention that all lists end with z: the input list must end with z (and therefore so does the b list); and the explicit instruction $c↑.next:=z$ puts z at the end of the a list. This program is quite simple to understand in a recursive formulation even though it actually is a rather sophisticated algorithm.

The running time of the program fits the standard "divide-and-conquer" recurrence $M(N) = 2M(N/2) + N$. The program is thus guaranteed to run in time proportional to $N \log N$. (See Chapter 4).

Our file of sample sorting keys is processed as shown in the following table. Each line in the table shows the result of a call on *merge*. First we merge O and S to get O S, then we merge this with A to get A O S. Eventually we merge R T with I N to get I N R T, then this is merged with A O S to get A I N O R S T, etc.:

1	2	3	4	5	6	7	8	9	10	11	12	13	14	15
A	S	O	R	T	I	N	G	E	X	A	M	P	L	E
	O	S												
A	O	S												
			R	T										
					I	N								
			I	N	R	T								
A	I	N	O	R	S	T								
							E	G						
									A	X				
							A	E	G	X				
											M	P		
													E	L
											E	L	M	P
							A	E	E	G	L	M	P	X
A	A	E	E	G	I	L	M	N	O	P	R	S	T	X

Thus, this method recursively builds up small sorted files into larger ones. Another version of mergesort processes the files in a slightly different order: first scan through the list performing 1-by-1 merges to produce sorted sublists of size 2, then scan through the list performing 2-by-2 merges to produce sorted sublists of size 4, then do 4-by-4 merges to get sorted sublists of size 8, etc., until the whole list is sorted. Our sample file is sorted in four passes using this "bottom-up" mergesort:

1	2	3	4	5	6	7	8	9	10	11	12	13	14	15
A	S	O	R	T	I	N	G	E	X	A	M	P	L	E
A	S |	O	R |	I	T |	G	N |	E	X |	A	M |	L	P |	E
A	O	R	S |	G	I	N	T |	A	E	M	X |	E	L	P
A	G	I	N	O	R	S	T |	A	E	E	L	M	P	X
A	A	E	E	G	I	L	M	N	O	P	R	S	T	X

In general, $\log N$ passes are required to sort a file of N elements, since each pass doubles the size of the sorted subfiles. A detailed implementation of this idea is given below.

```
function mergesort(c: link): link;
  var a, b, head, todo, t: link;
      i, N: integer;
  begin
  N:=1;  new(head);  head↑.next:=c;
  repeat
    todo:=head↑.next;  c:=head;
    repeat
      t:=todo;
      a:=t;  for i:=1 to N−1 do t:=t↑.next;
      b:=t↑.next;  t↑.next:=z;
      t:=b;  for i:=1 to N−1 do t:=t↑.next;
      todo:=t↑.next;  t↑.next:=z;
      c↑.next:=merge(a, b);
      for i:=1 to N+N do c:=c↑.next
    until todo=z;
    N:=N+N;
  until a=head↑.next;
  mergesort:=head↑.next
  end;
```

This program uses a "list header" node (pointed to by *head*) whose link field points to the file being sorted. Each iteration of the outer repeat loop passes through the file, producing a linked list comprised of sorted subfiles twice as long as for the previous pass. This is done by maintaining two pointers, one to the part of the list not yet seen (*todo*) and one to the end of the part of the list for which the subfiles have already been merged (*c*). The inner repeat loop merges the two subfiles of length N starting at the node pointed to by *todo* producing a subfile of length $N+N$ which is linked onto the *c* result list. The actual merge is accomplished by saving a link to the first subfile to be merged in *a*, then skipping N nodes (using the temporary link *t*), linking *z* onto the end of *a*'s list, then doing the same to get another list of N nodes pointed to by *b* (updating *todo* with the link of the last node visited), then calling *merge*. (Then *c* is updated by simply chasing down to the end of the list just merged. This is a simpler (but slightly less efficient) method than various alternatives which are available, such as having *merge* return pointers to both the beginning and the end, or maintaining multiple pointers in each list node.)

Like Heapsort, mergesort has a guaranteed $N \log N$ running time; like Quicksort, it has a short inner loop. Thus it combines the virtues of these methods, and will perform well for all inputs (though it won't be as quick as Quicksort for random files). The main advantage of mergesort over these methods is that it is stable; the main disadvantage of mergesort over these methods is that extra space proportional to N (for the links) is required. It is also possible to develop a nonrecursive implementation of mergesort using arrays, switching to a different array for each pass in the same way that we discussed in Chapter 10 for straight radix sort.

Recursion Revisited

The programs of this chapter (together with Quicksort) are typical of implementations of divide-and-conquer algorithms. We'll see several algorithms with similar structure in later chapters, so it's worthwhile to take a more detailed look at some basic characteristics of these implementations.

Quicksort is actually a "conquer-and-divide" algorithm: in a recursive implementation, most of the work is done *before* the recursive calls. On the other hand, the recursive mergesort is more in the spirit of divide-and-conquer: first the file is divided into two parts, then each part is conquered individually. The first problem for which mergesort does actual processing is a small one; at the finish the largest subfile is processed. Quicksort starts with actual processing on the largest subfile, finishes up with the small ones.

This difference manifests itself in the non-recursive implementations of the two methods. Quicksort must maintain a stack, since it has to save large subproblems which are divided up in a data-dependent manner. Mergesort admits to a simple non-recursive version because the way in which it divides the file is independent of the data, so the order in which it processes subproblems can be rearranged somewhat to give a simpler program.

Another practical difference which manifests itself is that mergesort is stable (if properly implemented); Quicksort is not (without going to extra trouble). For mergesort, if we assume (inductively) that the subfiles have been sorted stably, then we need only be sure that the merge is done in a stable manner, which is easily arranged. But for Quicksort, no easy way of doing the partitioning in a stable manner suggests itself, so the possibility of being stable is foreclosed even before the recursion comes into play.

Many algorithms are quite simply expressed in a recursive formulation. In modern programming environments, recursive programs implementing such algorithms can be quite useful. However, it is always worthwhile to study the nature of the recursive structure of the program and the possibility of removing the recursion. If the result is not a simpler, more efficient implementation of the algorithm, such study will at least lead to better understanding of the method.

Exercises

1. For $N = 1000$, empirically determine the value of k for which the Quick-sort-based partitioning procedure becomes faster than using heaps to find the k th smallest element in a random file.

2. Describe how you would rearrange an array of $4N$ elements so that the N smallest keys fall in the first N positions, the next N keys fall in the next N positions, the next N in the next N positions, and the N largest in the last N positions.

3. Show the recursive calls made when *select* is used to find the median of the keys EASYQUESTION.

4. Write a program to rearrange a file so that *all* the elements with keys equal to the median are in place, with smaller elements to the left and larger elements to the right.

5. What method would be best for an application that requires selection of the kth largest element (for various arbitrary k) a large number of times on the same file?

6. True or false: the running time of mergesort does not depend on the order of the keys in the input file. Explain your answer.

7. What is the smallest number of steps mergesort could use (to within a constant factor)?

8. Implement a bottom-up non-recursive mergesort that uses two arrays instead of linked lists.

9. Show the contents of the linked lists passed as arguments to each call when the recursive mergesort is used to sort the keys EASYQUESTION.

10. Show the contents of the linked list at each iteration when the non-recursive mergesort is used to sort the keys EASYQUESTION.

13. External Sorting

Many important sorting applications involve processing very large files, much too large to fit into the primary memory of any computer. Methods appropriate for such applications are called *external* methods, since they involve a large amount of processing external to the central processing unit (as opposed to the *internal* methods that we've been studying).

There are two major factors which make external algorithms quite different from those we've seen until now. First, the cost of accessing an item is orders of magnitude greater than any bookkeeping or calculating costs. Second, even with this higher cost, there are severe restrictions on access, depending on the external storage medium used: for example, items on a magnetic tape can be accessed only in a sequential manner.

The wide variety of external storage device types and costs make the development of external sorting methods very dependent on current technology. The methods can be complicated, and many parameters affect their performance: that a clever method might go unappreciated or unused because of a simple change in the technology is a definite possibility in external sorting. For this reason, we'll concentrate on general methods in this chapter rather than on developing specific implementations.

In short, for external sorting, the "systems" aspect of the problem is certainly as important as the "algorithms" aspect. Both areas must be carefully considered if an effective external sort is to be developed. The primary costs in external sorting are for input-output. A good exercise for someone planning to implement an efficient program to sort a very large file is first to implement an efficient program to copy a large file, then (if that was too easy) implement an efficient program to reverse the order of the elements in a large file. The systems problems that arise in trying to solve these problems efficiently are similar to those that arise in external sorts. Permuting a large external file in any non-trivial way is about as difficult as sorting it, even though no key comparisons, etc. are required. In external sorting, we are mainly concerned with limiting the number of times each piece of data is moved between the external storage medium and the primary memory, and being sure that such transfers are done as efficiently as allowed by the available hardware.

External sorting methods have been developed which are suitable for the punched cards and paper tape of the past, the magnetic tapes and disks of the present, and the bubble memories and videodisks of the future. The essential differences among the various devices are the relative size and speed of available storage and the types of data access restrictions. We'll concentrate on basic methods for sorting on magnetic tape and disk because these devices are likely to remain in widespread use and illustrate the two fundamentally different modes of access that characterize many external storage systems. Often, modern computer systems have a "storage hierarchy" of several progressively slower, cheaper, and larger memories. Many of the algorithms that we will consider can be adapted to run well in such an environment, but we'll deal exclusively with "two-level" memory hierarchies consisting of main memory and disk or tape.

Sort-Merge

Most external sorting methods use the following general strategy: make a first pass through the file to be sorted, breaking it up into blocks about the size of the internal memory, and *sort* these blocks. Then *merge* the sorted blocks together, by making several passes through the file, making successively larger sorted blocks until the whole file is sorted. The data is most often accessed in a sequential manner, which makes this method appropriate for most external devices. Algorithms for external sorting strive to reduce the number of passes through the file and to reduce the cost of a single pass to be as close to the cost of a copy as possible.

Since most of the cost of an external sorting method is for input-output, we can get a rough measure of the cost of a sort-merge by counting the number of times each word in the file is read or written, (the number of passes over all the data). For many applications, the methods that we consider will involve on the order of ten or less such passes. Note that this implies that we're interested in methods that can eliminate even a single pass. Also, the running time of the whole external sort can be easily estimated from the running time of something like the "reverse file copy" exercise suggested above.

Balanced Multiway Merging

To begin, we'll trace through the various steps of the simplest sort-merge procedure for an example file. Suppose that we have records with the keys A S O R T I N G A N D M E R G I N G E X A M P L E on an input tape; these are to be sorted and put onto an output tape. Using a "tape" simply means that we're restricted to read the records sequentially: the second record can't be read until the first has been, etc. Assume further that we have only enough room for three records in our computer memory but that we have plenty of tapes available.

The first step is to read in the file three records at a time, sort them to make three-record blocks, and output the sorted blocks. Thus, first we read in A S O and output the block A O S, next we read in R T I and output the block I R T, and so forth. Now, in order for these blocks to be merged together, they must be on different tapes. If we want to do a three-way merge, then we would use three tapes, ending up with the following configuration after the sorting pass:

Tape 1: A O S   D M N   A E X
Tape 2: I R T   E G R   L M P
Tape 3: A G N   G I N   E

Now we're ready to merge the sorted blocks of size three together. We read the first record off each input tape (there's just enough room in the memory) and output the one with the smallest key. Then the next record from the same tape as the record just output is read in and, again, the record in memory with the smallest key is output. When the end of a three-word block in the input is encountered, then that tape is ignored until the blocks from the other two tapes have been processed, and nine records have been output. Then the process is repeated to merge the second three-word block on each tape into a nine-word block (which is output on a different tape, to get ready for the next merge). Continuing, we get three long blocks configured as follows:

Tape 4: A A G I N O R S T
Tape 5: D E G G I M N N R
Tape 6: A E E L M P X

Now one more three-way merge completes the sort. If we had a much longer file with many blocks of size 9 on each tape, then we would finish the second pass with blocks of size 27 on tapes 1, 2, and 3, then a third pass would produce blocks of size 81 on tapes 4, 5, and 6, and so forth. We need six tapes to sort an arbitrarily large file: three for the input and three for the output of each three-way merge. Actually, we could get by with just four tapes: the output could be put on just one tape, then the blocks from that tape distributed to the three input tapes in between merging passes.

This method is called the *balanced multiway merge*: it is a reasonable algorithm for external sorting and a good starting point for the implementation of an external sort. The more sophisticated algorithms below could make the sort run perhaps 50% faster, but not much more. (On the other hand, when execution times are measured in hours, which is not uncommon in external sorting, even a small percentage decrease in running time can be helpful and 50% can be quite significant.)

Suppose that we have N words to be manipulated by the sort and an internal memory of size M. Then the "sort" pass produces about N/M sorted blocks. (This estimate assumes 1-word records: for larger records, the number of sorted blocks is computed by multiplying further by the record size.) If we do P-way merges on each subsequent pass, then the number of subsequent passes is about $\log_P(N/M)$, since each pass reduces the number of sorted blocks by a factor of P.

Though small examples can help one understand the details of the algorithm, it is best to think in terms of very large files when working with external sorts. For example, the formula above says that using a 4-way merge to sort a 200-million-word file on a computer with 1 million words of memory should take a total of about five passes. A very rough estimate of the running time can be found by multiplying by five the running time for the reverse file copy implementation suggested above.

Replacement Selection

It turns out that the details of the implementation can be developed in an elegant and efficient way using priority queues. First, we'll see that priority queues provide a natural way to implement a multiway merge. More important, it turns out that we can use priority queues for the initial sorting pass in such a way that they can produce sorted blocks much longer than could fit into internal memory.

The basic operation needed to do P-way merging is to repeatedly output the smallest of the smallest elements not yet output from each of the P blocks to be merged. That element should be replaced with the next element from the block from which it came. The *replace* operation on a priority queue of size P is exactly what is needed. (Actually, the "indirect" verions of the priority queue routines, as described in Chapter 11, are more appropriate for this application.) Specifically, to do a P-way merge we begin by filling up a priority queue of size P with the smallest element from each of the P inputs using the *pqinsert* procedure from Chapter 11 (appropriately modified so that the smallest element rather than the largest is at the top of the heap). Then, using the *pqreplace* procedure from Chapter 11 (modified in the same way) we output the smallest element and replace it in the priority queue with the next element from its block.

For example, the following table shows the result of merging A O S with I R T and A G N (the first merge from our example above):

1	2	3
A	I	A
A	I	O
G	I	O
I	N	O
N	R	O
O	R	
R	S	
S	T	
T		

The lines in the table represent the contents of a heap of size three used in the merging process. We begin with the first three keys in each block. (The "heap condition" is that the first key must be smaller than the second and third.) Then the first A is output and replaced with the O (the next key in its block). This violates the heap condition, so the O is exchanged with the other A. Then that A is output and replaced with the next key in its block, the G. This does not violate the heap condition, so no further change is necessary. Continuing in this way, we produce the sorted file (read down in the table to see the keys in the order in which they appear in the first heap position and are output). When a block is exhausted, a sentinel is put on the heap and considered to be larger than all the other keys. When the heap consists of all sentinels, the merge is completed. This way of using priority queues is sometimes called *replacement selection*.

Thus to do a *P*-way merge, we can use replacement selection on a priority queue of size P to find each element to be output in $\log P$ steps. This performance difference is not of particular practical relevance, since a brute-force implementation can find each element to output in P steps, and P is normally so small that this cost is dwarfed by the cost of actually outputting the element. The real importance of replacement selection is the way that it can be used in the first part of the sort-merge process: to form the initial sorted blocks which provide the basis for the merging passes.

The idea is to pass the (unordered) input through a large priority queue, always writing out the smallest element on the priority queue as above, and always replacing it with the next element from the input, with one additional proviso: if the new element is smaller than the last one put out, then, since it could not possibly become part of the current sorted block, it should be marked as a member of the next block and treated as greater than all elements in the current block. When a marked element makes it to the top of the priority queue, the old block is ended and a new block started. Again, this is easily implemented with *pqinsert* and *pqreplace* from Chapter 11, again appropriately modified so that the smallest element is at the top of the heap, and with *pqreplace* changed to treat marked elements as always greater than unmarked elements.

Our example file clearly demonstrates the value of replacement selection. With an internal memory capable of holding only three records, we can produce sorted blocks of size 5, 4, 9, 6, and 1, as illustrated in the following table. Each step in the diagram below shows the next key to be input (boxed) and the contents of the heap just before that key is input. (As before, the order in which the keys occupy the first position in the heap is the order in which they are output.) Asterisks are used to indicate which keys in the heap belong to different blocks: an element marked the same way as the element at the root belongs to the current sorted block, others belong to the next sorted block. Always, the heap condition (first key less than the second and third) is maintained, with elements in the next sorted block considered to be greater than elements in the current sorted block.

R	A	S	O		E	A	M	D		M	A*	G*	E*
T	O	S	R		R	D	M	E		P	E*	G*	M*
I	R	S	T		G	E	M	R		L	G*	P*	M*
N	S	I*	T		I	G	M	R		E	L*	P*	M*
G	T	I*	N*		N	I	M	R			M*	P*	E
A	G*	I*	N*		G	M	N	R			P*		E
N	I*	A	N*		E	N	G*	R			E		
D	N*	A	N*		X	R	G*	E*					
M	N*	A	D		A	X	G*	E*					

For example, when *pqreplace* is called for M, it returns N for output (A and D are considered greater) and then sifts down M to make the heap A M D.

It can be shown that, if the keys are random, the runs produced using replacement selection are about twice the size of what could be produced using an internal method. The practical effect of this is to save one merging pass: rather than starting with sorted runs about the size of the internal memory and then taking a merging pass to produce runs about twice the size of the internal memory, we can start right off with runs about twice the size of the internal memory, by using replacement selection with a priority queue of size M. If there is some order in the keys, then the runs will be much, much longer. For example, if no key has more than M larger keys before it in the file, the file will be completely sorted by the replacement selection pass, and no merging will be necessary! This is the most important practical reason to use the method.

In summary, the replacement selection technique can be used for both the "sort" and the "merge" steps of a balanced multiway merge. To sort N 1-word records using an internal memory of size M and $P + 1$ tapes, first use replacement selection with a priority queue of size M to produce initial runs of size about $2M$ (in a random situation) or longer (if the file is partially ordered) then use replacement selection with a priority queue of size P for about $\log_P(N/2M)$ (or fewer) merge passes.

Practical Considerations

To complete an implementation of the sorting method outlined above, it is necessary to implement the input-output functions which actually transfer data between the processor and the external devices. These functions are obviously the key to good performance for the external sort, and they just as obviously require careful consideration of some systems (as opposed to algorithm) issues. (Readers unfamiliar with computers at the "systems" level may wish to skim the next few paragraphs.)

A major goal in the implementation should be to overlap reading, writing, and computing as much as possible. Most large computer systems have independent processing units for controlling the large-scale input/output (I/O) devices which make this overlapping possible. The efficiency achievable by an external sorting method depends on the number of such devices available.

For each file being read or written, there is a standard systems programming technique called *double-buffering* which can be used to maximize the overlap of I/O with computing. The idea is to maintain two "buffers," one for use by the main processor, one for use by the I/O device (or the processor which controls the I/O device). For input, the processor uses one buffer while the input device is filling the other. When the processor has finished using its buffer, it waits until the input device has filled its buffer, then the buffers switch roles: the processor uses the new data in the just-filled buffer while the input device refills the buffer with the data already used by the processor. The same technique works for output, with the roles of the processor and the device reversed. Usually the I/O time is far greater than the processing time and so the effect of double-buffering is to overlap the computation time entirely; thus the buffers should be as large as possible.

A difficulty with double-buffering is that it really uses only about half the available memory space. This can lead to inefficiency if a large number of buffers are involved, as is the case in P-way merging when P is not small. This problem can be dealt with using a technique called *forecasting*, which requires the use of only one extra buffer (not P) during the merging process. Forecasting works as follows. Certainly the best way to overlap input with computation during the replacement selection process is to overlap the input of the buffer that needs to be filled next with the processing part of the algorithm. But it is easy to determine which buffer this is: the next input buffer to be emptied is the one whose *last* item is smallest. For example, when merging A O S with I R T and A G N we know that the third buffer will be the first to empty, then the first. A simple way to overlap processing with input for multiway merging is therefore to keep one extra buffer which is filled by the input device according to this rule. When the processor encounters an empty buffer, it waits until the input buffer is filled (if it hasn't been filled already), then switches to begin using that buffer and directs the input device to begin filling the buffer just emptied according to the forecasting rule.

The most important decision to be made in the implementation of the multiway merge is the choice of the value of P, the "order" of the merge. For tape sorting, when only sequential access is allowed, this choice is easy: P must be chosen to be one less than the number of tape units available: the multiway merge uses P input tapes and one output tape. Obviously, there should be at least two input tapes, so it doesn't make sense to try to do tape sorting with less than three tapes.

For disk sorting, when access to arbitrary positions is allowed but is somewhat more expensive than sequential access, it is also reasonable to choose P to be one less than the number of disks available, to avoid the higher cost of non-sequential access that would be involved, for example, if two different input files were on the same disk. Another alternative commonly used is to pick P large enough so that the sort will be complete in two merging phases: it is usually unreasonable to try to do the sort in one pass, but a two-pass sort can often be done with a reasonably small P. Since replacement selection produces about $N/2M$ runs and each merging pass divides the number of runs by P, this means P should be chosen to be the smallest integer with $P^2 > N/2M$. For our example of sorting a 200-million-word file on a computer with a 1-million-word memory, this implies that $P = 11$ would be a safe choice to ensure a two-pass sort. (The right value of P could be computed exactly after the sort phase is completed.) The best choice between these two alternatives of the lowest reasonable value of P and the highest reasonable value of P is obviously very dependent on many systems parameters: both alternatives (and some in between) should be considered.

Polyphase Merging

One problem with balanced multiway merging for tape sorting is that it requires either an excessive number of tape units or excessive copying. For P-way merging either we must use $2P$ tapes (P for input and P for output) or we must copy almost all of the file from a single output tape to P input tapes between merging passes, which effectively doubles the number of passes to be about $2\log_P(N/2M)$. Several clever tape-sorting algorithms have been invented which eliminate virtually all of this copying by changing the way in which the small sorted blocks are merged together. The most prominent of these methods is called *polyphase merging*.

The basic idea behind polyphase merging is to distribute the sorted blocks produced by replacement selection somewhat unevenly among the available tape units (leaving one empty) and then to apply a "merge until empty" strategy, at which point one of the output tapes and the input tape switch roles.

For example, suppose that we have just three tapes, and we start out with the following initial configuration of sorted blocks on the tapes. (This comes from applying replacement selection to our example file with an internal memory that can only hold two records.)

> *Tape 1*: A O R S T   I N   A G N   D E M R   G I N
> *Tape 2*: E G X   A M P   E L
> *Tape 3*:

After three 2-way merges from tapes 1 and 2 to tape 3, the second tape becomes empty and we are left with the configuration:

> *Tape 1*: D E M R   G I N
> *Tape 2*:
> *Tape 3*: A E G O R S T X   A I M N P   A E G L N

Then, after two 2-way merges from tapes 1 and 3 to tape 2, the first tape becomes empty, leaving:

> *Tape 1*:
> *Tape 2*: A D E E G M O R R S T X   A G I I M N N P
> *Tape 3*: A E G L N

The sort is completed in two more steps. First, a two-way merge from tapes 2 and 3 to tape 1 leaves one file on tape 2, one file on tape 1. Then a two-way merge from tapes 1 and 2 to tape 3 leaves the entire sorted file on tape 3.

This "merge until empty" strategy can be extended to work for an arbitrary number of tapes. For example, if we have four tape units T1, T2, T3, and T4 and we start out with T1 being the output tape, T2 having 13 initial runs, T3 having 11 initial runs, and T4 having 7 initial runs, then after running a 3-way "merge until empty," we have T4 empty, T1 with 7 (long) runs, T2 with 6 runs, and T3 with 4 runs. At this point, we can rewind T1 and make it an input tape, and rewind T4 and make it an output tape. Continuing in this way, we eventually get the whole sorted file onto T1:

T1	T2	T3	T4
0	13	11	7
7	6	4	0
3	2	0	4
1	0	2	2
0	1	1	1
1	0	0	0

The merge is broken up into many *phases* which don't involve all the data, but no direct copying is involved.

The main difficulty in implementing a polyphase merge is to determine how to distribute the initial runs. It is not difficult to see how to build the table above by working backwards: take the largest number on each line, make it zero, and add it to each of the other numbers to get the previous line. This corresponds to defining the highest-order merge for the previous line which could give the present line. This technique works for any number of tapes (at least three): the numbers which arise are "generalized Fibonacci numbers" which have many interesting properties. Of course, the number of initial runs may not be known in advance, and it probably won't be exactly a generalized Fibonacci number. Thus a number of "dummy" runs must be added to make the number of initial runs exactly what is needed for the table.

The analysis of polyphase merging is complicated, interesting, and yields surprising results. For example, it turns out that the very best method for distributing dummy runs among the tapes involves using extra phases and more dummy runs than would seem to be needed. The reason for this is that some runs are used in merges much more often than others.

There are many other factors to be taken into consideration in implementing a most efficient tape-sorting method. For example, a major factor which we have not considered at all is the time that it takes to rewind a tape. This subject has been studied extensively, and many fascinating methods have been defined. However, as mentioned above, the savings achievable over the simple multiway balanced merge are quite limited. Even polyphase merging is only better than balanced merging for small P, and then not substantially. For $P > 8$, balanced merging is likely to run faster than polyphase, and for smaller P the effect of polyphase is basically to save two tapes (a balanced merge with two extra tapes will run faster).

An Easier Way

Many modern computer systems provide a large *virtual memory* capability which should not be overlooked in implementing a method for sorting very large files. In a good virtual memory system, the programmer has the ability to address a very large amount of data, leaving to the system the responsibility of making sure that addressed data is transferred from external to internal storage when needed. This strategy relies on the fact that many programs have a relatively small "locality of reference": each reference to memory is likely to be to an area of memory that is relatively close to other recently referenced areas. This implies that transfers from external to internal storage are needed infrequently. An internal sorting method with a small locality of reference can work very well on a virtual memory system. (For example, Quicksort has two "localities": most references are near one of the two partitioning pointers.) But check with your systems programmer before trying it on a very large file: a method such as radix sorting, which has no locality of reference whatsoever, would be disastrous on a virtual memory system, and even Quicksort could cause problems, depending on how well the available virtual memory system is implemented. On the other hand, the strategy of using a simple internal sorting method for sorting disk files deserves serious consideration in a good virtual memory environment.

Exercises

1. Describe how you would do external *selection*: find the kth largest in a file of N elements, where N is much too large for the file to fit in main memory.

2. Implement the replacement selection algorithm, then use it to test the claim that the runs produced are about twice the internal memory size.

3. What is the *worst* that can happen when replacement selection is used to produce initial runs in a file of N records, using a priority queue of size M, with $M < N$.

4. How would you sort the contents of a disk if no other storage (except main memory) were available for use?

5. How would you sort the contents of a disk if only one tape (and main memory) were available for use?

6. Compare the 4-tape and 6-tape multiway balanced merge to polyphase merge with the same number of tapes, for 31 initial runs.

7. How many phases does 5-tape polyphase merge use when started up with four tapes containing 26,15,22,28 runs?

8. Suppose the 31 initial runs in a 4-tape polyphase merge are each one record long (distributed 0, 13, 11, 7 initially). How many records are there in each of the files involved in the last three-way merge?

9. How should small files be handled in a Quicksort implementation to be run on a very large file within a virtual memory environment?

10. How would you organize an external priority queue? (Specifically, design a way to support the *insert* and *remove* operations of Chapter 11, when the number of elements in the priority queue could grow to be much to large for the queue to fit in main memory.)

SOURCES for Sorting

The primary reference for this section is volume three of D. E. Knuth's series on sorting and searching. Further information on virtually every topic that we've touched upon can be found in that book. In particular, the results that we've quoted on performance characteristics of the various algorithms are backed up by complete mathematical analyses in Knuth's book.

There is a vast amount of literature on sorting. Knuth and Rivest's 1973 bibliography contains hundreds of entries, and this doesn't include the treatment of sorting in countless books and articles on other subjects (not to mention work since 1973).

For Quicksort, the best reference is Hoare's original 1962 paper, which suggests all the important variants, including the use for selection discussed in Chapter 12. Many more details on the mathematical analysis and the practical effects of many of the modifications and embellishments which have been suggested over the years may be found in this author's 1975 Ph.D. thesis.

A good example of an advanced priority queue structure, as mentioned in Chapter 11, is J. Vuillemin's "binomial queues" as implemented and analyzed by M. R. Brown. This data structure supports all of the priority queue operations in an elegant and efficient manner.

To get an impression of the myriad details of reducing algorithms like those we have discussed to general-purpose practical implementations, a reader would be advised to study the reference material for his particular computer system's sort utility. Such material necessarily deals primarily with formats of keys, records and files as well as many other details, and it is often interesting to identify how the algorithms themselves are brought into play.

M. R. Brown, "Implementation and analysis of binomial queue algorithms," *SIAM Journal of Computing*, **7**, 3, (August, 1978).

C. A. R. Hoare, "Quicksort," *Computer Journal*, **5**, 1 (1962).

D. E. Knuth, *The Art of Computer Programming. Volume 3: Sorting and Searching*, Addison-Wesley, Reading, MA, second printing, 1975.

R. L. Rivest and D. E. Knuth, "Bibliography 26: Computing Sorting," *Computing Reviews*, **13**, 6 (June, 1972).

R. Sedgewick, *Quicksort*, Garland, New York, 1978. (Also appeared as the author's Ph.D. dissertation, Stanford University, 1975).

SEARCHING

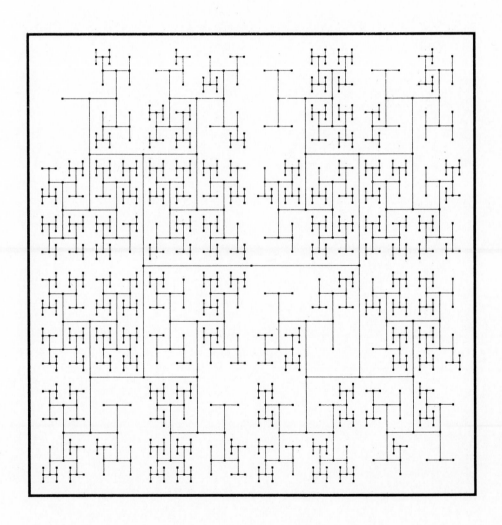

14. Elementary Searching Methods

A fundamental operation intrinsic to a great many computational tasks is *searching*: retrieving some particular information from a large amount of previously stored information. Normally we think of the information as divided up into *records*, each record having a *key* for use in searching. The goal of the search is to find all records with keys matching a given *search key*. The purpose of the search is usually to access information within the record (not merely the key) for processing.

Two common terms often used to describe data structures for searching are *dictionaries* and *symbol tables*. For example, in an English language dictionary, the "keys" are the words and the "records" the entries associated with the words which contain the definition, pronunciation, and other associated information. (One can prepare for learning and appreciating searching methods by thinking about how one would implement a system allowing access to an English language dictionary.) A symbol table is the dictionary for a program: the "keys" are the symbolic names used in the program, and the "records" contain information describing the object named.

In searching (as in sorting) we have programs which are in widespread use on a very frequent basis, so that it will be worthwhile to study a variety of methods in some detail. As with sorting, we'll begin by looking at some elementary methods which are very useful for small tables and in other special situations and illustrate fundamental techniques exploited by more advanced methods. We'll look at methods which store records in arrays which are either searched with key comparisons or indexed by key value, and we'll look at a fundamental method which builds structures defined by the key values.

As with priority queues, it is best to think of search algorithms as belonging to packages implementing a variety of generic operations which can be separated from particular implementations, so that alternate implementations could be substituted easily. The operations of interest include:

Initialize the data structure.

Search for a record (or records) having a given key.

Insert a new record.

Delete a specified record.

Join two dictionaries to make a large one.

Sort the dictionary; output all the records in sorted order.

As with priority queues, it is sometimes convenient to combine some of these operations. For example, a *search and insert* operation is often included for efficiency in situations where records with duplicate keys are not to be kept within the data structure. In many methods, once it has been determined that a key does not appear in the data structure, then the internal state of the search procedure contains precisely the information needed to insert a new record with the given key.

Records with duplicate keys can be handled in one of several ways, depending on the application. First, we could insist that the primary searching data structure contain only records with distinct keys. Then each "record" in this data structure might contain, for example, a link to a list of all records having that key. This is the most convenient arrangement from the point of view of the design of searching algorithms, and it is convenient in some applications since *all* records with a given search key are returned with one *search*. The second possibility is to leave records with equal keys in the primary searching data structure and return *any* record with the given key for a *search*. This is simpler for applications that process one record at a time, where the order in which records with duplicate keys are processed is not important. It is inconvenient from the algorithm design point of view because some mechanism for retrieving all records with a given key must still be provided. A third possibility is to assume that each record has a unique identifier (apart from the key), and require that a *search* find the record with a given identifier, given the key. Or, some more complicated mechanism could be used to distinguish among records with equal keys.

Each of the fundamental operations listed above has important applications, and quite a large number of basic organizations have been suggested to support efficient use of various combinations of the operations. In this and the next few chapters, we'll concentrate on implementations of the fundamental functions *search* and *insert* (and, of course, *initialize*), with some comment on *delete* and *sort* when appropriate. As with priority queues, the *join* operation normally requires advanced techniques which we won't be able to consider here.

Sequential Searching

The simplest method for searching is simply to store the records in an array, then look through the array sequentially each time a record is sought. The following code shows an implementation of the basic functions using this simple organization, and illustrates some of the conventions that we'll use in implementing searching methods.

```
type node=record key, info: integer end;
var a: array [0..maxN] of node;
    N: integer;
procedure initialize;
  begin N:=0 end;
function seqsearch(v: integer; x: integer): integer;
  begin
  a[N+1].key:=v;
  if (x>=0) and (x<=N) then
     repeat x:=x+1 until v=a[x].key;
  seqsearch:=x
  end;
function seqinsert(v: integer): integer;
  begin
  N:=N+1;  a[N].key:=v;
  seqinsert:=N;
  end;
```

The code above processes records that have integer keys (*key*) and "associated information" (*info*). As with sorting, it will be necessary in many applications to extend the programs to handle more complicated records and keys, but this won't fundamentally change the algorithms. For example, *info* could be made into a pointer to an arbitrarily complicated record structure. In such a case, this field can serve as the unique identifier for the record for use in distinguishing among records with equal keys.

The *search* procedure takes two arguments in this implementation: the key value being sought and an index (x) into the array. The index is included to handle the case where several records have the same key value: by successively executing $t:= search(v, t)$ starting at $t=0$ we can successively set t to the index of each record with key value v.

A sentinel record containing the key value being sought is used, which ensures that the search will always terminate, and therefore involves only one completion test within the inner loop. After the inner loop has finished, testing whether the index returned is greater than N will tell whether the search found the sentinel or a key from the table. This is analogous to our use of a sentinel record containing the smallest or largest key value to simplify the coding of the inner loop of various sorting algorithms.

This method takes about N steps for an *unsuccessful* search (every record must be examined to decide that a record with any particular key is absent) and about $N/2$ steps, on the average, for a *successful* search (a "random" search for a record in the table will require examining about half the entries, on the average).

Sequential List Searching

The *seqsearch* program above uses purely sequential access to the records, and thus can be naturally adapted to use a linked list representation for the records. One advantage of doing so is that it becomes easy to keep the list sorted, as shown in the following implementation:

```
type  link=↑node;
      node=record key, info: integer;  next: link end;
var head, t, z: link;
    i: integer;
procedure initialize;
  begin
  new(z);  z↑.next:=z;
  new(head);  head↑.next:=z;
  end;
function listsearch(v: integer;  t: link): link;
  begin
  z↑.key:=v;
  repeat t:=t↑.next until v<=t↑.key;
  if v=t↑.key then listsearch:=t
              else  listsearch:=z
  end;
function listinsert(v: integer;  t: link): link;
  var x: link;
  begin
  z↑.key:=v;
  while t↑.next↑.key<v do t:=t↑.next;
  new(x);  x↑.next:=t↑.next;  t↑.next:=x;
  x↑.key:=v;
  listinsert:=x;
  end;
```

With a sorted list, a search can be terminated unsuccessfully when a record with a key larger than the search key is found. Thus only about half the records (not all) need to be examined for an unsuccessful search. The sorted order is easy to maintain because a new record can simply be inserted into the list at the point at which the unsuccessful search terminates. As usual with linked lists, a dummy header node *head* and a tail node *z* allow the code to be substantially simpler than without them. Thus, the call *listinsert*(*v, head*) will put a new node with key *v* into the list pointed to by the *next* field of the *head*, and *listsearch* is similar. Repeated calls on *listsearch* using the links returned will return records with duplicate keys. The tail node *z* is used as a sentinel in the same way as above. If *listsearch* returns *z*, then the search was unsuccessful.

If something is known about the relative frequency of access for various records, then substantial savings can often be realized simply by ordering the records intelligently. The "optimal" arrangement is to put the most frequently accessed record at the beginning, the second most frequently accessed record in the second position, etc. This technique can be very effective, especially if only a small set of records is frequently accessed.

If information is not available about the frequency of access, then an approximation to the optimal arrangement can be achieved with a "self-organizing" search: each time a record is accessed, move it to the beginning of the list. This method is more conveniently implemented when a linked-list implementation is used. Of course the running time for the method depends on the record access distributions, so it is difficult to predict how it will do in general. However, it is well suited to the quite common situation when most of the accesses to each record tend to happen close together.

Binary Search

If the set of records is large, then the total search time can be significantly reduced by using a search procedure based on applying the "divide-and-conquer" paradigm: divide the set of records into two parts, determine which of the two parts the key being sought belongs to, then concentrate on that part. A reasonable way to divide the sets of records into parts is to keep the records sorted, then use indices into the sorted array to delimit the part of the array being worked on. To find if a given key *v* is in the table, first compare it with the element at the middle position of the table. If *v* is smaller, then it must be in the first half of the table; if *v* is greater, then it must be in the second half of the table. Then apply the method recursively. (Since only one recursive call is involved, it is simpler to express the method iteratively.) This brings us directly to the following implementation, which assumes that the array *a* is sorted.

```
function binarysearch(v: integer): integer;
   var x, l, r: integer;
   begin
   l:=1; r:=N;
   repeat
      x:=(l+r) div 2;
      if v<a[x].key then r:=x−1 else l:=x+1
   until (v=a[x].key) or (l>r);
   if v=a[x].key then binarysearch:=x
                 else   binarysearch:=N+1
   end;
```

Like Quicksort and radix exchange sort, this method uses the pointers l and r to delimit the subfile currently being worked on. Each time through the loop, the variable x is set to point to the midpoint of the current interval, and the loop terminates successfully, or the left pointer is changed to $x+1$, or the right pointer is changed to $x−1$, depending on whether the search value v is equal to, less than, or greater than the key value of the record stored at $a[x]$.

The following table shows the subfiles examined by this method when searching for S in a table built by inserting the keys A S E A R C H I N G E X A M P L E:

1	2	3	4	5	6	7	8	9	10	11	12	13	14	15	16	17
A	A	A	C	E	E	E	G	H	I	L	M	N	P	R	S	X
A	A	A	C	E	E	E	G	H	I	L	M	N	P	R	S	X
									I	L	M	N	P	R	S	X
												P	R	S	X
														S	X

The interval size is at least halved at each step, so the total number of times through the loop is only about $\lg N$. However, the time required to insert new records is high: the array must be kept sorted, so records must be moved to make room for new records. For example, if the new record has a smaller key than any record in the table, then every entry must be moved over one position. A random insertion requires that $N/2$ records be moved, on the average. Thus, this method should not be used for applications which involve many insertions.

Some care must be exercised to properly handle records with equal keys for this algorithm: the index returned could fall in the middle of a block of records with key v, so loops which scan in both directions from that index should be used to pick up all the records. Of course, in this case the running time for the search is proportional to $\lg N$ plus the number of records found.

The sequence of comparisons made by the binary search algorithm is predetermined: the specific sequence used is based on the value of the key being sought and the value of N. The comparison structure can be simply described by a binary tree structure. The following binary tree describes the comparison structure for our example set of keys:

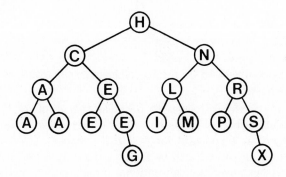

In searching for the key S for instance, it is first compared to H. Since it is greater, it is next compared to N; otherwise it would have been compared to C), etc. Below we will see algorithms that use an explicitly constructed binary tree structure to guide the search.

One improvement suggested for binary search is to try to guess more precisely where the key being sought falls within the current interval of interest (rather than blindly using the middle element at each step). This mimics the way one looks up a number in the telephone directory, for example: if the name sought begins with B, one looks near the beginning, but if it begins with Y, one looks near the end. This method, called *interpolation search*, requires only a simple modification to the program above. In the program above, the new place to search (the midpoint of the interval) is computed with the statement $x:=(l+r)$ **div** 2. This is derived from the computation $x = l + \frac{1}{2}(r - l)$: the middle of the interval is computed by adding half the size of the interval to the left endpoint. Interpolation search simply amounts to replacing $\frac{1}{2}$ in this formula by an estimate of where the key might be based on the values available: $\frac{1}{2}$ would be appropriate if v were in the middle of the interval between $a[l].key$ and $a[r].key$, but we might have better luck trying

$x:=l+(v-a[l].key)*(r-l)$ **div** $(a[r].key-a[l].key)$. Of course, this assumes numerical key values. Suppose in our example that the ith letter in the alphabet is represented by the number i. Then, in a search for S, the first table position examined would be $x = 1 + (19-1)*(17-1)/(24-1) = 13$. The search is completed in just three steps:

1	2	3	4	5	6	7	8	9	10	11	12	13	14	15	16	17
A	A	A	C	E	E	E	G	H	I	L	M	N	P	R	S	X
													P	R	S	X	
														S	X		

Other search keys are found even more efficiently: for example X and A are found in the first step.

Interpolation search manages to decrease the number of elements examined to about $\log \log N$. This is a very slowly growing function which can be thought of as a constant for practical purposes: if N is one billion, $\lg \lg N < 5$. Thus, any record can be found using only a few accesses, a substantial improvement over the conventional binary search method. But this assumes that the keys are rather well distributed over the interval, and it does require some computation: for small N, the $\log N$ cost of straight binary search is close enough to $\log \log N$ that the cost of interpolating is not likely to be worthwhile. But interpolation search certainly should be considered for large files, for applications where comparisons are particularly expensive, or for external methods where very high access costs are involved.

Binary Tree Search

Binary tree search is a simple, efficient dynamic searching method which qualifies as one of the most fundamental algorithms in computer science. It's classified here as an "elementary" method because it is so simple; but in fact it is the method of choice in many situations.

The idea is to build up an explicit structure consisting of *nodes*, each node consisting of a record containing a key and *left* and *right* links. The left and right links are either null, or they point to nodes called the *left son* and the *right son*. The sons are themselves the roots of trees, called the *left subtree* and the *right subtree* respectively. For example, consider the following diagram, where nodes are represented as encircled key values and the links by lines connected to nodes:

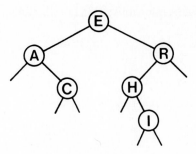

The links in this diagram all point down. Thus, for example, E's right link points to R, but H's left link is null.

The defining property of a *tree* is ~~the~~ that every node is pointed to by only one other node called its *father*. (We assume the existence of an imaginary node which points to the root.) The defining property of a *binary tree* is that each node has left and right links. For searching, each node also has a record with a key value; in a *binary search tree* we insist that all records with smaller keys are in the left subtree and that all records in the right subtree have larger (or equal) key values. We'll soon see that it is quite simple to ensure that binary search trees built by successively inserting new nodes satisfy this defining property.

A search procedure like *binarysearch* immediately suggests itself for this structure. To find a record with a given key v, first compare it against the root. If it is smaller, go to the left subtree; if it is equal, stop; and if it is greater, go to the right subtree. Apply the method recursively. At each step, we're guaranteed that no parts of the tree other than the current subtree could contain records with key v, and, just as the size of the interval in binary search shrinks, the "current subtree" always gets smaller. The procedure stops either when a record with key v is found or, if there is no such record, when the "current subtree" becomes empty. (The words "binary," "search," and "tree" are admittedly somewhat overused at this point, and the reader should be sure to understand the difference between the *binarysearch* function given above and the binary search trees described here. Above, we used a binary tree to describe the sequence of comparisons made by a function searching in an array; here we actually construct a data structure of records connected with links which is used for the search.)

```
type link=↑node;
      node=record key, info: integer; l, r: link end;
var t, head, z: link;
function treesearch(v: integer; x: link): link;
   begin
   z↑.key:=v;
   repeat
      if v<x↑.key then x:=x↑.l else x:=x↑.r
   until v=x↑.key;
   treesearch:=x
   end;
```

As with sequential list searching, the coding in this program is simplified by the use of a "tail" node z. Similarly, the insertion code given below is simplified by the use of a tree header node *head* whose right link points to the root. To search for a record with key v we set x:=*search*(v, *head*).

If a node has no left (right) subtree then its left (right) link is set to point to z. As in sequential search, we put the value sought in z to stop an unsuccessful search. Thus, the "current subtree" pointed to by x never becomes empty and all searches are "successful": the calling program can check whether the link returned points to z to determine whether the search was successful. It is sometimes convenient to think of links which point to z as pointing to imaginary *external nodes* with all unsuccessful searches ending at external nodes. The normal nodes which contain our keys are called *internal nodes*; by introducing external nodes we can say that every internal nodes points to two other nodes in the tree, even though, in our implementation, all of the external nodes are represented by the single node z.

For example, if D is sought in the tree above, first it is compared against E, the key at the root. Since D is less, it is next compared against A, the key in the left son of the node containing E. Continuing in this way, D is compared next against the C to the right of that node. The links in the node containing C are pointers to z so the search terminates with D being compared to itself in z and the search is unsuccessful.

To insert a node into the tree, we just do an unsuccessful search for it, then hook it on in place of z at the point where the search terminated, as in the following code:

```
function treeinsert(v: integer; x:link): link;
    var f: link;
    begin
    repeat
        f:=x;
        if v<x↑.key then x:=x↑.l else x:=x↑.r
    until x=z;
    new(x); x↑.key:=v; x↑.l:=z; x↑.r:=z;
    if v<f↑.key then f↑.l:=x else f↑.r:=x;
    treeinsert:=x
    end;
```

To insert a new key in a tree with a tree header node pointed to by *head*, we call *treeinsert*(*v, head*). To be able to do the insertion, we must keep track of the father *f* of *x*, as it proceeds down the tree. When the bottom of the tree (*x=z*) is reached, *f* points to the node whose link must be changed to point to the new node inserted. The function returns a link to the newly created node so that the calling routine can fill in the *info* field as appropriate.

When a new node whose key is equal to some key already in the tree is inserted, it will be inserted to the right of the node already in the tree. All records with key equal to *v* can be processed by successively setting *t* to *search*(*v, t*) as we did for sequential searching.

As mentioned above, it is convenient to use a tree header node *head* whose right link points to the actual root node of the tree, and whose key is smaller than all other key values (for simplicity, we use 0 assuming the keys are all positive integers). The left link of *head* is not used. The empty tree is represented by having the right link of *head* point to *z*, as constructed by the following code:

```
procedure treeinitialize;
    begin
    new(z); new(head);
    head↑.key:=0; head↑.r:=z;
    end;
```

To see the need for *head*, consider what happens when the first node is inserted into an empty tree constructed by *treeinitialize*.

The diagram below shows the tree constructed when our sample keys are inserted into an initially empty tree.

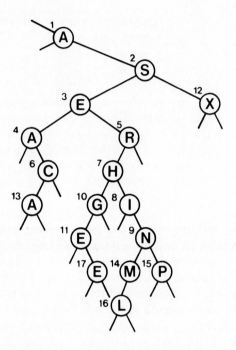

The nodes in this tree are numbered in the order in which they were inserted. Since new nodes are added at the bottom of the tree, the construction process can be traced out from this diagram: the tree as it was after k records had been inserted is simply the part of the tree consisting only of nodes with numbers less than k (and keys from the first k letters of A S E A R C H I N G E X A M P L E).

The *sort* function comes almost for free when binary search trees are used, since a binary search tree represents a sorted file if you look at it the right way. Consider the following recursive program:

inorder

```
procedure treeprint(x: link);
    begin
    if x<>z then
        begin
        treeprint(x↑.l);
        printnode(x);
        treeprint(x↑.r)
        end
    end;
```

The call *treeprint*(*head*↑.*r*) will print out the keys of the tree in order. This defines a sorting method which is remarkably similar to Quicksort, with the node at the root of the tree playing a role similar to that of the partitioning element in Quicksort. A major difference is that the tree-sorting method must use extra memory for the links, while Quicksort sorts with only a little extra memory.

The running times of algorithms on binary search trees are quite dependent on the shapes of the trees. In the best case, the tree could be shaped like that given above for describing the comparison structure for binary search, with about lg N nodes between the root and each external node. We might, roughly, expect logarithmic search times on the average because the first element inserted becomes the root of the tree; if N keys are to be inserted at random, then this element would divide the keys in half (on the average), leading to logarithmic search times (using the same argument on the subtrees). Indeed, were it not for the equal keys, it could happen that the tree given above for describing the comparison structure for binary search would be built. This would be the best case of the algorithm, with guaranteed logarithmic running time for all searches. Actually, the root is equally likely to be any key in a truly random situation, so such a perfectly balanced tree would be extremely rare. But if random keys are inserted, it turns out that the trees are nicely balanced. The average number of steps for a *treesearch* in a tree built by successive insertion of N random keys is proportional to $2 \ln N$.

On the other hand, binary tree searching is susceptible to the same worst-case problems as Quicksort. For example, when the keys are inserted in order (or in reverse order) the binary tree search method is no better than the sequential search method that we saw at the beginning of this chapter. In the next chapter, we'll examine a technique for eliminating this worst case and making all trees look more like the best-case tree.

The implementations given above for the fundamental *search, insert,* and *sort* functions using binary tree structures are quite straightforward. However, binary trees also provide a good example of a recurrent theme in the study of searching algorithms: the *delete* function is often quite cumbersome to implement. To delete a node from a binary tree is easy if the node has no sons, like L or P in the tree above (lop it off by making the appropriate link in its father null); or if it has just one son, like G or R in the tree above (move the link in the son to the appropriate father link); but what about nodes with two sons, such as H or S in the tree above? Suppose that x is a link to such a node. One way to delete the node pointed to by x is to first set y to the node with the *next* highest key. By examining the *treeprint* routine, one can become convinced that this node must have a null left link, and that it can be found by y:=x↑.r; **while** y↑.l<>z **do** y:=y↑.l. Now the deletion can be accomplished by copying y↑.key and y↑.info into x↑.key and x↑.info, then deleting the node pointed to by y. Thus, we delete H in the example above by copying I into H, then deleting I; and we delete the E at node 3 by copying the E at node 11 into node 3, then deleting node 11. A full implementation of a *treedelete* procedure according to this description involves a fair amount of code to cover all the cases: we'll forego the details because we'll be doing similar, but more complicated manipulations on trees in the next chapter. It is quite typical for searching algorithms to require significantly more complicated implementations for deletion: the keys themselves tend to be integral to the structure, and removal of a key can involve complicated repairs.

Indirect Binary Search Trees

As we saw with heaps in Chapter 11, for many applications we want a searching structure to simply help us find records, without moving them around. For example, we might have an array $a[1..N]$ of records with keys, and we would like the *search* routine to give us the index into that array of the record matching a certain key. Or we might want to remove the record with a given index from the searching structure, but still keep it in the array for some other use.

To adapt binary search trees to such a situation, we simply make the *info* field of the nodes the array index. Then we could eliminate the *key* field by having the search routines access the keys in the records directly, e.g. via an instruction like **if** $v<a[x\uparrow.info]$ **then** .... However, it is often better to make a copy of the key, and use the code above just as it is given. We'll use the function name *bstinsert*($v, info$: integer; x: *link*) to refer to a function just like *treeinsert*, except that it also sets the *info* field to the value given in the argument. Similarly, a function *bstdelete*($v, info$: integer; x: *link*) to delete the node with key v and array index *info* from the binary search tree rooted at x will refer to an implementation of the *delete* function as described above. These functions use an extra copy of the keys (one in the array, one in the tree), but this allows the same function to be used for more than one array, or as we'll see in Chapter 27, for more than one key field in the same array. (There are other ways to achieve this: for example, a procedure could be associated with each tree which extracts keys from records.)

Another direct way to achieve "indirection" for binary search trees is to simply do away with the linked implementation entirely. That is, all links just become indices into an array $a[1..N]$ of records which contain a *key* field and l and r index fields. Then link references such as **if** $v<x\uparrow.key$ **then** $x:=x\uparrow.l$ **else** ... become array references such as **if** $v<a[x].key$ **then** $x:=a[x].l$ **else** .... No calls to *new* are used, since the tree exists within the record array: *new*(*head*) becomes *head*:=0, *new*(*z*) becomes $z:=N+1$, and to insert the Mth node, we would pass M, not v, to *treeinsert*, and then simply refer to $a[M].key$ instead of v and replace the line containing *new*(x) in *treeinsert* with $x:=M$. This

way of implementing binary search trees to aid in searching large arrays of records is preferred for many applications, since it avoids the extra expense of copying keys as in the previous paragraph, and it avoids the overhead of the storage allocation mechanism implied by *new*. The disadvantage is that space is reserved with the record array for links which may not be in use, which could lead to problems with large arrays in a dynamic situation.

Exercises

1. Implement a sequential searching algorithm which averages about $N/2$ steps for both successful and unsuccessful search, keeping the records in a sorted array.

2. Give the order of the keys after records with the keys E A S Y Q U E S T I O N have been put into an intially empty table with *search and insert* using the self-organizing search heuristic.

3. Give a recursive implementation of binary search.

4. Suppose $a[i]=2i$ for $1 \leq i \leq N$. How many table positions are examined by interpolation search during the unsuccessful search for $2k - 1$?

5. Draw the binary search tree that results from inserting records with the keys E A S Y Q U E S T I O N into an initially empty tree.

6. Write a recursive program to compute the *height* of a binary tree: the longest distance from the root to an external node.

7. Suppose that we have an estimate ahead of time of how often search keys are to be accessed in a binary tree. Should the keys be inserted into the tree in increasing or decreasing order of likely frequency of access? Why?

8. Give a way to modify binary tree search so that it would keep equal keys together in the tree. (If there are any other nodes in the tree with the same key as any given node, then either its father or one of its sons should have an equal key.)

9. Write a nonrecursive program to print out the keys from a binary search tree in order.

10. Use a least-squares curvefitter to find values of a and b that give the best formula of the form $aN \ln N + bN$ for describing the total number of instructions executed when a binary search tree is built from N random keys.

15. Balanced Trees

The binary tree algorithms of the previous section work very well for a wide variety of applications, but they do have the problem of bad worst-case performance. What's more, as with Quicksort, it's embarrassingly true that the bad worst case is one that's likely to occur in practice if the person using the algorithm is not watching for it. Files already in order, files in reverse order, files with alternating large and small keys, or files with any large segment having a simple structure can cause the binary tree search algorithm to perform very badly.

With Quicksort, our only recourse for improving the situation was to resort to randomness: by choosing a random partitioning element, we could rely on the laws of probability to save us from the worst case. Fortunately, for binary tree searching, we can do much better: there is a general technique that will enable us to *guarantee* that this worst case will not occur. This technique, called *balancing*, has been used as the basis for several different "balanced tree" algorithms. We'll look closely at one such algorithm and discuss briefly how it relates to some of the other methods that are used.

As will become apparent below, the implementation of balanced tree algorithms is certainly a case of "easier said than done." Often, the general concept behind an algorithm is easily described, but an implementation is a morass of special and symmetric cases. Not only is the program developed in this chapter an important searching method, but also it is a nice illustration of the relationship between a "high-level" algorithm description and a "low-level" Pascal program to implement the algorithm.

Top-Down 2-3-4 Trees

To eliminate the worst case for binary search trees, we'll need some flexibility in the data structures that we use. To get this flexibility, let's assume that we can have nodes in our trees that can hold more than one key. Specifically, we'll allow *3-nodes* and *4-nodes*, which can hold two and three keys respectively. A 3-node has three links coming out of it, one for all records with keys smaller than both its keys, one for all records with keys in between its two keys, and one for all records with keys larger than both its keys. Similarly, a 4-node has four links coming out of it, one for each of the intervals defined by its three keys. (The nodes in a standard binary search tree could thus be called *2-nodes*: one key, two links.) We'll see below some efficient ways of defining and implementing the basic operations on these extended nodes; for now, let's assume we can manipulate them conveniently and see how they can be put together to form trees.

For example, below is a *2-3-4 tree* which contains some keys from our searching example.

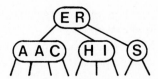

It is easy to see how to search in such a tree. For example, to search for O in the tree above, we would follow the middle link from the root, since O is between E and R then terminate the unsuccessful search at the right link from the node containing H and I.

To insert a new node in a 2-3-4 tree, we would like to do an unsuccessful search and then hook the node on, as before. It is easy to see what to if the node at which the search terminates is a 2-node: just turn it into a 3-node. Similarly, a 3-node can easily be turned into a 4-node. But what should we do if we need to insert our new node into a 4-node? The answer is that we should first split the 4-node into two 2-nodes and pass one of its keys further up in the tree. To see exactly how to do this, let's consider what happens when the keys from A S E A R C H I N G E X A M P L E are inserted into an initially empty tree. We start out with a 2-node, then a 3-node, then a 4-node:

Now we need to put a second A into the 4-node. But notice that as far as the search procedure is concerned, the 4-node at the right above is exactly equivalent to the binary tree:

If our algorithm "splits" the 4-node to make this binary tree before trying to insert the A, then there will be room for A at the bottom:

Now R, C, and the H can be inserted, but when it's time for I to be inserted, there's no room in the 4-node at the right:

Again, this 4-node must be split into two 2-nodes to make room for the I, but this time the extra key needs to be inserted into the father, changing it from a 2-node to a 3-node. Then the N can be inserted with no splits, then the G causes another split, turning the root into a 4-node:

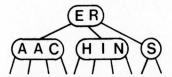   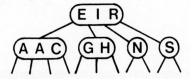

But what if we were to need to split a 4-node whose father is also a 4-node? One method would be to split the father also, but this could keep happening all the way back up the tree. An easier way is to make sure that the father of any node we see won't be a 4-node by splitting any 4-node we see on the way down the tree. For example, when E is inserted, the tree above first becomes

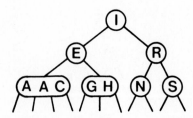

This ensures that we could handle the situation at the bottom even if E were to go into a 4-node (for example, if we were inserting another A instead). Now, the insertion of E, X, A, M, P, L, and E finally leads to the tree:

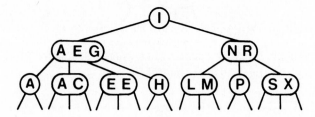

The above example shows that we can easily insert new nodes into 2-3-4 trees by doing a search and splitting 4-nodes on the way down the tree. Specifically, every time we encounter a 2-node connected to a 4-node, we should transform it into a 3-node connected to two 2-nodes:

and every time we encounter a 3-node connected to a 4-node, we should transform it into a 4-node connected to two 2-nodes:

These transformations are purely "local": no part of the tree need be examined or modified other than what is diagrammed. Each of the transformations passes up one of the keys from a 4-node to its father in the tree, restructuring links accordingly. Note that we don't have to worry explicitly about the father being a 4-node since our transformations ensure that as we pass through each node in the tree, we come out on a node that is not a 4-node. In particular, when we come out the bottom of the tree, we are not on a 4-node, and we can directly insert the new node either by transforming a 2-node to a 3-node or a 3-node to a 4-node. Actually, it is convenient to treat the insertion as a split of an imaginary 4-node at the bottom which passes up the new key to be inserted. Whenever the root of the tree becomes a 4-node, we'll split it into three 2-nodes, as we did for our first node split in the example above. This (and only this) makes the tree grow one level "higher."

The algorithm sketched in the previous paragraph gives a way to do searches and insertions in 2-3-4 trees; since the 4-nodes are split up on the way from the top down, the trees are called *top-down 2-3-4 trees*. What's interesting is that, even though we haven't been worrying about balancing at all, the resulting trees are perfectly balanced! The distance from the root to every external node is the same, which implies that the time required by a search or an insertion is always proportional to $\log N$. The proof that the trees are always perfectly balanced is simple: the transformations that we perform have no effect on the distance from any node to the root, except when we split the root, and in this case the distance from all nodes to the root is increased by one.

The description given above is sufficient to define an algorithm for searching using binary trees which has guaranteed worst-case performance. However, we are only halfway towards an actual implementation. While it would be possible to write algorithms which actually perform transformations on distinct data types representing 2-, 3-, and 4-nodes, most of the things that need to be done are very inconvenient in this direct representation. (One can become convinced of this by trying to implement even the simpler of the two node transformations.) Furthermore, the overhead incurred in manipulating the more complex node structures is likely to make the algorithms slower than standard binary tree search. The primary purpose of balancing is to provide "insurance" against a bad worst case, but it would be unfortunate to have to pay the overhead cost for that insurance on every run of the algorithm. Fortunately, as we'll see below, there is a relatively simple representation of 2-, 3-, and 4-nodes that allows the transformations to be done in a uniform way with very little overhead beyond the costs incurred by standard binary tree search.

Red-Black Trees

Remarkably, it is possible to represent 2-3-4 trees as standard binary trees (2-nodes only) by using only one extra bit per node. The idea is to represent 3-nodes and 4-nodes as small binary trees bound together by "red" links which contrast with the "black" links which bind the 2-3-4 tree together. The representation is simple: 4-nodes are represented as three 2-nodes connected by red links and 3-nodes are represented as two 2-nodes connected by a red link (red links are drawn as double lines):

(Either orientation for a 3-node is legal.) The binary tree drawn below is one way to represent the final tree from the example above. If we eliminate the red links and collapse the nodes they connect together, the result is the 2-3-4 tree from above. The extra bit per node is used to store the color of the link pointing to that node: we'll refer to 2-3-4 trees represented in this way as *red-black trees*.

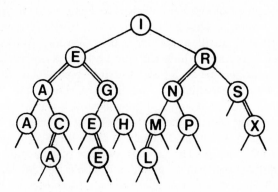

The "slant" of each 3-node is determined by the dynamics of the algorithm to be described below. There are many red-black trees corresponding to each 2-3-4 tree. It would be possible to enforce a rule that 3-nodes all slant the same way, but there is no reason to do so.

These trees have many structural properties that follow directly from the way in which they are defined. For example, there are never two red links in a row along any path from the root to an external node, and all such paths have an equal number of black links. Note that it is possible that one path (alternating black-red) be twice as long as another (all black), but that all path lengths are still proportional to $\log N$.

A striking feature of the tree above is the positioning of duplicate keys. On reflection, it is clear that any balanced tree algorithm must allow records with keys equal to a given node to fall on both sides of that node: otherwise, severe imbalance could result from long strings of duplicate keys. This implies that we can't find all nodes with a given key by repeated calls to the searching procedure, as in the previous chapter. However, this does not present a real problem, because all nodes in the subtree rooted at a given node with the same key as that node can be found with a simple recursive procedure like the *treeprint* procedure of the previous chapter. Or, the option of requiring distinct keys in the data structure (with linked lists of records with duplicate keys) could be used.

One very nice property of red-black trees is that the *treesearch* procedure for standard binary tree search works without modification (except for the problem with duplicate keys discussed in the previous paragraph). We'll implement the link colors by adding a boolean field *red* to each node which is true if the link pointing to the node is red, false if it is black; the *treesearch* procedure simply never examines that field. That is, no "overhead" is added by the balancing mechanism to the time taken by the fundamental searching procedure. Each key is inserted just once, but might be searched for many times in a typical application, so the end result is that we get improved search times (because the trees are balanced) at relatively little cost (because no work for balancing is done during the searches).

Moreover, the overhead for insertion is very small: we have to do something different only when we see 4-nodes, and there aren't many 4-nodes in the tree because we're always breaking them up. The inner loop needs only one extra test (if a node has two red sons, it's a part of a 4-node), as shown in the following implementation of the insert procedure:

```
function rbtreeinsert(v: integer; x:link): link;
   var gg, g, f: link;
   begin
   f:=x; g:=x;
   repeat
      gg:=g; g:=f; f:=x;
      if v<x↑.key then x:=x↑.l else x:=x↑.r;
      if x↑.l↑.red and x↑.r↑.red then x:=split(v, gg, g, f, x);
   until x=z;
   new(x); x↑.key:=v; x↑.l:=z; x↑.r:=z;
   if v<f↑.key then f↑.l:=x else f↑.r:=x;
   rbtreeinsert:=x;
   x:=split(v, gg, g, f, x);
   end;
```

In this program, x moves down the tree as before, with *gg*, *g*, and *f* kept pointing to *x*'s great-grandfather, grandfather, and father in the tree. To see why all these links are needed, consider the addition of Y to the tree above. When the external node at the right of the 3-node containing S and X is reached, *gg* is R, *g* is S, and *f* is X. Now, Y must be added to make a 4-node containing S, X, and Y, resulting in the following tree:

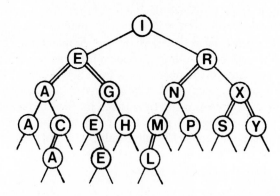

We need a pointer to R (*gg*) because R's right link must be changed to point to X, not S. To see exactly how this comes about, we need to look at the operation of the *split* procedure.

To understand how to implement the *split* operation, let's consider the red-black representation for the two transformations we must perform: if we have a 2-node connected to a 4-node, then we should convert them into a

3-node connected to two 2-nodes; if we have a 3-node connected to a 4-node, we should convert them into a 4-node connected to two 2-nodes. When a new node is added at the bottom, it is considered to be the middle node of an imaginary 4-node (that is, think of z as being *red*, though this never gets explicitly tested).

The transformation required when we encounter a 2-node connected to a 4-node is easy:

This same transformation works if we have a 3-node connected to a 4-node in the "right" way:

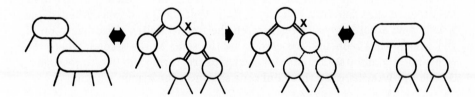

Thus, *split* will begin by marking x to be red and the sons of x to be black. This leaves the two other situations that can arise if we encounter a 3-node connected to a 4-node:

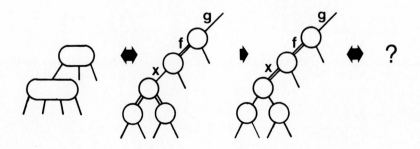

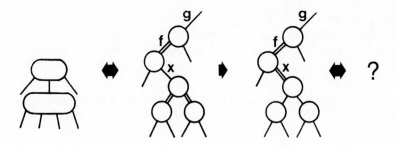

(Actually, there are four situations, since the mirror images of these two can
also occur for 3-nodes of the other orientation.) In these cases, the split-up of
the 4-node has left two red links in a row, an illegal situation which must be
corrected. This is easily tested for in the code: we just marked x red; if x's
father f is also red, we must take further action. The situation is not too bad
because we do have three nodes connected by red links: all we need to do is
transform the tree so that the red links point down from the same node.

Fortunately, there is a simple operation which achieves the desired effect.
Let's begin with the easier of the two, the third case, where the red links
are oriented the same way. The problem is that the 3-node was oriented the
wrong way: accordingly, we restructure the tree to switch the orientation of
the 3-node, thus reducing this case to be the same as the second, where the
color flip of x and its sons was sufficient. Restructuring the tree to reorient a
3-node involves changing three links, as shown in the example below:

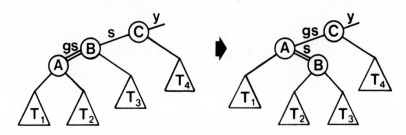

In this diagram, T_1 represents the tree containing all the records with keys less than A, T_2, contains all the records with keys between A and B, and so forth. The transformation switches the orientation of the 3-node containing A and B without disturbing the rest of the tree. Thus none of the keys in T_1, T_2, T_3, and T_4 are touched. In this case, the transformation is effected by the link changes $s\uparrow.l:=gs\uparrow.r; gs\uparrow.r:=s; y\uparrow.l:=gs$. Also, note carefully that the colors of A and B are switched. There are three analogous cases: the 3-node could be oriented the other way, or it could be on the right side of y (oriented either way).

Disregarding the colors, this *single rotation* operation is defined on any binary search tree and is the basis for several balanced tree algorithms. It is important to note, however, that doing a single rotation doesn't necessarily improve the balance of the tree. In the diagram above, the rotation brings all the nodes in T_1 one step closer to the root, but all the nodes in T_3 are *lowered* one step. If T_3 were to have more nodes than T_1, then the tree after the rotation would become less balanced, not more balanced. Top-down 2-3-4 trees may be viewed as simply a convenient way to identify single rotations which *are* likely to improve the balance.

Doing a single rotation involves structurally modifying the tree, something that should be done with caution. A convenient way to handle the four different cases outlined above is to use the search key v to "rediscover" the relevant son (s) and grandson (gs) of the node y. (We know that we'll only be reorienting a 3-node if the search took us to its bottom node.) This leads to somewhat simpler code that the alternative of remembering during the search not only the two links corresponding to s and gs but also whether they are right or left links. We have the following function for reorienting a 3-node along the search path for v whose father is y:

```
function rotate(v: integer; y: link): link;
   var s, gs: link;
   begin
   if v<y↑.key then s:=y↑.l else s:=y↑.r;
   if v<s↑.key
      then begin gs:=s↑.l; s↑.l:=gs↑.r; gs↑.r:=s end
      else begin gs:=s↑.r; s↑.r:=gs↑.l; gs↑.l:=s end;
   if v<y↑.key then y↑.l:=gs else y↑.r:=gs;
   rotate:=gs
   end;
```

If s is the left link of y and gs is the left link of s, this makes exactly the link transformations for the diagram above. The reader may wish to check the other cases. This function returns the link to the top of the 3-node, but does not do the color switch itself.

Thus, to handle the third case for *split*, we can make g red, then set x to *rotate*(v, gg), then make x black. This reorients the 3-node consisting of the two nodes pointed to by g and f and reduces this case to be the same as the second case, when the 3-node was oriented the right way.

Finally, to handle the case when the two red links are oriented in different directions, we simply set f to *rotate*(v, g). This reorients the "illegal" 3-node consisting of the two nodes pointed to by f and x. These nodes are the same color, so no color change is necessary, and we are immediately reduced to the third case. Combining this and the rotation for the third case is called a *double rotation* for obvious reasons.

This completes the description of the operations which must be performed by *split*. It must switch the colors of x and its sons, do the bottom part of a double rotation if necessary, then do the single rotation if necessary:

```
function split(v: integer; gg, g, f, x: link): link;
  begin
  x↑.red:=true; x↑.l↑.red:=false; x↑.r↑.red:=false;
  if f↑.red then
    begin
    g↑.red:=true;
    if (v<g↑.key)<>(v<f↑.key) then f:=rotate(v, g);
    x:=rotate(v, gg);
    x↑.red:=false
    end;
  head↑.r↑.red:=false;
  split:=x
  end;
```

This procedure takes care of fixing the colors after a rotation and also restarts x high enough in the tree to ensure that the search doesn't get lost due to all the link changes. The long argument list is included for clarity; this procedure should more properly be declared local to *rbtreeinsert*, with access to its variables.

If the root is a 4-node then the *split* procedure will make the root red, corresponding to transforming it, along with the dummy node above it into a 3-node. Of course, there is no reason to do this, so a statement is included at the end of *split* to keep the root black.

Assembling the code fragments above gives a very efficient, relatively simple algorithm for insertion using a binary tree structure that is guaranteed to take a logarithmic number of steps for all searches and insertions. This is one of the few searching algorithms with that property, and its use is justified whenever bad worst-case performance simply cannot be tolerated. Furthermore, this is achieved at very little cost. Searching is done just as quickly as if the balanced tree were constructed by the elementary algorithm, and insertion involves only one extra bit test and an occasional *split*. For random keys the height of the tree seems to be quite close to $\lg N$ (and only one or two *splits* are done for the average insertion) but no one has been able to analyze this statistic for any balanced tree algorithm. Thus a key in a file of, say, half a million records can be found by comparing it against only about twenty other keys.

Other Algorithms

The "top-down 2-3-4 tree" implementation using the "red-black" framework given in the previous section is one of several similar strategies than have been proposed for implementing balanced binary trees. As we saw above, it is actually the "rotate" operations that balance the trees: we've been looking at a particular view of the trees that makes it easy to decide when to rotate. Other views of the trees lead to other algorithms, a few of which we'll mention briefly in this section.

The oldest and most well-known data structure for balanced trees is the *AVL tree*. These trees have the property that the heights of the two subtrees of each node differ by at most one. If this condition is violated because of an insertion, it turns out that it can be reinstated using rotations. But this requires an extra loop: the basic algorithm is to search for the value being inserted, then proceed *up* the tree along the path just travelled adjusting the heights of nodes using rotations. Also, it is necessary to know whether each node has a height that is one less than, the same, or one greater than the height of its brother. This requires two bits if encoded in a straightforward way, though there is a way to get by with just one bit per node.

A second well-known balanced tree structure is the *2-3 tree*, where only 2-nodes and 3-nodes are allowed. It is possible to implement *insert* using an "extra loop" involving rotations as with AVL trees, but there is not quite enough flexibility to give a convenient top-down version.

In Chapter 18, we'll study the most important type of balanced tree, an extension of 2-3-4 trees called *B-trees*. These allow up to M keys per node for large M, and are widely used for searching applications involving very large files.

Exercises

1. Draw the top-down 2-3-4 tree that is built when the keys E A S Y Q U E S T I O N are inserted into an initially empty tree (in that order).

2. Draw a red-black representation of the tree from the previous question.

3. Exactly what links are modified by *split* and *rotate* when Z is inserted (after Y) into the example tree for this chapter?

4. Draw the red-black tree that results when the letters A to K are inserted in order, and describe what happens in general when keys are inserted into the trees in ascending order.

5. How many tree links actually must be changed for a double rotation, and how many are actually changed in the given implementation?

6. Generate two random 32-node red-black trees, draw them (either by hand or with a program), and compare them with the unbalanced binary search trees built with the same keys.

7. Generate ten random 1000-node red-black trees. Compute the number of rotations required to build the trees and the average distance from the root to an external node for the trees that you generate. Discuss the results.

8. With 1 bit per node for "color," we can represent 2-, 3-, and 4-nodes. How many different types of nodes could we represent if we used 2 bits per node for "color"?

9. Rotations are required in red-black trees when 3-nodes are made into 4-nodes in an "unbalanced" way. Why not eliminate rotations by allowing 4-nodes to be represented as any three nodes connected by two red links (perfectly balanced or not)?

10. Use a least-squares curvefitter to find values of a and b that give the best formula of the form $aN \lg N + bN$ for describing the total number of instructions executed when a red-black tree is built from N random keys.

16. Hashing

A completely different approach to searching from the comparison-based tree structures of the last section is provided by *hashing*: directly referencing records in a table by doing arithmetic transformations on keys into table addresses. If we were to know that the keys are distinct integers from 1 to N, then we could store the record with key i in table position i, ready for immediate access with the key value. Hashing is a generalization of this trivial method for typical searching applications when we don't have such specialized knowledge about the key values.

The first step in a search using hashing is to compute a *hash function* which transforms the search key into a table address. No hash function is perfect, and two or more different keys might hash to the same table address: the second part of a hashing search is a *collision resolution* process which deals with such keys. One of the collision resolution methods that we'll study uses linked lists, and is appropriate in a highly dynamic situation where the number of search keys can not be predicted in advance. The other two collision resolution methods that we'll examine achieve fast search times on records stored within a fixed array.

Hashing is a good example of a "time-space tradeoff." If there were no memory limitation, then we could do any search with only one memory access by simply using the key as a memory address. If there were no time limitation, then we could get by with only a minimum amount of memory by using a sequential search method. Hashing provides a way to use a reasonable amount of memory and time to strike a balance between these two extremes. Efficient use of available memory and fast access to the memory are prime concerns of any hashing method.

Hashing is a "classical" computer science problem in the sense that the various algorithms have been studied in some depth and are very widely used. There is a great deal of empirical and analytic evidence to support the utility of hashing for a broad variety of applications.

Hash Functions

The first problem we must address is the computation of the hash function which transforms keys into table addresses. This is an arithmetic computation with properties similar to the random number generators that we have studied. What is needed is a function which transforms keys (usually integers or short character strings) into integers in the range $[0..M-1]$, where M is the amount of memory available. An ideal hash function is one which is easy to compute and approximates a "random" function: for each input, every output should be "equally likely."

Since the methods that we will use are arithmetic, the first step is to transform keys into numbers which can be operated on (and are as large as possible). For example, this could involve removing bits from character strings and packing them together in a machine word. From now on, we'll assume that such an operation has been performed and that our keys are integers which fit into a machine word.

One commonly used method is to take M to be prime and, for any key k, compute $h(k) = k \bmod M$. This is a straightforward method which is easy to compute in many environments and spreads the key values out well.

A second commonly used method is similar to the linear congruential random number generator: take $M = 2^m$ and $h(k)$ to be the leading m bits of $(bk \bmod w)$, where w is the word size of the computer and b is chosen as for the random number generator. This can be more efficiently computed than the method above on some computers, and it has the advantage that it can spread out key values which are close to one another (e. g., *temp1, temp2, temp3*). As we've noted before, languages like Pascal are not well-suited to such operations.

Separate Chaining

The hash functions above will convert keys into table addresses: we still need to decide how to handle the case when two keys hash to the same address. The most straightforward method is to simply build a linked list, for each table address, of the records whose keys hash to that address. Since the keys which hash to the same table position are kept in a linked list, they might as well be kept in order. This leads directly to a generalization of the elementary list searching method that we discussed in Chapter 14. Rather than maintaining a single list with a single list header node *head* as discussed there, we maintain M lists with M list header nodes, initialized as follows:

```
type  link=↑node;
        node=record key, info: integer; next: link end;
var heads: array [0..M] of link;
    t, z: link;
procedure initialize;
    var i: integer;
    begin
    new(z);  z↑.next:=z;
    for i:=0 to M−1 do
        begin new(heads[i]);  heads[i]↑.next:=z end;
    end;
```

Now the procedures from Chapter 14 can be used as is, with a hash function used to choose among the lists. For example, *listinsert(v, heads[v mod M])* can be used to add something to the table, *t:=listsearch(v, heads[v mod M])* to find the first record with key *v*, and successively set *t:=listsearch(v, t)* until *t=z* to find subsequent records with key *v*.

For example if the *i*th letter in the alphabet is represented with the number *i* and we use the hash function $h(k) = k \bmod M$, then we get the following hash values for our sample set of keys with $M = 11$:

Key:	A	S	E	A	R	C	H	I	N	G	E	X	A	M	P	L	E
Hash:	1	8	5	1	7	3	8	9	3	7	5	2	1	2	5	1	5

if these keys are successively inserted into an initially empty table, the following set of lists would result:

0	1	2	3	4	5	6	7	8	9	10
	A	M	C		E		G	H	I	
	A	X	N		E		R	S		
	A				E					
	L				P					

Obviously, the amount of time required for a search depends on the length of the lists (and the relative positions of the keys in them). The lists could be left unordered: maintaining sorted lists is not as important for this application as it was for the elementary sequential search because the lists are quite short.

For an "unsuccessful search" (for a record with a key not in the table) we can assume that the hash function scrambles things enough so that each of the M lists is equally likely to be searched and, as in sequential list searching, that the list searched is only traversed halfway (on the average). The average length of the list examined (not counting z) in this example for unsuccessful search is $(0+4+2+2+0+4+0+2+2+1+0)/11 \approx 1.545$. This would be the average time for an unsuccessful search were the lists unordered; by keeping them ordered we cut the time in half. For a "successful search" (for one of the records in the table), we assume that each record is equally likely to be sought: seven of the keys would be found as the first list item examined, six would be found as the second item examined, etc., so the average is $(7 \cdot 1 + 6 \cdot 2 + 2 \cdot 3 + 2 \cdot 4)/17 \approx 1.941$. (This count assumes that equal keys are distinguished with a unique identifier or some other mechanism, and the search routine modified appropriately to be able to search for each individual key.)

If N, the number of keys in the table, is much larger than M then a good approximation to the average length of the lists is N/M. As in Chapter 14, unsuccessful and successful searches would be expected to go about halfway down some list. Thus, hashing provides an easy way to cut down the time required for sequential search by a factor of M, on the average.

In a separate chaining implementation, M is typically chosen to be relatively small so as not to use up a large area of contiguous memory. But it's probably best to choose M sufficiently large that the lists are short enough to make sequential search the most efficient method for them: "hybrid" methods (such as using binary trees instead of linked lists) are probably not worth the trouble.

The implementation given above uses a hash table of links to headers of the lists containing the actual keys. Maintaining M list header nodes is somewhat wasteful of space; it is probably worthwhile to eliminate them and make *heads* be a table of links to the first keys in the lists. This leads to some complication in the algorithm. For example, adding a new record to the beginning of a list becomes a different operation than adding a new record anywhere else in a list, because it involves modifying an entry in the table of links, not a field of a record. An alternate implementation is to put the first key within the table. If space is at a premium, it is necessary to carefully analyze the tradeoff between wasting space for a table of links and wasting space for a key and a link for each empty list. If N is much bigger than M then the alternate method is probably better, though M is usually small enough that the extra convenience of using list header nodes is probably justified.

Open Addressing

If the number of elements to be put in the hash table can be estimated in advance, then it is probably not worthwhile to use any links at all in the hash table. Several methods have been devised which store N records in a table of size $M > N$, relying on empty places in the table to help with collision resolution. Such methods are called *open-addressing* hashing methods.

The simplest open-addressing method is called *linear probing*: when there is a collision (when we hash to a place in the table which is already occupied and whose key is not the same as the search key), then just *probe* the next position in the table: that is, compare the key in the record there against the search key. There are three possible outcomes of the probe: if the keys match, then the search terminates successfully; if there's no record there, then the search terminates unsuccessfully; otherwise probe the next position, continuing until either the search key or an empty table position is found. If a record containing the search key is to be inserted following an unsuccessful search, then it can simply be put into the empty table space which terminated the search. This method is easily implemented as follows:

```
type node=record key, info: integer end;
var a: array [0..M] of node;
function h(k: integer): integer;
   begin h:=k mod M end;
procedure hashinitialize;
   var i: integer;
   begin
   for i:=0 to M do a[i].key:=maxint;
   end;
function hashinsert(v: integer): integer;
   var x: integer;
   begin
   x:=h(v);
   while a[x].key<>maxint do x:=(x+1) mod M;
   a[x].key:=v;
   hashinsert:=x;
   end;
```

Linear probing requires a special key value to signal an empty spot in the table: this program uses *maxint* for that purpose. The computation $x:=(x+1)$ **mod** M corresponds to examining the next position (wrapping back to the beginning when the end of the table is reached). Note that unsuccessful search is signaled by returning M, an index outside the hash table. Also note that this program does not checked for the table being filled to capacity.

The implementation of *hashsearch* is similar to *hashinsert*: simply add the condition "$a[x].key<>v$" to the **while** loop, and delete the following instruction which stores v. This leaves the calling routine with the task of checking if the search was unsuccessful, by testing whether the table position returned actually contains v (successful) or *maxint* (unsuccessful). Other conventions could be used, for example *hashsearch* could return M for unsuccessful search. For reasons that will become obvious below, open addressing is not appropriate if large numbers of records with duplicate keys are to be processed, but *hashsearch* can easily be adapted to handle equal keys in the case where each record has a unique identifier.

For our example set of keys with $M = 19$, we might get the hash values:

Key:	A	S	E	A	R	C	H	I	N	G	E	X	A	M	P	L	E
Hash:	1	0	5	1	18	3	8	9	14	7	5	5	1	13	16	12	5

The following table shows the steps of successively inserting these into an initially empty hash table:

0	1	2	3	4	5	6	7	8	9	10	11	12	13	14	15	16	17	18
	A																	
S	A																	
S	A				E													
S	A	A			E													
S	A	A			E													R
S	A	A	C		E													R
S	A	A	C		E			H										R
S	A	A	C		E			H	I									R
S	A	A	C		E			H	I				N					R
S	A	A	C		E		G	H	I				N					R
S	A	A	C		E	E	G	H	I				N					R
S	A	A	C		E	E	G	H	I	X			N					R
S	A	A	C	A	E	E	G	H	I	X			N					R
S	A	A	C	A	E	E	G	H	I	X			M	N				R
S	A	A	C	A	E	E	G	H	I	X			M	N		P		R
S	A	A	C	A	E	E	G	H	I	X		L	M	N		P		R
S	A	A	C	A	E	E	G	H	I	X	E	L	M	N		P		R

The table size is greater than before, since we must have $M > N$, but the total amount of memory space used is less, since no links are used. The average number of items that must be examined for a successful search for this example is: $33/17 \approx 1.941$.

Linear probing (indeed, any hashing method) works because it guarantees that, when searching for a particular key, we look at every key that hashes to the same table address (in particular, the key itself if it's in the table). Unfortunately, in linear probing, other keys are also examined, especially when the table begins to fill up: in the example above, the search for X involved looking at G, H, and I which did not have the same hash value. What's worse, insertion of a key with one hash value can drastically increase the search times for keys with other hash values: in the example, an insertion at position 17 would cause greatly increased search times for position 16. This phenomenon, called *clustering*, can make linear probing run very slowly for nearly full tables.

Fortunately, there is an easy way to virtually eliminate the clustering problem: *double hashing*. The basic strategy is the same; the only difference is that, instead of examining each successive entry following a collided position, we use a second hash function to get a fixed increment to use for the "probe" sequence. This is easily implemented by inserting $u:=h2(v)$ at the beginning of the procedure and changing $x:=(x+1) \bmod M$ to $x:=(x+u) \bmod M$ within the **while** loop. The second hash function $h2$ must be chosen with some care, otherwise the program might not work at all.

First, we obviously don't want to have $u=0$, since that would lead to an infinite loop on collision. Second, it is important that M and u be relatively prime here, since otherwise some of the probe sequences could be very short (for example, consider the case $M=2u$). This is easily enforced by making M prime and $u<M$. Third, the second hash function should be "different" from the first, otherwise a slightly more complicated clustering could occur. A function such as $h_2(k) = M - 2 - k \bmod (M-2)$ will produce a good range of "second" hash values.

For our sample keys, we get the following hash values:

Key	A	S	E	A	R	C	H	I	N	G	E	X	A	M	P	L	E
Hash 1:	1	0	5	1	18	3	8	9	14	7	5	5	1	13	16	12	5
Hash 2:	16	15	12	16	16	14	9	8	3	10	12	10	16	4	1	5	12

The following table is produced by successively inserting our sample keys into an initially empty table using double hashing with these values.

0	1	2	3	4	5	6	7	8	9	10	11	12	13	14	15	16	17	18
	A																	
S	A																	
S	A				**E**													
S	**A**				E												**A**	
S	A				E												A	**R**
S	A		**C**		E												A	R
S	A		C		E			**H**									A	R
S	A		C		E			H	**I**								A	R
S	A		C		E			H	I					**N**			A	R
S	A		C		E		**G**	H	I					N			A	R
S	A		C		**E**		G	H	I	**E**				N			**A**	R
S	A		C		**E**		G	H	I	E				N	**X**		A	R
S	**A**		C		E		G	H	I	E	**A**			N	X		**A**	R
S	A		C		E		G	H	I	E	A		**M**	N	X		A	R
S	A		C		E		G	H	I	E	A		M	N	X	**P**	A	R
S	A		C		E		G	H	I	E	A	**L**	M	N	X	P	A	R
S	**A**		**C**		**E**	**E**	G	**H**	I	**E**	A	L	**M**	N	**X**	P	**A**	R

This technique uses the same amount of space as linear probing but the average number of items examined for successful search is slightly smaller: $32/17 \approx 1.882$. Still, such a full table is to be avoided as shown by the 9 probes used to insert the last E in this example.

Open addressing methods can be inconvenient in a dynamic situation, when an unpredictable number of insertions and deletions might have to be processed. First, how big should the table be? Some estimate must be made of how many insertions are expected but performance degrades drastically as the table starts to get full. A common solution to this problem is to rehash everything into a larger table on a (very) infrequent basis. Second, a word of caution is necessary about deletion: a record can't simply be removed from a table built with linear probing or double hashing. The reason is that later insertions into the table might have skipped over that record, and searches for those records will terminate at the hole left by the deleted record. A way to solve this problem is to have another special key which can serve as a placeholder for searches but can be identified and remembered as an

empty position for insertions. Note that neither table size nor deletion are a
particular problem with separate chaining.

Analytic Results

The methods discussed above have been analyzed completely and it is possible to compare their performance in some detail. The following formulas,
summarized from detailed analyses described by D. E. Knuth in his book on
sorting and searching, give the average number of items examined (probes) for
unsuccessful and successful searching using the methods we've studied. The
formulas are most conveniently expressed in terms of the "load factor" of the
hash table, $\alpha = N/M$. Note that for separate chaining we can have $\alpha > 1$,
but for the other methods we must have $\alpha < 1$.

	Unsuccessful	*Successful*
Separate Chaining:	$1 + \alpha/2$	$(\alpha + 1)/2$
Linear Probing:	$1/2 + 1/2(1 - \alpha)^2$	$1/2 + 1/2(1 - \alpha)$
Double Hashing:	$1/(1 - \alpha)$	$-\ln(1 - \alpha)/\alpha$

For small α, it turns out that all the formulas reduce to the basic result that
unsuccessful search takes about $1 + N/M$ probes and successful search takes
about $1 + N/2M$ probes. (Except, as we've noted, the cost of an unsuccessful
search for separate chaining is reduced by about half by ordering the lists.)
The formulas indicate how badly performance degrades for open addressing
as α gets close to 1. For large M and N, with a table about 90% full, linear
probing will take about 50 probes for an unsuccessful search, compared to
10 for double hashing. Comparing linear probing and double hashing against
separate chaining is more complicated, because there is more memory available
in the open addressing methods (since there are no links). The value of α used
should be modified to take this into account, based on the relative size of keys
and links. This means that it is not normally justified to choose separate
chaining over double hashing on the basis of performance.

   The choice of the very best hashing method for a particular application can be very difficult. However, the very best method is rarely needed
for a given situation, and the various methods do have similar performance
characteristics as long as the memory resource is not being severely strained.
Generally, the best course of action is to use the simple separate chaining
method to reduce search times drastically when the number of records to be
processed is not known in advance (and a good storage allocator is available)
and to use double hashing to search among a set of keys whose size can be
roughly predicted ahead of time.

   Many other hashing methods have been developed which have application
in some special situations. Although we can't go into details, we'll briefly

consider two examples to illustrate the nature of specially adapted hashing methods. These and many other methods are fully described in Knuth's book.

The first, called *ordered hashing*, is a method for making use of ordering within an open addressing table: in standard linear probing, we stop the search when we find an empty table position or a record with a key equal to the search key; in ordered hashing, we stop the search when we find a record with a key greater than or equal to the search key (the table must be cleverly constructed to make this work). This method turns out to reduce the time for unsuccessful search to approximately that for successful search. (This is the same kind of improvement that comes in separate chaining.) This method is useful for applications where unsuccessful searching is frequently used. For example, a text processing system might have an algorithm for hyphenating words that works well for most words, but not for bizarre cases (such as "bizarre"). The situation could be handled by looking up all words in a relatively small *exception dictionary* of words which must be handled in a special way, with most searches likely to be unsuccessful.

Similarly, there are methods for moving some records around during unsuccessful search to make successful searching more efficient. In fact, R. P. Brent developed a method for which the average time for a successful search can be bounded by a constant, giving a very useful method for applications with frequent successful searching in very large tables such as dictionaries.

These are only two examples of a large number of algorithmic improvements which have been suggested for hashing. Many of these improvements are interesting and have important applications. However, our usual cautions must be raised against premature use of advanced methods except by experts with serious searching applications, because separate chaining and double hashing are simple, efficient, and quite acceptable for most applications.

Hashing is preferred to the binary tree structures of the previous two chapters for many applications because it is somewhat simpler and it can provide very fast (constant) searching times, if space is available for a large enough table. Binary tree structures have the advantages that they are dynamic (no advance information on the number of insertions is needed), they can provide guaranteed worst-case performance (everything could hash to the same place even in the best hashing method), and they support a wider range of operations (most important, the *sort* function). When these factors are not important, hashing is certainly the searching method of choice.

Exercises

1. Describe how you might implement a hash function by making use of a good random number generator. Would it make sense to implement a random number generator by making use of a hash function?

2. How long could it take in the worst case to insert N keys into an initially empty table, using separate chaining with unordered lists? Answer the same question for sorted lists.

3. Give the contents of the hash table that results when the keys E A S Y Q U E S T I O N are inserted in that order into an initially empty table of size 13 using linear probing. (Use $h_1(k) = k \bmod 13$ for the hash function for the kth letter of the alphabet.)

4. Give the contents of the hash table that results when the keys E A S Y Q U E S T I O N are inserted in that order into an initially empty table of size 13 using double hashing. (Use $h_1(k)$ from the previous question, $h_2(k) = 1 + (k \bmod 11)$ for the second hash function.)

5. About how many probes are involved when double hashing is used to build a table consisting of N equal keys?

6. Which hashing method would you use for an application in which many equal keys are likely to be present?

7. Suppose that the number of items to be put into a hash table is known in advance. Under what condition will separate chaining be preferable to double hashing?

8. Suppose a programmer has a bug in his double hashing code so that one of the hash functions always returns the same value (not 0). Describe what happens in each situation (when the first one is wrong and when the second one is wrong).

9. What hash function should be used if it is known in advance that the key values fall into a relatively small range?

10. Criticize the following algorithm for deletion from a hash table built with linear probing. Scan right from the element to be deleted (wrapping as necessary) to find an empty position, then scan left to find an element with the same hash value. Then replace the element to be deleted with that element, leaving its table position empty.

17. Radix Searching

Several searching methods proceed by examining the search keys one bit at a time (rather than using full comparisons between keys at each step). These methods, called *radix searching methods*, work with the bits of the keys themselves, as opposed to the transformed version of the keys used in hashing. As with radix sorting methods, these methods can be useful when the bits of the search keys are easily accessible and the values of the search keys are well distributed.

The principal advantages of radix searching methods are that they provide reasonable worst-case performance without the complication of balanced trees; they provide an easy way to handle variable-length keys; some allow some savings in space by storing part of the key within the search structure; and they can provide very fast access to data, competitive with both binary search trees and hashing. The disadvantages are that biased data can lead to degenerate trees with bad performance (and data comprised of characters is biased) and that some of the methods can make very inefficient use of space. Also, as with radix sorting, these methods are designed to take advantage of particular characteristics of the computer's architecture: since they use digital properties of the keys, it's difficult or impossible to do efficient implementations in languages such as Pascal.

We'll examine a series of methods, each one correcting a problem inherent in the previous one, culminating in an important method which is quite useful for searching applications where very long keys are involved. In addition, we'll see the analogue to the "linear-time sort" of Chapter 10, a "constant-time" search which is based on the same principle.

Digital Search Trees

The simplest radix search method is digital tree searching: the algorithm is precisely the same as that for binary tree searching, except that rather than

213

branching in the tree based on the result of the comparison between the keys, we branch according to the key's bits. At the first level the leading bit is used, at the second level the second leading bit, and so on until an external node is encountered. The code for this is virtually the same as the code for binary tree search. The only difference is that the key comparisons are replaced by calls on the *bits* function that we used in radix sorting. (Recall from Chapter 10 that $bits(x, k, j)$ is the j bits which appear k from the right and can be efficiently implemented in machine language by shifting right k bits then setting to 0 all but the rightmost j bits.)

```
function digitalsearch(v: integer; x: link): link;
  var b: integer;
  begin
  z↑.key:=v;  b:=maxb;
  repeat
    if bits(v, b, 1)=0 then x:=x↑.l else x:=x↑.r;
    b:=b−1;
  until v=x↑.key;
  digitalsearch:=x
  end;
```

The data structures for this program are the same as those that we used for elementary binary search trees. The constant *maxb* is the number of bits in the keys to be sorted. The program assumes that the first bit in each key (the $(maxb+1)$st from the right) is 0 (perhaps the key is the result of a call to *bits* with a third argument of *maxb*), so that searching is done by setting $x:=$ *digitalsearch*$(v, head)$, where *head* is a link to a tree header node with 0 key and a left link pointing to the search tree. Thus the initialization procedure for this program is the same as for binary tree search, except that we begin with $head↑.l:=z$ instead of $head↑.r:=z$.

We saw in Chapter 10 that equal keys are anathema in radix sorting; the same is true in radix searching, not in this particular algorithm, but in the ones that we'll be examining later. Thus we'll assume in this chapter that all the keys to appear in the data structure are distinct: if necessary, a linked list could be maintained for each key value of the records whose keys have that value. As in previous chapters, we'll assume that the ith letter of the alphabet is represented by the five-bit binary representation of i. That is, we'll use the following sample keys in this chapter:

A	00001
S	10011
E	00101
R	10010
C	00011
H	01000
I	01001
N	01110
G	00111
X	11000
M	01101
P	10000
L	01100

To be consistent with *bits*, we consider the bits to be numbered 0–4, from right to left. Thus bit 0 is A's only nonzero bit and bit 4 is P's only nonzero bit.

The insert procedure for digital search trees also derives directly from the corresponding procedure for binary search trees:

```
function digitalinsert(v: integer; x: link): link;
  var f: link; b: integer;
  begin
  b:=maxb;
  repeat
    f:=x;
    if bits(v, b, 1)=0 then x:=x↑.l else x:=x↑.r;
    b:=b−1;
  until x=z;
  new(x); x↑.key:=v; x↑.l:=z; x↑.r:=z;
  if bits(v, b+1, 1)=0 then f↑.l:=x else f↑.r:=x;
  digitalinsert:=x
  end;
```

To see how the algorithm works, consider what happens when a new key Z= 11010 is added to the tree below. We go right twice because the leading two bits of Z are 1, then we go left, where we hit the external node at the left of X, where Z would be inserted.

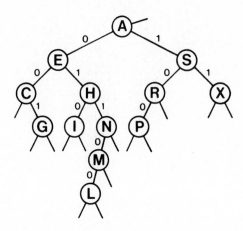

The worst case for trees built with digital searching will be much better than for binary search trees. The length of the longest path in a digital search tree is the length of the longest match in the leading bits between any two keys in the tree, and this is likely to be relatively short. And it is obvious that no path will ever be any longer than the number of bits in the keys: for example, a digital search tree built from eight-character keys with, say, six bits per character will have no path longer than 48, even if there are hundreds of thousands of keys. For random keys, digital search trees are nearly perfectly balanced (the height is about $\lg N$). Thus, they provide an attractive alternative to standard binary search trees, *provided* that bit extraction can be done as easily as key comparison (which is not really the case in Pascal).

Radix Search Tries

It is quite often the case that search keys are very long, perhaps consisting of twenty characters or more. In such a situation, the cost of comparing a search key for equality with a key from the data structure can be a dominant cost which cannot be neglected. Digital tree searching uses such a comparison at each tree node: in this section we'll see that it is possible to get by with only one comparison per search in most cases.

The idea is to not store keys in tree nodes at all, but rather to put all the keys in external nodes of the tree. That is, instead of using z for external nodes of the structure, we put nodes which contain the search keys. Thus, we have two types of nodes: internal nodes, which just contain links to other nodes, and external nodes, which contain keys and no links. (E. Fredkin named this method "trie" because it is useful for re*trie*val; in conversation it's usually pronounced "try-ee" or just "try" for obvious reasons.) To search for a key in such a structure, we just branch according to its bits, as above, but we don't compare it to anything until we get to an external node. Each key in the tree is stored in an external node on the path described by the leading bit pattern of the key and each search key winds up at one external node, so one full key comparison completes the search.

After an unsuccessful search, we can insert the key sought by replacing the external node which terminated the search by an internal node which will have the key sought and the key which terminated the search in external nodes below it. Unfortunately, if these keys agree in more bit positions, it is necessary to add some external nodes which do not correspond to any keys in the tree (or put another way, some internal nodes which have an empty external node as a son). The following is the (binary) radix search trie for our sample keys:

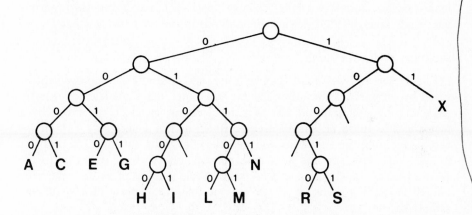

Now inserting Z=11010 into this tree involves replacing X with a new internal node whose left son is another new internal node whose sons are X and Z.

The implementation of this method in Pascal is actually relatively complicated because of the necessity to maintain two types of nodes, both of which could be pointed to by links in internal nodes. This is an example of an algorithm for which a low-level implementation might be simpler than a high-level implementation. We'll omit the code for this because we'll see an improvement below which avoids this problem.

The left subtree of a binary radix search trie has all the keys which have 0 for the leading bit; the right subtree has all the keys which have 1 for the leading bit. This leads to an immediate correspondence with radix sorting: binary trie searching partitions the file in exactly the same way as radix exchange sorting. (Compare the trie above with the partitioning diagram we examined for radix exchange sorting, after noting that the keys are slightly different.) This correspondence is analogous to that between binary tree searching and Quicksort.

An annoying feature of radix tries is the "one-way" branching required for keys with a large number of bits in common. For example, keys which differ only in the last bit require a path whose length is equal to the key length, no matter how many keys there are in the tree. The number of internal nodes can be somewhat larger than the number of keys. The height of such trees is still limited by the number of bits in the keys, but we would like to consider the possibility of processing records with very long keys (say 1000 bits or more) which perhaps have some uniformity, as might occur in character encoded data. One way to shorten the paths in the trees is to use many more than two links per node (though this exacerbates the "space" problem of using too many nodes); another way is to "collapse" paths containing one-way branches into single links. We'll discuss these methods in the next two sections.

Multiway Radix Searching

For radix sorting, we found that we could get a significant improvement in speed by considering more than one bit at a time. The same is true for radix searching: by examining m bits at a time, we can speed up the search by a factor of 2^m. However, there's a catch which makes it necessary to be more careful applying this idea than was necessary for radix sorting. The problem is that considering m bits at a time corresponds to using tree nodes with $M = 2^m$ links, which can lead to a considerable amount of wasted space for unused links. For example, if $M = 4$ the following tree is formed for our sample keys:

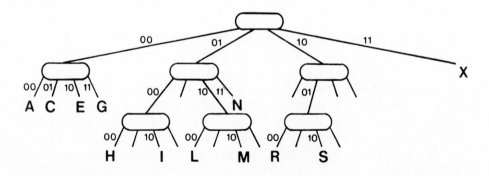

Note that there is some wasted space in this tree because of the large number of unused external links. As M gets larger, this effect gets worse: it turns out that the number of links used is about $MN/\ln M$ for random keys. On the other hand this provides a very efficient searching method: the running time is about $\log_M N$. A reasonable compromise can be struck between the time efficiency of multiway tries and the space efficiency of other methods by using a "hybrid" method with a large value of M at the top (say the first two levels) and a small value of M (or some elementary method) at the bottom. Again, efficient implementations of such methods can be quite complicated because of multiple node types.

For example, a two-level 32-way tree will divide the keys into 1024 categories, each accessible in two steps down the tree. This would be quite useful for files of thousands of keys, because there are likely to be (only) a few keys per category. On the other hand, a smaller M would be appropriate for files of hundreds of keys, because otherwise most categories would be empty and too much space would be wasted, and a larger M would be appropriate for files with millions of keys, because otherwise most categories would have too many keys and too much time would be wasted.

It is amusing to note that "hybrid" searching corresponds quite closely to the way humans search for things, for example, names in a telephone book. The first step is a multiway decision ("Let's see, it starts with 'A'"), followed perhaps by some two way decisions ("It's before 'Andrews', but after 'Aitken'") followed by sequential search (" 'Algonquin' ... 'Algren' ... No, 'Algorithms' isn't listed!"). Of course computers are likely to be somewhat better than humans at multiway search, so two levels are appropriate. Also, 26-way branching (with even more levels) is a quite reasonable alternative to consider for keys which are composed simply of letters (for example, in a dictionary).

In the next chapter, we'll see a systematic way to adapt the structure to take advantage of multiway radix searching for arbitrary file sizes.

Patricia

The radix trie searching method as outlined above has two annoying flaws: there is "one-way branching" which leads to the creation of extra nodes in the tree, and there are two different types of nodes in the tree, which complicates the code somewhat (especially the insertion code). D. R. Morrison discovered a way to avoid both of these problems in a method which he named *Patricia* ("Practical Algorithm To Retrieve Information Coded In Alphanumeric"). The algorithm given below is not in precisely the same form as presented by Morrison, because he was interested in "string searching" applications of the type that we'll see in Chapter 19. In the present context, Patricia allows searching for N arbitrarily long keys in a tree with just N nodes, but requires only one full key comparison per search.

One-way branching is avoided by a simple device: each node contains the index of the bit to be tested to decide which path to take out of that node. External nodes are avoided by replacing links to external nodes with links that point upwards in the tree, back to our normal type of tree node with a key and two links. But in Patricia, the keys in the nodes are not used on the way down the tree to control the search; they are merely stored there for reference when the bottom of the tree is reached. To see how Patrica works, we'll first look at the search algorithm operating on a typical tree, then we'll examine how the tree is constructed in the first place. For our example keys, the following Patricia tree is constructed when the keys are successively inserted.

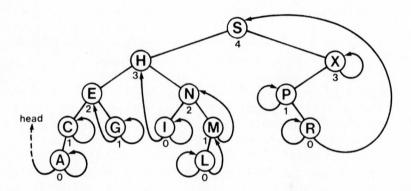

To search in this tree, we start at the root and proceed down the tree, using the bit index in each node to tell us which bit to examine in the search key, going right if that bit is 1, left if it is 0. The keys in the nodes are not examined at all on the way down the tree. Eventually, an upwards link is encountered: each upward link points to the unique key in the tree that has the bits that would cause a search to take that link. For example, S is the only key in the tree that matches the bit pattern 10x11. Thus if the key at the node pointed to by the first upward link encountered is equal to the search key, then the search is successful, otherwise it is unsuccessful. For tries, all searches terminate at external nodes, whereupon one full key comparison is done to determine whether the search was successful or not; for Patricia all searches terminate at upwards links, whereupon one full key comparison is done to determine whether the search was successful or not. Futhermore, it's easy to test whether a link points up, because the bit indices in the nodes (by

definition) decrease as we travel down the tree. This leads to the following
search code for Patricia, which is as simple as the code for radix tree or trie
searching:

```
type  link=↑node;
       node=record key, info, b: integer; l, r: link end;
var head: link;
function patriciasearch(v: integer; x: link): link;
  var f: link;
  begin
  repeat
    f:=x;
    if bits(v, x↑.b, 1)=0 then x:=x↑.l else x:=x↑.r;
  until f↑.b<=x↑.b;
  patriciasearch:=x
  end;
```

This function returns a link to the unique node which could contain the record
with key v. The calling routine then can test whether the search was successful
or not. Thus to search for Z=11010 in the above tree we go right, then up at
the right link of X. The key there is not Z so the search is unsuccessful.

The following diagram shows the transformations made on the right
subtree of the tree above if Z, then T are added.

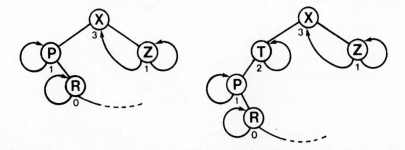

The search for Z=11010 ends at the node containing X=11000. By the defining
property of the tree, X is the only key in the tree for which a search would
terminate at that node. If Z is inserted, there would be two such nodes, so
the upward link that was followed into the node containing X should be made
to point to a new node containing Z, with a bit index corresponding to the
leftmost point where X and Z differ, and with two upward links: one pointing
to X and the other pointing to Z. This corresponds precisely to replacing the

external node containing X with a new internal node with X and Z as sons in radix trie insertion, with one-way branching eliminated by including the bit index.

The insertion of T=10100 illustrates a more complicated case. The search for T ends at P=10000, indicating that P is the only key in the tree with the pattern 10x0x. Now, T and P differ at bit 2, a position that was skipped during the search. The requirement that the bit indices decrease as we go down the tree dictates that T be inserted between X and P, with an upward self pointer corresponding to its own bit 2. Note carefully that the fact that bit 2 was skipped before the insertion of T implies that P and R have the same bit 2 value.

The examples above illustrate the only two cases that arise in insertion for Patricia. The following implementation gives the details:

```
function patriciainsert(v: integer;  x: link): link;
   var t, f: link;  i: integer;
   begin
   t:=patriciasearch(v,  x);
   i:=maxb;
   while bits(v, i, 1)=bits(t↑.key, i, 1) do i:=i−1;
   repeat
       f:=x;
       if bits(v, x↑.b, 1)=0 then x:=x↑.l else x:=x↑.r;
   until (x↑.b<=i) or (f↑.b<=x↑.b);
   new(t);  t↑.key:=v;  t↑.b:=i;
   if bits(v, t↑.b, 1)=0
       then begin t↑.l:=t;  t↑.r:=x end
       else  begin t↑.r:=t;  t↑.l:=x end;
   if bits(v, f↑.b, 1)=0 then f↑.l:=t else f↑.r:=t;
   patriciainsert:=t
   end;
```

(This code assumes that *head* is initialized with key field of 0, a bit index of *maxb* and both links upward self pointers.) First, we do a search to find the key which must be distinguished from *v*, then we determine the leftmost bit position at which they differ, travel down the tree to that point, and insert a new node containing *v* at that point.

Patricia is the quintessential radix searching method: it manages to identify the bits which distinguish the search keys and build them into a data structure (with no surplus nodes) that quickly leads from any search key to the only key in the data structure that could be equal. Clearly, the same technique as used in Patricia can be used in binary radix trie searching to eliminate one-way branching, but this only exacerbates the multiple node type problem.

Unlike standard binary tree search, the radix methods are insensitive to the order in which keys are inserted; they depend only upon the structure of the keys themselves. For Patricia the placement of the upwards links depend on the order of insertion, but the tree structure depends only on the bits in the keys, as for the other methods. Thus, even Patricia would have trouble with a set of keys like 001, 0001, 00001, 000001, etc., but for normal key sets, the tree should be relatively well-balanced so the number of bit inspections, even for very long keys, will be roughly proportional to $\lg N$ when there are N nodes in the tree.

The most useful feature of radix trie searching is that it can be done efficiently with keys of varying length. In all of the other searching methods we have seen the length of the key is "built into" the searching procedure in some way, so that the running time is dependent on the length of the keys as well as the number of keys. The specific savings available depends on the method of bit access used. For example, suppose we have a computer which can efficiently access 8-bit "bytes" of data, and we have to search among hundreds of 1000-bit keys. Then Patricia would require access of only about 9 or 10 bytes of the search key for the search, plus one 125-byte equality comparison while hashing requires access of all 125-bytes of the search key for computing the hash function plus a few equality comparisons, and comparison-based methods require several long comparisons. This effect makes Patricia (or radix trie searching with one-way branching removed) the search method of choice when very long keys are involved.

Exercises

1. Draw the digital search tree that results when the keys E A S Y Q U E S T I O N are inserted into an initially empty tree (in that order).

2. Generate a 1000 node digital search tree and compare its height and the number of nodes at each level against a standard binary search tree and a red-black tree (Chapter 15) built from the same keys.

3. Find a set of 12 keys that make a particulary badly balanced digital search trie.

4. Draw the radix search trie that results when the keys E A S Y Q U E S T I O N are inserted into an initially empty tree (in that order).

5. A problem with 26-way multiway radix search tries is that some letters of the alphabet are very infrequently used. Suggest a way to fix this problem.

6. Describe how you would delete an element from a multiway radix search tree.

7. Draw the Patricia tree that results when the keys E A S Y Q U E S T I O N are inserted into an initially empty tree (in that order).

8. Find a set of 12 keys that make a particulary badly balanced Patricia tree.

9. Write a program that prints out all keys in a Patricia tree having the same initial t bits as a given search key.

10. Use a least-squares curvefitter to find values of a and b that give the best formula of the form $aN \lg N + bN$ for describing the total number of instructions executed when a Patricia tree is built from N random keys.

18. External Searching

Searching algorithms appropriate for accessing items from very large files are of immense practical importance. Searching is the fundamental operation on large data files, and certainly consumes a very significant fraction of the resources used in many computer installations.

We'll be concerned mainly with methods for searching on large disk files, since disk searching is of the most practical interest. With sequential devices such as tapes, searching quickly degenerates to the trivially slow method: to search a tape for an item, one can't do much better than to mount the tape and read until the item is found. Remarkably, the methods that we'll study can find an item from a disk as large as a billion words with only two or three disk accesses.

As with external sorting, the "systems" aspect of using complex I/O hardware is a primary factor in the performance of external searching methods that we won't be able to study in detail. However, unlike sorting, where the external methods are really quite different from the internal methods, we'll see that external searching methods are logical extensions of the internal methods that we've studied.

Searching is a fundamental operation for disk devices. Files are typically organized to take advantage of particular device characteristics to make access of information as efficient as possible. As we did with sorting, we'll work with a rather simple and imprecise model of "disk" devices in order to explain the principal characteristics of the fundamental methods. Determining the best external searching method for a particular application is extremely complicated and very dependent on characteristics of the hardware (and systems software), and so it is quite beyond the scope of this book. However, we can suggest some general approaches to use.

For many applications we would like to frequently change, add, delete or (most important) quickly access small bits of information inside very, very large files. In this chapter, we'll examine some methods for such dynamic situations which offer the same kinds of advantages over the straightforward methods that binary search trees and hashing offer over binary search and sequential search.

A very large collection of information to be processed using a computer is called a *database*. A great deal of study has gone into methods of building, maintaining and using databases. However, large databases have very high inertia: once a very large database has been built around a particular searching strategy, it can be very expensive to rebuild it around another. For this reason, the older, static methods are in widespread use and likely to remain so, though the newer, dynamic methods are beginning to be used for new databases.

Database applications systems typically support much more complicated operations than a simple search for an item based on a single key. Searches are often based on criteria involving more than one key and are expected to return a large number of records. In later chapters we'll see some examples of algorithms which are appropriate for some search requests of this type, but general search requests are sufficiently complicated that it is typical to do a sequential search over the entire database, testing each record to see if it meets the criteria.

The methods that we will discuss are of practical importance in the implementation of large file systems in which every file has a unique identifier and the purpose of the file system is to support efficient access, insertion and deletion based on that identifier. Our model will consider the disk storage to be divided up into *pages*, contiguous blocks of information that can be efficiently accessed by the disk hardware. Each page will hold many records; our task is to organize the records within the pages in such a way that any record can be accessed by reading only a few pages. We assume that the I/O time required to read a page completely dominates the processing time required to do any computing involving that page. As mentioned above, this is an oversimplified model for many reasons, but it retains enough characteristics of actual external storage devices to allow us to consider some of the fundamental methods which are used.

Indexed Sequential Access

Sequential disk searching is the natural extension of the elementary sequential searching methods that we considered in Chapter 14: the records are stored in increasing order of their keys, and searches are done by simply reading in the records one after the other until one containing a key greater than or equal to the search key is found. For example, if our search keys come from E X T E R N A L S E A R C H I N G E X A M P L E and we have disks capable of holding three pages of four records each, then we would have the configuration:

```
Disk 1:  A  A  A  C     E  E  E     E  G  H  I
Disk 2:  L  L  M  N     N  P  R  R     S  T  X  X
```

As with external sorting, we must consider very small examples to understand the algorithms but think about very large examples to appreciate their performance. Obviously, pure sequential searching is unattractive because, for example, searching for W in the example above would require reading all the pages.

To vastly improve the speed of a search, we can keep, for each disk, an "index" of which keys belong to which pages on that disk, as in the following example:

```
Disk 1:  * 1 c 2 e    A  A  A  C     E  E  E  E
Disk 2:  e 1 i 2 n    E  G  H  I     L  L  M  N
Disk 3:  n 1 r 2 x    N  P  R  R     S  T  X  X
```

The first page of each disk is its index: lower case letters indicate that only the key value is stored, not the full record; numbers are page indices. In the index, each page number is followed by the value of its last key and preceded by the value of the last key on the previous page. (The "*" is a sentinel key, smaller than all the others.) Thus, for example, the index for disk 2 says that its first page contains records with keys between E and I inclusive and its second page contains records with keys between I and N inclusive. Normally, it should be possible to fit many more keys and page indices on an index page than records on a "data" page; in fact, the index for a whole disk should require only a few pages. These indices are coupled with a "master index" which tells which keys are on which disk. For our example, the master index would be "* **1** e **2** n **3** x," where boldface integers are disk numbers. The master index is likely to be small enough that it can be kept in memory, so that most records can be found with only two pages accessed, one for the index on the appropriate disk and one for the page containing the approriate record. For example, a search for W would involve first reading the index page from disk 3, then reading the second page (from disk 3) which is the only one that could contain W. Searches for keys which appear in the index require reading three pages the index plus the two pages flanking the key value in the index. If no duplicate keys are in the file, then the extra page access can be avoided. On the other hand, if there are many equal keys in the file, several page accesses might be called for (records with equal keys might fill several pages).

Because it combines a sequential key organization with indexed access, this organization is called *indexed sequential.* It is the method of choice for applications where changes to the database are likely to be made infrequently. The disadvantage of using indexed sequential access is that it is very inflexible. For example, adding B to the configuration above requires that virtually the whole database be rebuilt, with new positions for many of the keys and new values for the indices.

B-Trees

A better way to handle searching in a dynamic situation is to use balanced trees. In order to reduce the number of (relatively expensive) disk accesses, it is reasonable to allow a large number of keys per node so that the nodes have a large branching factor. Such trees were named B-trees by R. Bayer and E. McCreight, who were the first to consider the use of multiway balanced trees for external searching. (Many people reserve the term "B-tree" to describe the exact data structure built by the algorithm suggested by Bayer and McCreight; we'll use it as a generic term to mean "external balanced trees.")

The top-down algorithm that we used for 2-3-4 trees extends readily to handle more keys per node: assume that there are anywhere from 1 to $M - 1$ keys per node (and so anywhere from 2 to M links per node). Searching proceeds in a way analogous to 2-3-4 trees: to move from one node to the next, first find the proper interval for the search key in the current node and then exit through the corresponding link to get to the next node. Continue in this way until an external node is reached, then insert the new key into the last internal node reached. As with top-down 2-3-4 trees, it is necessary to "split" nodes that are "full" on the way down the tree: any time we see a k-node attached to an M node, we replace it by a $(k + 1)$-node attached to two $M/2$ nodes. This guarantees that when the bottom is reached there is room to insert the new node. The B-tree constructed for $M = 4$ and our sample keys is diagrammed below:

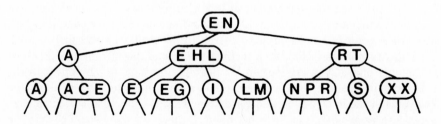

This tree has 13 nodes, each corresponding to a disk page. Each node must contain links as well as records. The choice $M = 4$ (even though it leaves us with familiar 2-3-4 trees) is meant to emphasize this point: before we could fit four records per page, now only three will fit, to leave room for the links. The actual amount of space used up depends on the relative size of records and links. We'll see a method below which avoids this mixing of records and links.

For example, the root node might be stored as "10 E 11 N 12", indicating that the root of the subtree containing records with keys less than or equal to E is on page 0 of disk 1, etc. Just as we kept the master index for indexed sequential search in memory, it's reasonable to keep the root node of the B-tree in memory. The other nodes for our example might be stored as follows:

Disk 1:	20 A 21	22 E 30 H 31 L 32	40 R 41 T 42
Disk 2:	0 A 0	0 A 0 C 0 E 0	0 E 0
Disk 3:	0 E 0 G 0	0 I 0	0 L 0 M 0
Disk 4:	0 N 0 P 0 R 0	0 S 0	0 X 0 X 0

The assignment of nodes to disk pages in this example is simply to proceed down the tree, working from right to left at each level, assigning nodes to disk 1, then disk 2, etc. In an actual application, other assignments might be indicated. For example, it might be better to avoid having all searches going through disk 1 by assigning first to page 0 of all the disks, etc. In truth, more sophisticated strategies are needed because of the dynamics of the tree construction (consider the difficulty of implementing a *split* routine that respects either of the above strategies).

The nodes at the bottom level in the B-trees described above all contain many 0 links which can be eliminated by marking such nodes in some way. Furthermore, a much larger value of M can be used at the higher levels of the tree if we store just keys (not full records) in the interior nodes as in indexed sequential access. To see how to take advantage of these observations in our example, suppose that we can fit up to seven keys and eight links on a page, so that we can use $M = 8$ for the interior nodes and $M = 5$ for the bottom-level nodes (*not* $M = 4$ because no space for links need be reserved at the bottom). A bottom node splits when a fifth record is added to it; the split involves "inserting" the key of the middle record into the tree above, which operates as a normal B-tree from $M = 8$ (on stored keys, not records). This leads to the following tree:

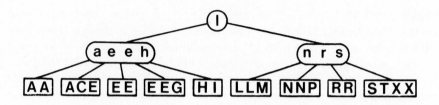

The effect for a typical application is likely to be much more dramatic since the branching factor of the tree is increased by roughly the ratio of the record size to key size, which is likely to be large. Also, with this type of organization, the "index" (which contains keys and links) can be separated from the actual records, as in indexed sequential search:

Disk 1:	11 l 12	20 a 21 e 22 e 30 h 31	32 n 40 r 41 s 42
Disk 2:	A A	A C E	E E
Disk 3:	E E G	H I	L L M
Disk 4:	N N P	R R	S T X X

As before, the root node is kept in memory. Also the same issues as discussed above regarding node placement on the disks arise.

Now we have two values of M, one for the interior nodes which determines the branching factor of the tree (M_I) and one for the bottom-level nodes which determines the allocation of records to pages (M_B). To minimize the number of disk accesses, we want to make both M_I and M_B as large as possible, even at the expense of some extra computation. On the other hand, we don't want to make M_I huge, because then most tree nodes would be largely empty and space would be wasted and we don't want to make M_B huge because this would reduce to sequential search of the bottom-level nodes. Usually, it is best to relate both M_I and M_B to the page size. The obvious choice for M_B is the number of records that can fit on a page: the goal of the search is to find the page containing the record sought. If M_I is taken to be the number of keys that can fit on two to four pages, then the B-tree is likely to be only be three levels deep, even for very large files (a three-level tree with $M_I = 1024$ can handle up to 1024^3, or over a billion, entries). But recall that the root node of the tree, which is accessed for every operation on the tree, is kept in memory, so that only two disk accesses are required to find any element in the file.

As discussed in Chapter 15, a more complicated "bottom-up" insertion method is commonly used for B-trees, though the distinction between top-down and bottom up methods loses importance for three level trees. Other variations of balanced trees are more important for external searching. For example, when a node becomes full, splitting (and the resultant half-empty nodes) can be forestalled by dumping some of the contents of the node into its "brother" node (if it's not too full). This leads to better space utilization within the nodes, which is likely to be of central concern in a large-scale disk searching application.

Extendible Hashing

An alternative to B-trees which extends digital searching algorithms to apply to external searching was developed in 1978 by R. Fagin, J. Nievergelt, N. Pippenger, and R. Strong. This method, called *extendible hashing*, guarantees that no more than two disk accesses will be used for any search. As with B-trees, our records are stored on pages which are split into two pieces when they fill up; as with indexed sequential access, we maintain an index which we access to find the page containing the records which match our search key. Extendible hashing combines these approaches by using digital properties of the search keys.

To see how extendible hashing works, we'll consider how it handles successive insertions of keys from E X T E R N A L S E A R C H I N G E X A M P L E, using pages with a capacity of up to four records.

We start with an "index" with just one entry, a pointer to the page which is to hold the records. The first four records fit on the page, leaving the following trivial structure:

$$
\begin{array}{ll}
00101\ \text{E} & \\
00101\ \text{E} & \textit{Disk 1}\text{: } \mathbf{20} \\
10100\ \text{T} & \textit{Disk 2}\text{: E E T X} \\
11000\ \text{X} & \\
\end{array}
$$

The directory on disk 1 says that all records are on page 0 of disk 2, where they are kept in sorted order of their keys. For reference, we also give the binary value of the keys, using our standard encoding of the five-bit binary representation fo i for the ith letter of the alphabet. Now the page is full, and must be split in order to add the key R=10010. The strategy is simple: put records with keys that begin with 0 on one page and records with keys that begin with 1 on another page. This necessitates doubling the size of the directory, and moving half the keys from page 0 of disk 2 to a new page, leaving the following structure:

```
0:   0|0101 E
     0|0101 E              Disk 1: 20 21
1:   1|0010 R              Disk 2: E E          R T X
     1|0100 T
     1|1000 X
```

Now N=01110 and A=00001 can be added, but another split is needed before L=01100 can be added:

```
0:   0|0001 A
     0|0101 E
     0|0101 E              Disk 1: 20 21
     0|1110 N              Disk 2: A E E N    R T X
1:   1|0010 R
     1|0100 T
     1|1000 X
```

Recall our basic assumption that we do disk I/O in page units, and that processing time is negligible compared to the time to input or output a page. Thus, keeping the records in sorted order of their keys is not a real expense: to add a record to a page, we must read the page into memory, modify it, and write it back out. The extra time required to insert the new record to maintain sorted order is not likely to be noticeable in the typical case when the pages are small.

Proceeding in the same way, as for the first split, we make room for L= 01100 by splitting the first page into two pieces, one for keys that begin with 00 and one for keys that begin with 01. What's not immediately clear is what to do with the directory. One alternative would be to simply add another entry, one pointer to each page. This is unattractive because it essentially reduces to indexed sequential search (albeit a radix version): the directory has to be scanned sequentially to find the proper page during a search. Alternatively, we can just double the size of the directory again, giving the structure:

```
00:   00|001 A
      00|101 E
      00|101 E
01:   01|110 L              Disk 1: 20 21 22 22
      01|110 N              Disk 2: A E E      L N        R T X
10:   1|0010 R
11:   1|0100 T
      1|1000 X
```

Now we can access any record by using the first two bits of its key to access directly the directory entry that contains the address of the page containing the record.

Continuing a little further, we can add S=10011 and E=00101 before another split is necessary to add A=00001. This split also requires doubling the directory, leaving the structure:

000: 000|01 A
     000|01 A
001: 001|01 E
     001|01 E
     001|01 E   *Disk 1*: 20 21 22 22 30 30 30 30
010: 01|100 L   *Disk 2*: A A   E E E   L N
011: 01|110 N   *Disk 3*: R S T X
100: 10|010 R
101: 1|0011 S
110: 1|0100 T
111: 1|1000 X

In general, the structure built by extendible hashing consists of a *directory* of 2^d words (one for each d-bit pattern) and a set of *leaf pages* which contain all records with keys beginning with a specific bit pattern (of less than or equal to d bits). A search entails using the leading d bits of the key to index into the directory, which contains pointers to leaf pages. Then the referenced leaf page is accessed and searched (using any strategy) for the proper record. A leaf page can be pointed to by more than one directory entry: to be precise, if a leaf page contains all the records with keys that begin with a specific k bits (those marked with a vertical line in the pages on the diagram above), then it will have 2^{d-k} directory entries pointing to it. In the example above, we have $d = 3$, and page 0 of disk 3 contains all the records with keys that begin with a 1 bit, so there are four directory entries pointing to it.

The directory contains only pointers to pages. These are likely to be smaller than keys or records, so more directory entries will fit on each page. For our example, we'll assume that we can fit twice as many directory entries as records per page, though this ratio is likely to be much higher in practice. When the directory spans more than one page, we keep a "root node" in memory which tells where the directory pages are, using the same indexing scheme. For example, if the directory spans two pages, the root node might contain the two entries "10 11," indicating that the directory for all the records with keys beginning with 0 are on page 0 of disk 1, and the directory for all keys beginning with 1 are on page 1 of disk 1. For our example, this split occurs when the E is inserted. Continuing up until the last E (see below), we get the following disk storage structure:

Disk 1:	**20 20 21 22 30 30 31 32 40 40 41 41 42 42 42 42**
Disk 2:	A A A C E E E G
Disk 3:	H I L L M N N
Disk 4:	P R R S T X X

As illustrated in the above example, insertion into an extendible hashing structure can involve one of three operations, after the leaf page which could contain the search key is accessed. If there's room in the leaf page, the new record is simply inserted there; otherwise the leaf page is split in two (half the records are moved to a new page). If the directory has more than one entry pointing to that leaf page, then the directory entries can be split as the page is. If not, the size of the directory must be doubled.

As described so far, this algorithm is very susceptible to a bad input key distribution: the value of d is the largest number of bits required to separate the keys into sets small enough to fit on leaf pages, and thus if a large number of keys agree in a large number of leading bits, then the directory could get unacceptably large. For actual large-scale applications, this problem can be headed off by *hashing* the keys to make the leading bits (pseudo-)random. To search for a record, we hash its key to get a bit sequence which we use to access the directory, which tells us which page to search for a record with the same key. From a hashing standpoint, we can think of the algorithm as splitting nodes to take care of hash value collisions: hence the name "extendible hashing." This method presents a very attractive alternative to B-trees and indexed sequential access because it always uses exactly two disk accesses for each search (like indexed sequential), while still retaining the capability for efficient insertion (like B-trees).

Even with hashing, extraordinary steps must be taken if large numbers of equal keys are present. They can make the directory artificially large; and the algorithm breaks down entirely if there are more equal keys than fit in one leaf page. (This actually occurs in our example, since we have five E's.) If many equal keys are present then we could (for example) assume distinct keys in the data structure and put pointers to linked lists of records containing equal keys in the leaf pages. To see the complication involved, consider the insertion of the last E into the structure above.

Virtual Memory

The "easier way" discussed at the end of Chapter 13 for external sorting applies directly and trivially to the searching problem. A virtual memory is actually nothing more than a general-purpose external searching method: given an address (key), return the information associated with that address.

However, direct use of the virtual memory is *not* recommended as an easy searching application. As mentioned in Chapter 13, virtual memories perform best when most accesses are relatively close to previous accesses. Sorting algorithms can be adapted to this, but the very nature of searching is that requests are for information from arbitrary parts of the database.

Exercises

1. Give the contents of the B-tree that results when the keys E A S Y Q U E S T I O N are inserted in that order into an initially empty tree, with $M = 5$.

2. Give the contents of the B-tree that results when the keys E A S Y Q U E S T I O N are inserted in that order into an initially empty tree, with $M = 6$, using the variant of the method where all the records are kept in external nodes.

3. Draw the B-tree that is built when sixteen equal keys are inserted into an initially empty tree, with $M = 5$.

4. Suppose that one page from the database is destroyed. Describe how you would handle this event for each of the B-tree structures described in the text.

5. Give the contents of the extendible hashing table that results when the keys E A S Y Q U E S T I O N are inserted in that order into an initially empty table, with a page capacity of four records. (Following the example in the text, don't hash, but use the five-bit binary representation of i as the key for the ith letter.)

6. Give a sequence of as few distinct keys as possible which make an extendible hashing directory grow to size 16, from an initially empty table, with a page capacity of three records.

7. Outline a method for *deleting* an item from an extendible hashing table.

8. Why are "top-down" B-trees better than "bottom-up" B-trees for concurrent access to data? (For example, suppose two programs are trying to insert a new node at the same time.)

9. Implement *search* and *insert* for *internal* searching using the extendible hashing method.

10. Discuss how the program of the previous exercise compares with double hashing and radix trie searching for internal searching applications.

SOURCES for Searching

Again, the primary reference for this section is Knuth's volume three. Most of the algorithms that we've studied are treated in great detail in that book, including mathematical analyses and suggestions for practical applications.

The material in Chapter 15 comes from Guibas and Sedgewick's 1978 paper, which shows how to fit many classical balanced tree algorithms into the "red-black" framework, and which gives several other implementations. There is actually quite a large literature on balanced trees. Comer's 1979 survey gives many references on the subject of B-trees.

The extendible hashing algorithm presented in Chapter 18 comes from Fagin, Nievergelt, Pippenger and Strong's 1979 paper. This paper is a must for anyone wishing further information on external searching methods: it ties together material from our Chapters 16 and 17 to bring out the algorithm in Chapter 18.

Trees and binary trees as purely mathematical objects have been studied extensively, quite apart from computer science. A great deal is known about the combinatorial properties of these objects. A reader interested in studying this type of material might begin with Knuth's volume 1.

Many practical applications of the methods discussed here, especially Chapter 18, arise within the context of database systems. An introduction to this field is given in Ullman's 1980 book.

D. Comer, "The ubquitous B-tree," *Computing Surveys*, **11** (1979).

R. Fagin, J. Nievergelt, N. Pippenger and H. R. Strong, "Extendible Hashing - a fast access method for dynamic files," *ACM transactions on Database Systems*, **4**, 3 (September, 1979).

L. Guibas and R. Sedgewick, "A dichromatic framework for balanced trees," in *19th Annual Symposium on Foundations of Computer Science*, IEEE, 1978. Also in *A Decade of Progress 1970–1980*, Xerox PARC, Palo Alto, CA.

D. E. Knuth, *The Art of Computer Programming. Volume 1: Fundamental Algorithms*, Addison-Wesley, Reading, MA, 1968.

D. E. Knuth, *The Art of Computer Programming. Volume 3: Sorting and Searching*, Addison-Wesley, Reading, MA, second printing, 1975.

J. D. Ullman, *Principles of Database Systems*, Computer Science Press, Rockville, MD, 1982.

STRING PROCESSING

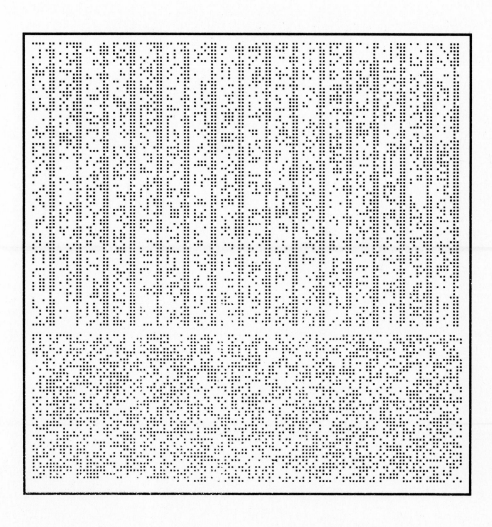

19. String Searching

Data to be processed often does not decompose logically into independent records with small identifiable pieces. This type of data is characterized only by the fact that it can be written down as a *string*: a linear (typically very long) sequence of characters.

Strings are obviously central in "word processing" systems, which provide a variety of capabilities for the manipulation of text. Such systems process *text strings*, which might be loosely defined as sequences of letters, numbers, and special characters. These objects can be quite large (for example, this book contains over a million characters), and efficient algorithms play an important role in manipulating them.

Another type of string is the *binary string*, a simple sequence of 0 and 1 values. This is in a sense merely a special type of text string, but it is worth making the distinction not only because different algorithms are appropriate but also binary strings arise naturally in many applications. For example, some computer graphics systems represent pictures as binary strings. (This book was printed on such a system: the present page was represented at one time as a binary string consisting of millions of bits.)

In one sense, text strings are quite different objects than binary strings, since they are made up of characters from a large alphabet. In another, the two types of strings are equivalent, since each text character can be represented by (say) eight binary bits and a binary string can be viewed as a text string by treating eight-bit chunks as characters. We'll see that the size of the alphabet from which the characters are taken to form a string is an important factor in the design of string processing algorithms.

A fundamental operation on strings is *pattern matching*: given a *text* string of length N and a *pattern* of length M, find an occurrence of the pattern within the text. (We will use the term "text" even when referring to a sequence of 0-1 values or some other special type of string.) Most algorithms for this problem can easily be extended to find *all* occurrences of the pattern in the text, since they scan sequentially through the text and can be restarted at the point directly after the beginning of a match to find the next match.

The pattern-matching problem can be characterized as a searching problem with the pattern as the key, but the searching algorithms that we have studied do not apply directly because the pattern can be long and because it "lines up" with the text in an unknown way. It is an interesting problem to study because several very different (and surprising) algorithms have only recently been discovered which not only provide a spectrum of useful practical methods but also illustrate some fundamental algorithm design techniques.

A Short History

The development of the algorithms that we'll be examining has an interesting history: we'll summarize it here to place the various methods into perspective.

There is an obvious brute-force algorithm for string processing which is in widespread use. While it has a worst-case running time proportional to MN, the strings which arise in many applications lead to a running time which is virtually always proportional to $M + N$. Furthermore, it is well suited to good architectural features on most computer systems, so an optimized version provides a "standard" which is difficult to beat with a clever algorithm.

In 1970, S. A. Cook proved a theoretical result about a particular type of abstract machine implying that an algorithm exists which solves the pattern-matching problem in time proportional to $M + N$ in the worst case. D. E. Knuth and V. R. Pratt laboriously followed through the construction Cook used to prove his theorem (which was not intended at all to be practical) to get an algorithm which they were then able to refine to be a relatively simple practical algorithm. This seemed a rare and satisfying example of a theoretical result having immediate (and unexpected) practical applicability. But it turned out that J. H. Morris had discovered virtually the same algorithm as a solution to an annoying practical problem that confronted him when implementing a text editor (he didn't want to ever "back up" in the text string). However, the fact that the same algorithm arose from two such different approaches lends it credibility as a fundamental solution to the problem.

Knuth, Morris, and Pratt didn't get around to publishing their algorithm until 1976, and in the meantime R. S. Boyer and J. S. Moore (and, independently, R. W. Gosper) discovered an algorithm which is much faster in many applications, since it often examines only a fraction of the characters in the text string. Many text editors use this algorithm to achieve a noticeable decrease in response time for string searches.

Both the Knuth-Morris-Pratt and the Boyer-Moore algorithms require some complicated preprocessing on the pattern that is difficult to understand

and has limited the extent to which they are used. (In fact, the story goes that an unknown systems programmer found Morris' algorithm too difficult to understand and replaced it with a brute-force implementation.)

In 1980, R. M. Karp and M. O. Rabin observed that the problem is not as different from the standard searching problem as it had seemed, and came up with an algorithm almost as simple as the brute-force algorithm which virtually always runs in time proportional to $M + N$. Furthermore, their algorithm extends easily to two-dimensional patterns and text, which makes it more useful than the others for picture processing.

This story illustrates that the search for a "better algorithm" is still very often justified: one suspects that there are still more developments on the horizon even for this problem.

Brute-Force Algorithm

The obvious method for pattern matching that immediately comes to mind is just to check, for each possible position in the text at which the pattern could match, whether it does in fact match. The following program searches in this way for the first occurrence of a pattern $p[1..M]$ in a text string $a[1..N]$:

```
function brutesearch: integer;
   var i, j: integer;
   begin
   i:=1; j:=1;
   repeat
      if a[i]=p[j]
         then begin i:=i+1; j:=j+1 end
         else begin i:=i-j+2; j:=1 end;
   until (j>M) or (i>N);
   if j>M then brutesearch:=i-M else brutesearch:=i
   end;
```

The program keeps one pointer (i) into the text and another pointer (j) into the pattern. As long as they point to matching characters, both pointers are incremented. If the end of the pattern is reached ($j>M$), then a match has been found. If i and j point to mismatching characters, then j is reset to point to the beginning of the pattern and i is reset to correspond to moving the pattern to the right one position for matching against the text. If the end of the text is reached ($i>N$), then there is no match. If the pattern does not occur in the text, the value $N+1$ is returned.

In a text-editing application, the inner loop of this program is seldom iterated, and the running time is very nearly proportional to the number of text characters examined. For example, suppose that we are looking for the pattern STING in the text string

A STRING SEARCHING EXAMPLE CONSISTING OF SIMPLE TEXT

Then the statement $j:=j+1$ is executed only four times (once for each S, but twice for the first ST) before the actual match is encountered. On the other hand, this program can be very slow for some patterns. For example, if the pattern is 00000001 and the text string is:

00000000000000000000000000000000000000000000000001

then j is incremented 7*45 (315) times before the match is encountered. Such degenerate strings are not likely in English (or Pascal) text, but the algorithm does run more slowly when used on binary (two-character) text, as might occur in picture processing and systems programming applications. The following table shows what happens when the algorithm is used to search for 10010111 in the following binary string:

```
          10011101001001001001 0111000111
          1001
             1
              1
              10
                 10010
                  10010
                    10010
                      10010111
```

There is one line in this table for each time the body of the **repeat** loop is entered, and one character for each time j is incremented. These are the "false starts" that occur when trying to find the pattern: an obvious goal is to try to limit the number and length of these.

Knuth-Morris-Pratt Algorithm

The basic idea behind the algorithm discovered by Knuth, Morris, and Pratt is this: when a mismatch is detected, our "false start" consists of characters that we know in advance (since they're in the pattern). Somehow we should be able to take advantage of this information instead of backing up the i pointer over all those known characters.

For a simple example of this, suppose that the first character in the pattern doesn't appear again in the pattern (say the pattern is 10000000). Then, suppose we have a false start j characters long at some position in the text. When the mismatch is detected, we know, by dint of the fact that j characters have matched, that we don't have to "back up" the text pointer i, since none of the previous $j-1$ characters in the text can match the first character in the pattern. This change could be implemented by replacing $i:=i-j+2$ in the program above by $i:=i+1$. The practical effect of this change is limited because such a specialized pattern is not particularly likely to occur, but the idea is worth thinking about because the Knuth-Morris-Pratt algorithm is a generalization. Surprisingly, it is always possible to arrange things so that the i pointer is never decremented.

Fully skipping past the pattern on detecting a mismatch as described in the previous paragraph won't work when the pattern could match itself at the point of the mismatch. For example, when searching for 10100111 in 1010100111 we first detect the mismatch at the fifth character, but we had better back up to the third character to continue the search, since otherwise we would miss the match. But we can figure out ahead of time exactly what to do, because it depends only on the pattern, as shown by the following table:

j	$p[1..j-1]$	$next[j]$
2	1│	1
3	10│	1
4	10│1	2
5	10│10	3
6	10100│	1
7	10100│1	2
8	101001│1	2

The array $next[1..M]$ will be used to determine how far to back up when a mismatch is detected. In the table, imagine that we slide a copy of the first $j-1$ characters of the pattern over itself, from left to right starting with the first character of the copy over the second character of the pattern, stopping when all overlapping characters match (or there are none). These overlapping characters define the next possible place that the pattern could match, if a mismatch is detected at $p[j]$. The distance to back up ($next[j]$) is exactly one plus the number of the overlapping characters. Specifically, for $j>1$, the value of $next[j]$ is the maximum $k<j$ for which the first $k-1$ characters of the pattern match the last $k-1$ characters of the first $j-1$ characters of the pattern. A vertical line is drawn just after $p[j-next[j]]$ on each line of the

table. As we'll soon see, it is convenient to define $next[1]$ to be 0.

This $next$ array immediately gives a way to limit (in fact, as we'll see, eliminate) the "backup" of the text pointer i: a generalization of the method above. When i and j point to mismatching characters (testing for a pattern match beginning at position $i-j+1$ in the text string), then the next possible position for a pattern match is beginning at position $i-next[j]+1$. But by definition of the $next$ table, the first $next[j]-1$ characters at that position match the first $next[j]-1$ characters of the pattern, so there's no need to back up the i pointer that far: we can simply leave the i pointer unchanged and set the j pointer to $next[j]$, as in the following program:

```
function kmpsearch: integer;
  var i, j: integer;
  begin
  i:=1; j:=1;
  repeat
    if (j=0) or (a[i]=p[j])
      then begin i:=i+1; j:=j+1 end
      else begin j:=next[j] end;
  until (j>M) or (i>N);
  if j>M then kmpsearch:=i-M else kmpsearch:=i;
  end;
```

When $j=1$ and $a[i]$ does not match the pattern, there is no overlap, so we want to increment i and set j to the beginning of the pattern. This is achieved by defining $next[1]$ to be 0, which results in j being set to 0, then i is incremented and j set to 1 next time through the loop. (For this trick to work, the pattern array must be declared to start at 0, otherwise standard Pascal will complain about subscript out of range when $j=0$ even though it doesn't really have to access $p[0]$ to determine the truth of the **or**.) Functionally, this program is the same as $brutesearch$, but it is likely to run faster for patterns which are highly self-repetitive.

It remains to compute the $next$ table. The program for this is short but tricky: it is basically the same program as above, except that it is used to match the pattern against itself.

```
procedure initnext;
  var i, j: integer;
  begin
  i:=1; j:=0; next[1]:=0;
  repeat
    if (j=0) or (p[i]=p[j])
      then begin i:=i+1; j:=j+1; next[i]:=j end
      else begin j:=next[j] end;
    until i>M;
  end;
```

Just after i and j are incremented, it has been determined that the first $j-1$ characters of the pattern match the characters in positions $p[i-j-1..i-1]$, the last $j-1$ characters in the first $i-1$ characters of the pattern. And this is the largest j with this property, since otherwise a "possible match" of the pattern with itself would have been missed. Thus, j is exactly the value to be assigned to $next[i]$.

An interesting way to view this algorithm is to consider the pattern as fixed, so that the $next$ table can be "wired in" to the program. For example, the following program is exactly equivalent to the program above for the pattern that we've been considering, but it's likely to be much more efficient.

```
     i:=0;
0: i:=i+1;
1: if a[i]<>'1' then goto 0; i:=i+1;
2: if a[i]<>'0' then goto 1; i:=i+1;
3: if a[i]<>'1' then goto 1; i:=i+1;
4: if a[i]<>'0' then goto 2; i:=i+1;
5: if a[i]<>'0' then goto 3; i:=i+1;
6: if a[i]<>'1' then goto 1; i:=i+1;
7: if a[i]<>'1' then goto 2; i:=i+1;
8: if a[i]<>'1' then goto 2; i:=i+1;
     search:=i-8;
```

The **goto** labels in this program correspond precisely to the $next$ table. In fact, the $initnext$ program above which computes the $next$ table could easily be modified to output this program! To avoid checking whether $i>N$ each time i is incremented, we assume that the pattern itself is stored at the end of the text as a sentinel, in $a[N+1..N+M]$. (This optimization could also be applied to the standard implementation.) This is a simple example of a "string-searching compiler": given a pattern, we can produce a very efficient program which can scan for that pattern in an arbitrarily long text string. We'll see generalizations of this concept in the next two chapters.

The program above uses just a few very basic operations to solve the string searching problem. This means that it can easily be described in terms of a very simple machine model, called a *finite-state machine*. The following diagram shows the finite-state machine for the program above:

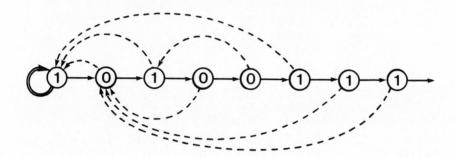

The machine consists of *states* (indicated by circled letters) and *transitions* (indicated by arrows). Each state has two transitions leaving it: a *match* transition (solid line) and a *non-match* transition (dotted line). The states are where the machine executes instructions; the transitions are the **goto** instructions. When in the state labeled "x," the machine can perform just one instruction: "if the current character is x then scan past it and take the match transition, otherwise take the non-match transition." To "scan past" a character means to take the next character in the string as the "current character"; the machine scans past characters as it matches them. There is one exception to this: the non-match transition in the first state (marked with a double line) also requires that the machine scan to the next character. (Essentially this corresponds to scanning for the first occurrence of the first character in the pattern.) In the next chapter we'll see how to use a similar (but more powerful) machine to help develop a much more powerful pattern-matching algorithm.

The alert reader may have noticed that there's still some room for improvement in this algorithm, because it doesn't take into account the character which caused the mismatch. For example, suppose that we encounter 1011 when searching for our sample pattern 10100111. After matching 101, we find a mismatch on the fourth character, at which point the *next* table says to check the second character, since we already matched the 1 in the third character. However, we could not have a match here: from the mismatch, we know that the next character in the text is not 0, as required by the pattern.

Another way to see this is to look at the version of the program with the next table "wired in": at label 4 we go to 2 if $a[i]$ is not 0, but at label 2 we go to 1 if $a[i]$ is not 0. Why not just go to 1 directly? Fortunately, it is easy to put this change into the algorithm. We need only replace the statement $next[i]:=j$ in the *initnext* program by

> **if** $p[j]<>p[i]$ **then** $next[i]:=j$ **else** $next[i]:=next[j]$;

With this change, we either increment j or reset it from the *next* table at most once for each value of i, so the algorithm is clearly linear.

The Knuth-Morris-Pratt algorithm is not likely to be significantly faster than the brute-force method in most actual applications, because few applications involve searching for highly self-repetitive patterns in highly self-repetitive text. However, the method does have a major virtue from a practical point of view: it proceeds sequentially through the input and never "backs up" in the input. This makes the method convenient for use on a large file being read in from some external device. (Algorithms which require backup require some complicated buffering in this situation.)

Boyer-Moore Algorithm

If "backing up" is not a problem, then a significantly faster string searching method can be developed by scanning the pattern from right to left when trying to match it against the text. When searching for our sample pattern 10100111, if we find matches on the eighth, seventh, and sixth character but not on the fifth, then we can immediately slide the pattern seven positions to the right, and check the fifteenth character next, because our partial match found 111, which might appear elsewhere in the pattern. Of course, the pattern at the end does appear elsewhere in general, so we need a *next* table as above. For example, the following is a right-to-left version of the *next* table for the pattern 10110101:

j	$p[M-j+2..M]$	$p[M-next[j]+1..M]$	$next[j]$
2	1|0101		4
3	01|0110101		7
4	101|01		2
5	0101|10101		5
6	10101|10101		5
7	110101|10101		5
8	0110101|10101		5

The number at the right on the jth line of the table gives the maximum number of character positions that the pattern can be shifted to the right given that a mismatch in a right-to-left scan occurred on the jth character from the right in the pattern. This is found in a similar manner as before, by sliding a copy of the pattern over the last $j-1$ characters of itself from left to right starting with the next-to-last character of the copy lined up with the last character of the pattern, stopping when all overlapping characters match (also taking into account the character which caused the mismatch).

This leads directly to a program which is quite similar to the above implementation of the Knuth-Morris-Pratt method. We won't go into this in more detail because there is a quite different way to skip over characters with right-to-left pattern scanning which is much better in many cases.

The idea is to decide what to do next based on the character that caused the mismatch in the *text* as well as the pattern. The simplest realization of this leads immediately to a quite useful program. Consider the first example that we studied, searching for the pattern STING in the text string

A STRING SEARCHING EXAMPLE CONSISTING OF SIMPLE TEXT

Proceeding from right to left to match the pattern, we first check the G in the pattern against the R (the fifth character) in the text. Not only do these not match, but also we can notice that R does not appear *anywhere* in the pattern, so we might as well slide it all the way past the R. The next comparison is of the G in the pattern against the fifth character following the R (the S in SEARCHING). This time, we can slide the pattern to the right until its S matches the S in the text. Then the G in the pattern is compared against the C in SEARCHING, which doesn't appear in the pattern, so it can be slid five more places to the right. After three more five-character skips, we arrive at the T in CONSISTING, at which point we align the pattern so that the its T matches the T in the text and find the full match. This method brings us right to the match position at a cost of examining only seven characters in the text (and five more to verify the match)! If the alphabet is not small and the pattern is not long, then this "mismatched character algorithm" will find a pattern of length M in a text string of length N in about N/M steps.

The mismatched character algorithm is quite easy to implement. It simply improves a brute-force right-to-left pattern scan by using an array *skip* which tells, for each character in the alphabet, how far to skip if that character appears in the text and causes a mismatch:

```
function mischarsearch: integer;
  var i, j: integer;
  begin
  i:=M; j:=M;
  repeat
    if a[i]=p[j]
      then begin i:=i−1; j:=j−1 end
      else begin i:=i+skip[index(a[i])]; j:=M; end;
  until (j<1) or (i>N);
  mischarsearch:=i+1
  end;
```

For simplicity, we assume that we have a **function** *index(c: char): integer;* that returns 0 for blanks and i for the ith letter of the alphabet. One left-to-right scan through the pattern will take care of initializing the *skip* array.

```
procedure initskip;
  var j: integer;
  begin
  for j:=0 to 26 do skip[j]:=M;
  for j:=1 to M do skip[index(p[j])]:=M−j
  end;
```

For example, for the pattern STING, the *skip* entry for G would be 0, the entry for N would be 1, the entry for I would be 2, the entry for T would be 3, the entry for S would be 4, and the entries for all other letters would be 5. Thus, for example, when an S is encountered during a right-to-left search, the i pointer is incremented by 4 so that the end of the pattern is aligned four positions to the right of the S (and consequently the S in the pattern lines up with the S in the text). If there were more than one S in the pattern, we would want to use the rightmost one for this calculation: hence the *skip* array is built by scanning from left to right.

Boyer and Moore suggested combining the two methods we have outlined for right-to-left pattern scanning, choosing the larger of the two skips called for.

The mismatched character algorithm obviously won't help much for binary strings, because there are only two possibilities for characters which cause the mismatch (and these are both likely to be in the pattern). However, the bits can be grouped together to make "characters" which can be used exactly as above. If we take b bits at a time, then we need a *skip* table with 2^b entries.

The value of b should be chosen small enough so that this table is not too large, but large enough that most b-bit sections of the text are not likely to be in the pattern. Specifically, there are $M - b + 1$ different b-bit sections in the pattern (one starting at each bit position from 1 through $M - b + 1$) so we want $M - b + 1$ to be significantly less than 2^b. For example, if we take b to be about $\lg(4M)$, then the *skip* table will be more than three-quarters filled with M entries. Also b must be less than $M/2$, otherwise we could miss the pattern entirely if it were split between two b-bit text sections.

Rabin-Karp Algorithm

A brute-force approach to string searching which we didn't examine above would be to use a large memory to advantage by treating each possible M-character section of the text as a key in a standard hash table. But it is not necessary to keep a whole hash table, since the problem is set up so that only one key is being sought: all that we need to do is to compute the hash function for each of the possible M-character sections of the text and check if it is equal to the hash function of the pattern. The problem with this method is that it seems at first to be just as hard to compute the hash function for M characters from the text as it is merely to check to see if they're equal to the pattern. Rabin and Karp found an easy way to get around this problem for the hash function $h(k) = k \bmod q$ where q (the table size) is a large prime. Their method is based on computing the hash function for position i in the text given its value for position $i - 1$. The method follows quite directly from the mathematical formulation. Let's assume that we translate our M characters to numbers by packing them together in a computer word, which we then treat as an integer. This corresponds to writing the characters as numbers in a base-d number system, where d is the number of possible characters. The number corresponding to $a[i..i + M - 1]$ is thus

$$x = a[i]d^{M-1} + a[i+1]d^{M-2} + \cdots + a[i + M - 1]$$

and we can assume that we know the value of $h(x) = x \bmod q$. But shifting one position right in the text simply corresponds to replacing x by

$$(x - a[i]d^{M-1})d + a[i + M].$$

A fundamental property of the mod operation is that we can perform it at any time during these operations and still get the same answer. Put another way, if we take the remainder when divided by q after each arithmetic operation (to keep the numbers that we're dealing with small) then we get the same answer that we would if we were to perform all of the arithmetic operations, then take the remainder when divided by q.

This leads to the very simple pattern-matching algorithm implemented below. The program assumes the same *index* function as above, but *d=32* is used for efficiency (the multiplications might be implemented as shifts).

```
function rksearch: integer;
    const q=33554393; d=32;
    var h1, h2, dM, i: integer;
    begin
    dM:=1; for i:=1 to M−1 do dM:=(d*dM) mod q;
    h1:=0; for i:=1 to M do h1:=(h1*d+index(p[i])) mod q;
    h2:=0; for i:=1 to M do h2:=(h2*d+index(a[i])) mod q;
    i:=1;
    while (h1<>h2) and (i<=N−M) do
      begin
      h2:=(h2+d*q−index(a[i])*dM) mod q;
      h2:=(h2*d+index(a[i+M])) mod q;
      i:=i+1;
      end;
    rksearch:=i;
    end;
```

The program first computes a hash value *h1* for the pattern, then a hash value *h2* for the first *M* characters of the text. (Also it computes the value of $d^{M-1} \bmod q$ in the variable *dM*.) Then it proceeds through the text string, using the technique above to compute the hash function for the *M* characters starting at position *i* for each *i*, comparing each new hash value to *h1*. The prime *q* is chosen to be as large as possible, but small enough that *(d+1)*q* doesn't cause overflow: this requires less **mod** operations then if we used the largest repesentable prime. (An extra *d*q* is added during the *h2* calculation to make sure that everything stays positive so that the **mod** operation works as it should.)

This algorithm obviously takes time proportional to $N + M$. Note that it really only finds a position in the text which has the same hash value as the pattern, so, to be sure, we really should do a direct comparison of that text with the pattern. However, the use of such a large value of *q*, made possible by the **mod** computations and by the fact that we don't have to keep the actual hash table around, makes it extremely unlikely that a collision will occur. Theoretically, this algorithm could still take NM steps in the (unbelievably) worst case, but in practice the algorithm can be relied upon to take about $N + M$ steps.

Multiple Searches

The algorithms that we've been discussing are all oriented towards a specific string searching problem: find an occurrence of a given pattern in a given text string. If the same text string is to be the object of many pattern searches, then it will be worthwhile to do some processing on the string to make subsequent searches efficient.

If there are a large number of searches, the string searching problem can be viewed as a special case of the general searching problem that we studied in the previous section. We simply treat the text string as N overlapping "keys," the ith key defined to be $a[1..N]$, the entire text string starting at position i. Of course, we don't manipulate the keys themselves, but pointers to them: when we need to compare keys i and j we do character-by-character compares starting at positions i and j of the text string. (If we use a "sentinel" character larger than all other characters at the end, then one of the keys will always be greater than the other.) Then the hashing, binary tree, and other algorithms of the previous section can be used directly. First, an entire structure is built up from the text string, and then efficient searches can be performed for particular patterns.

There are many details which need to be worked out in applying searching algorithms to string searching in this way; our intent is to point this out as a viable option for some string searching applications. Different methods will be appropriate in different situations. For example, if the searches will always be for patterns of the same length, a hash table constructed with a single scan as in the Rabin-Karp method will yield constant search times on the average. On the other hand, if the patterns are to be of varying length, then one of the tree-based methods might be appropriate. (Patricia is especially adaptable to such an application.)

Other variations in the problem can make it significantly more difficult and lead to drastically different methods, as we'll discover in the next two chapters.

Exercises

1. Implement a brute-force pattern matching algorithm that scans the pattern from right to left.

2. Give the *next* table for the Knuth-Morris-Pratt algorithm for the pattern AAAAAAAA.

3. Give the *next* table for the Knuth-Morris-Pratt algorithm for the pattern ABRACADABRA.

4. Draw a finite state machine which can search for the pattern ABRACAD ABRA.

5. How would you search a text file for a string of 50 consecutive blanks?

6. Give the right-to-left *skip* table for the right-left scan for the pattern ABRACADABRA.

7. Construct an example for which the right-to-left pattern scan with only the mismatch heuristic performs badly.

8. How would you modify the Rabin-Karp algorithm to search for a given pattern with the additional proviso that the middle character is a "wild card" (any text character at all can match it)?

9. Implement a version of the Rabin-Karp algorithm that can find a given two-dimensional pattern in a given two-dimensional text. Assume both pattern and text are rectangles of characters.

10. Write programs to generate a random 1000-bit text string, then find all occurrences of the last k bits elsewhere in the string, for $k = 5, 10, 15$. (Different methods might be appropriate for different values of k.)

20. Pattern Matching

It is often desirable to do string searching with somewhat less than complete information about the pattern to be found. For example, the user of a text editor may wish to specify only part of his pattern, or he may wish to specify a pattern which could match a few different words, or he might wish to specify that any number of occurrences of some specific characters should be ignored. In this chapter we'll consider how *pattern matching* of this type can be done efficiently.

The algorithms in the previous chapter have a rather fundamental dependence on complete specification of the pattern, so we have to consider different methods. The basic mechanisms that we will consider make possible a very powerful string searching facility which can match complicated M-character patterns in N-character text strings in time proportional to MN.

First, we have to develop a way to describe the patterns: a "language" that can be used to specify, in a rigorous way, the kinds of partial string searching problems suggested above. This language will involve more powerful primitive operations than the simple "check if the ith character of the text string matches the jth character of the pattern" operation used in the previous chapter. In this chapter, we consider three basic operations in terms of an imaginary type of machine that has the capability for searching for patterns in a text string. Our pattern-matching algorithm will be a way to simulate the operation of this type of machine. In the next chapter, we'll see how to translate from the pattern specification (which the user employs to describe his string searching task) to the machine specification (which the algorithm employs to actually carry out the search).

As we'll see, this solution to the pattern matching problem is intimately related to fundamental processes in computer science. For example, the method that we will use in our program to perform the string searching task implied by a given pattern description is akin to the method used by the

Pascal system to perform the computational task implied by a given Pascal program.

Describing Patterns

We'll consider pattern descriptions made up of symbols tied together with the following three fundamental operations.

(i)   *Concatenation*. This is the operation used in the last chapter. If two characters are adjacent in the pattern, then there is a match if and only if the same two characters are adjacent in the text. For example, AB means A followed by B.

(ii) *Or*. This is the operation that allows us to specify alternatives in the pattern. If we have an "or" between two characters, then there is a match if and only if either of the characters occurs in the text. We'll denote this operation by using the symbol + and use parentheses to allow it to be combined with concatenation in arbitrarily complicated ways. For example, A+B means "either A or B"; C(AC+B)D means "either CACD or CBD"; and (A+C)((B+C)D) means "either ABD or CBD or ACD or CCD."

(iii) *Closure*. This operation allows parts of the pattern to be repeated arbitrarily. If we have the closure of a symbol, then there is a match if and only if the symbol occurs any number of times (including 0). Closure will be denoted by placing a * after the character or parenthesized group to be repeated. For example, AB* matches strings consisting of an A followed by any number of B's, while (AB)* matches strings consisting of alternating A's and B's.

A string of symbols built up using these three operations is called a *regular expression*. Each regular expression describes many specific text patterns. Our goal is to develop an algorithm which will determine if any of the patterns described by a given regular expression occur in a given text string.

We'll concentrate on concatenation, *or*, and closure in order to show the basic principles in developing a regular-expression pattern matching algorithm. Various additions are commonly made in actual systems for convenience. For example, $-A$ might mean "match any character *except* A." This *not* operation is the same as an *or* involving all the characters except A but is much easier to use. Similarly, "?" might mean "match any letter." Again, this is obviously much more compact than a large *or*. Other examples of additional symbols which might make specification of large patterns easier are symbols which match the beginning or end of a line, any letter or any number, etc.

These operations can be remarkably descriptive. For example, the pattern description $?*(ie+ei)?*$ matches all words which have *ie* or *ei* in them (and so are likely to be misspelled!); and $(1+01)^*(0+1)$ describes all strings of 0's and 1's which do not have two consecutive 0's. Obviously there are many different pattern descriptions which describe the same strings: we must try to specify succinct pattern descriptions just as we try to write efficient algorithms.

The pattern matching algorithm that we'll examine may be viewed as a generalization of the brute force left-to-right string searching method (the first method that we looked at in Chapter 19). The algorithm looks for the leftmost substring in the text string which matches the pattern description by scanning the text string from left to right, testing, at each position whether there is a substring beginning at that position which matches the pattern description.

Pattern Matching Machines

Recall that we can view the Knuth-Morris-Pratt algorithm as a finite-state machine constructed from the search pattern which scans the text. The method we will use for regular-expression pattern matching is a generalization of this.

The finite-state machine for the Knuth-Morris-Pratt algorithm changes from state to state by looking at a character from the text string and then changing to one state if there's a match, to another state if not. A mismatch at any point means that the pattern couldn't occur in the text starting at that point. The algorithm itself can be thought of as a simulation of the machine. The characteristic of the machine that makes it easy to simulate is that it is *deterministic*: each state transition is completely determined by the next input character.

To handle regular expressions, it will be necessary to consider a more powerful abstract machine. Because of the *or* operation, the machine can't determine whether or not the pattern could occur at a given point by examining just one character; in fact, because of closure, it can't even determine how many characters might need to be examined before a mismatch is discovered. The most natural way to overcome these problems is to endow the machine with the power of *nondeterminism*: when faced with more than one way to try to match the pattern, the machine should "guess" the right one! This operation seems impossible to allow, but we will see that it is easy to write a program to simulate the actions of such a machine.

For example, the following diagram shows a nondeterministic finite-state machine that could be used to search for the pattern description $(A^*B+AC)D$ in a text string.

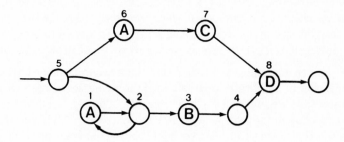

As in the deterministic machine of the previous chapter, the machine can travel from a state labeled with a character to the state "pointed to" by that state by matching (and scanning past) that character in the text string. What makes the machine nondeterministic is that there are some states (called *null* states) which not only are not labeled, but also can "point to" two different successor states. (Some null states, such as state 4 in the diagram, are "no-op" states with one exit, which don't affect the operation of the machine, but which make easier the implementation of the program which constructs the machine, as we'll see. State 9 is a null state with no exits, which stops the machine.) When in such a state, the machine can go to *either* successor state regardless of what's in the input (without scanning past anything). The machine has the power to guess which transition will lead to a match for the given text string (if any will). Note that there are no "non-match" transitions as in the previous chapter: the machine fails no find a match only if there is no way even to guess a sequence of transitions that leads to a match.

The machine has a unique *initial* state (which is pointed to by a "free" arrow) and a unique *final* state (which has no arrows going out). When started out in the initial state, the machine should be able to "recognize" any string described by the pattern by reading characters and changing state according to its rules, ending up in the "final state." Because it has the power of nondeterminism, the machine can guess the sequence of state changes that can lead to the solution. (But when we try to simulate the machine on a standard computer, we'll have to try all the possibilities.) For example, to determine if its pattern description (A*B+AC)D can occur in the text string

CDAABCAAABDDACDAAC

the machine would immediately report failure if started on the first or second character; it would work some to report failure on the next two characters; it would immediately report failure on the fifth or sixth characters; and it would guess the sequence of state transitions

5 2 1 2 1 2 1 2 3 4 8 9

to recognize AAABD if started on the seventh character.

We can construct the machine for a given regular expression by building partial machines for parts of the expression and defining the ways in which two partial machines can be composed into a larger machine for each of the three operations: concatenation, *or*, and closure.

We start with the trivial machine to recognize a particular character. It's convenient to write this as a two-state machine, with one initial state (which also recognizes the character) and one final state, as below:

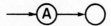

Now to build the machine for the concatenation of two expressions from the machines for the individual expressions, we simply merge the final state of the first with the initial state of the second:

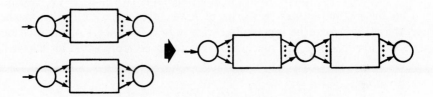

Similarly, the machine for the *or* operation is built by adding a new null state pointing to the two initial states, and making one final state point to the other, which becomes the final state of the combined machine.

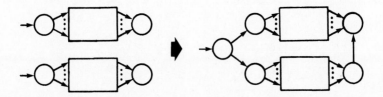

Finally, the machine for the closure operation is built by making the final state the initial state and making it point back to the old initial state and a new final state.

A machine can be built which corresponds to any regular expression by successively applying these rules. The numbers of the states for the example machine above are in order of creation as the machine is built by scanning the pattern from left to right, so the construction of the machine from the rules above can be easily traced. Note that we have a 2-state trivial machine for each letter in the regular expression, and each + and * causes one state to be created (concatenation causes one to be deleted) so the number of states is certainly less than twice the number of characters in the regular expression.

Representing the Machine

Our nondeterministic machines will all be constructed using only the three composition rules outlined above, and we can take advantage of their simple structure to manipulate them in a straightforward way. For example, no state has more than two arrows leaving it. In fact, there are only two types of states: those labeled by a character from the input alphabet (with one arrow leaving) and unlabeled (null) states (with two or fewer arrows leaving). This means that the machine can be represented with only a few pieces of information per node. For example, the machine above might be represented as follows:

State	Character	Next 1	Next 2
0		5	—
1	A	2	—
2		3	1
3	B	4	—
4		8	8
5		6	2
6	A	7	—
7	C	8	—
8	D	9	—
9		0	0

The rows in this table may be interpreted as instructions to the nondeterministic machine of the form "If you are in *State* and you see *Character* then scan the character and go to state *Next 1* (or *Next 2*)." State 9 is the final state in this example, and State 0 is a pseudo-initial state whose *Next 1* entry is the number of the actual initial state. (Note the special representation used for null states with 0 or 1 exits.)

Since we often will want to access states just by number, the most suitable organization for the machine is to use the array representation. We'll use the three arrays

ch: **array** [*0..Mmax*] **of** *char*;
next1, next2: **array** [*0..Mmax*] **of** *integer*;

Here *Mmax* is the maximum number of states (twice the maximum pattern length). It would be possible to get by with two-thirds this amount of space, since each state really uses only two meaningful pieces of information, but we'll forsake this improvement for the sake of clarity and also because pattern descriptions are not likely to be particularly long.

We've seen how to build up machines from regular expression pattern descriptions and how such machines might be represented as arrays. However, to write a program to do the translation from a regular expression to the corresponding nondeterministic machine representation automatically is quite another matter. In fact, even writing a program to determine if a given regular expression is legal is challenging for the uninitiated. In the next chapter, we'll study this operation, called *parsing*, in much more detail. For the moment, we'll assume that this translation has been done, so that we have available the *ch*, *next1*, and *next2* arrays representing a particular nondeterministic machine which corresponds to the regular expression pattern description of interest.

Simulating the Machine

The last step in the development of a general regular-expression pattern-matching algorithm is to write a program which somehow simulates the operation of a nondeterministic pattern-matching machine. The idea of writing a program which can "guess" the right answer seems ridiculous. However, in this case it turns out that we can keep track of *all possible* matches in a systematic way, so that we do eventually encounter the correct one.

One possibility would be to develop a recursive program which mimics the nondeterministic machine (but tries all possibilities rather than guessing the right one). Instead of using this approach, we'll look at a nonrecursive implementation which exposes the basic operating principles of the method by keeping the states under consideration in a rather peculiar data structure called a *deque*, described in some detail below.

The idea is to keep track of all states that could possibly be encountered while the machine is "looking at" the current input character. Each of these states are processed in turn: null states lead to two (or fewer) states, states for characters which do not match the current input are eliminated, and states for characters which do match the current input lead to new states for use when the machine is looking at the *next* input character. Thus, we maintain a list of all the states that the nondeterministic machine could possibly be in at a particular point in the text: the problem is to design an appropriate data structure for this list.

Processing null states seems to require a *stack*, since we are essentially postponing one of two things to be done, just as when we removed the recursion from Quicksort (so the new state should be put at the *beginning* of the current list, lest it get postponed indefinitely). Processing the other states seems to require a *queue*, since we don't want to examine states for the next input character until we've finished with the current character (so the new state should be put at the *end* of the current list). Rather than choosing between these two data structures, we'll use both! Deques ("double-ended queues") combine the features of stacks and queues: a deque is a list to which items can be added at either end. (Actually, we use an "output-restricted deque," since we always remove items from the beginning, not the end: that would be "dealing from the bottom of the deck.")

A crucial property of the machine is that there are no "loops" consisting of just null states: otherwise it could decide nondeterministically to loop forever. It turns out that this implies that the number of states on the deque at any time is less than the number of characters in the pattern description.

The program given below uses a deque to simulate the actions of a nondeterministic pattern-matching machine as described above. While examining a particular character in the input, the nondeterministic machine can be in any one of several possible states: the program keeps track of these in a deque *dq*. One pointer (*head*) to the head of the deque is maintained so that items can be inserted or removed at the beginning, and another pointer (*tail*) to the tail of the deque is maintained so that items can be inserted at the end. If the pattern description has M characters the deque can be implemented in a "circular" manner in an array of M integers. The contents of the deque are the elements "between" *head* and *tail* (inclusive): if *head*<=*tail*, the meaning is obvious; if *head*>*tail* we take the elements that would fall between *head* and *tail* if the elements of *dq* were arranged in a circle. $dq[head], dq[head+1], \ldots, dq[M-1], dq[0], dq[1], \ldots, dq[tail]$. This is quite simply implemented by using *head*:= *head*+1 **mod** M to increment *head* and similarly for *tail*. Similarly, *head*:= *head*+M-1 **mod** M refers to the element *before* head in the array: this is the position at which an element should be added to the beginning of the deque.

The main loop of the program removes a state from the deque (by incrementing *head* mod M and then referring to $dq[head]$) and performs the action required. If a character is to be matched, the input is checked for the required character: if it is found, the state transition is effected by putting the new state at the *end* of the deque (so that all states involving the current character are processed before those involving the next one). If the state is null, the two possible states to be simulated are put at the *beginning* of the deque. The states involving the current input character are kept separated from those involving the next by a marker *scan=−1* in the deque: when *scan* is encountered, the pointer into the input string is advanced. The loop terminates when the end of the input is reached (no match found), state 0 is reached (legal match found), or only one item, the *scan* marker is left on the deque (no match found). This leads directly to the following implementation:

```
function match(j: integer): integer;
  const scan=−1;
  var head, tail, n1, n2: integer;
      dq: array [0..Mmax] of integer;
  procedure addhead(x: integer);
    begin dq[head]:=x; head:=(head+M−1) mod M end;
  procedure addtail(x: integer);
    begin tail:=(tail+1) mod M; dq[tail]:=x end;
  begin
  head:=1; tail:=0;
  addtail(next1[0]); addtail(scan);
  match:=j−1;
  repeat
    if dq[head]=scan then
      begin j:=j+1; addtail(scan) end
    else if ch[dq[head]]=a[j] then
      addtail(next1[dq[head]])
    else if ch[dq[head]]=' ' then
      begin
      n1:=next1[dq[head]]; n2:=next2[dq[head]];
      addhead(n1); if n1<>n2 then addhead(n2)
      end;
    head:=(head+1) mod M
  until (j>N) or (dq[head]=0) or (head=tail);
  if dq[head]=0 then match:=j−1;
  end;
```

This function takes as its argument the position j in the text string a at which it should start trying to match. It returns the index of the last character in the match found (if any, otherwise it returns $j-1$).

The following table shows the contents of the deque each time a state is removed when our sample machine is run with the text string AABD. (For clarity, the details involving *head*, *tail*, and the maintenance of the circular deque are suppressed in this table: each line shows those elements in the deque between the *head* and *tail* pointers.) The characters appear in the lefthand column in the table at the point when the program has finished scanning them.

		5	*scan*			
		2	6	*scan*		
		1	3	6	*scan*	
		3	6	*scan*	2	
		6	*scan*	2		
A	*scan*	2	7			
		2	7	*scan*		
		1	3	7	*scan*	
		3	7	*scan*	2	
		7	*scan*	2		
A	*scan*	2				
		2	*scan*			
		1	3	*scan*		
		3	*scan*			
B	*scan*	4				
		4	*scan*			
		8	*scan*			
D	*scan*	9				
		9	*scan*			
		0	*scan*			

Thus, we start with State 5 while scanning the first character. First State 5 leads to States 2 and 6, then State 2 leads to States 1 and 3, all of which need to scan the same character and are on the beginning of the deque. Then State 1 leads to State 2, but at the end of the deque (for the next input character). State 3 only leads to another state while scanning a B, so it is ignored while an A is being scanned. When the "*scan*" sentinel finally reaches the front of the deque, we see that the machine could be either in State 2 or State 7 after scanning an A. Continuing, the program eventually ends up the final state, after considering all transitions consistent with the text string.

The running time of this program obviously depends very heavily on the pattern being matched. However, for each of the N input characters, it processes at most M states of the machine, so the worst case running time is proportional to MN. For sure, not all nondeterministic machines can be simulated so efficiently, as discussed in more detail in Chapter 40, but the use of a simple hypothetical pattern-matching machine in this application leads to a quite reasonable algorithm for a quite difficult problem. However, to complete the algorithm, we need a program which translates arbitrary regular expressions into "machines" for interpretation by the above code. In the next chapter, we'll look at the implementation of such a program in the context of a more general discussion of compilers and parsing techniques.

Exercises

1. Give a regular expression for recognizing all occurrences of four or fewer consecutive 1's in a binary string.

2. Draw the nondeterministic pattern matching machine for the pattern description $(A+B)* +C$.

3. Give the state transitions your machine from the previous exercise would make to recognize ABBAC.

4. Explain how you would modify the nondeterministic machine to handle the "not" function.

5. Explain how you would modify the nondeterministic machine to handle "don't-care" characters.

6. What would happen if *match* were to try to simulate the following machine?

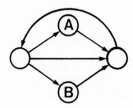

7. Modify *match* to handle regular expressions with the "not" function and "don't-care" characters.

8. Show how to construct a pattern description of length M and a text string of length N for which the running time of *match* is as large as possible.

9. Why must the deque in *match* have only one "*scan*" sentinel in it?

10. Show the contents of the deque each time a state is removed when *match* is used to simulate the example machine in the text with the text string ACD.

21. Parsing

Several fundamental algorithms have been developed to recognize legal computer programs and to decompose their structure into a form suitable for further processing. This operation, called *parsing*, has application beyond computer science, since it is directly related to the study of the structure of language in general. For example, parsing plays an important role in systems which try to "understand" natural (human) languages and in systems for translating from one language to another. One particular case of interest is translating from a "high-level" computer language like Pascal (suitable for human use) to a "low-level" assembly or machine language (suitable for machine execution). A program for doing such a translation is called a *compiler*.

Two general approaches are used for parsing. *Top-down* methods look for a legal program by first looking for parts of a legal program, then looking for parts of parts, etc. until the pieces are small enough to match the input directly. *Bottom-up* methods put pieces of the input together in a structured way making bigger and bigger pieces until a legal program is constructed. In general, top-down methods are recursive, bottom-up methods are iterative; top-down methods are thought to be easier to implement, bottom-up methods are thought to be more efficient.

A full treatment of the issues involved in parser and compiler construction would clearly be beyond the scope of this book. However, by building a simple "compiler" to complete the pattern-matching algorithm of the previous chapter, we will be able to consider some of the fundamental concepts involved. First we'll construct a top-down parser for a simple language for describing regular expressions. Then we'll modify the parser to make a program which translates regular expressions into pattern-matching machines for use by the *match* procedure of the previous chapter.

Our intent in this chapter is to give some feeling for the basic principles of parsing and compiling while at the same time developing a useful pattern matching algorithm. Certainly we cannot treat the issues involved at the level of depth that they deserve. The reader should be warned that subtle difficulties are likely to arise in applying the same approach to similar problems, and advised that compiler construction is a quite well-developed field with a variety of advanced methods available for serious applications.

Context-Free Grammars

Before we can write a program to determine whether a program written in a given language is legal, we need a description of exactly what constitutes a legal program. This description is called a *grammar*: to appreciate the terminology, think of the language as English and read "sentence" for "program" in the previous sentence (except for the first occurrence!). Programming languages are often described by a particular type of grammar called a *context-free grammar*. For example, the context-free grammar which defines the set of all legal regular expressions (as described in the previous chapter) is given below.

$$\langle expression \rangle ::= \langle term \rangle \mid \langle term \rangle + \langle expression \rangle$$
$$\langle term \rangle ::= \langle factor \rangle \mid \langle factor \rangle \langle term \rangle$$
$$\langle factor \rangle ::= (\langle expression \rangle) \mid v \mid \langle factor \rangle^*$$

This grammar describes regular expressions like those that we used in the last chapter, such as $(1+01)^*(0+1)$ or $(A^*B+AC)D$. Each line in the grammar is called a *production* or *replacement rule*. The productions consist of *terminal symbols* (,), + and * which are the symbols used in the language being described ("*v*," a special symbol, stands for any letter or digit); *nonterminal symbols* $\langle expression \rangle$, $\langle term \rangle$, and $\langle factor \rangle$ which are internal to the grammar; and *metasymbols* ::= and | which are used to describe the meaning of the productions. The ::= symbol, which may be read "*is a*," defines the left-hand side of the production in terms of the right-hand side; and the | symbol, which may be read as "*or*" indicates alternative choices. The various productions, though expressed in this concise symbolic notation, correspond in a simple way to an intuitive description of the grammar. For example, the second production in the example grammar might be read "a $\langle term \rangle$ is a $\langle factor \rangle$ or a $\langle factor \rangle$ followed by a $\langle term \rangle$." One nonterminal symbol, in this case $\langle expression \rangle$, is *distinguished* in the sense that a string of terminal symbols is in the language described by the grammar if and only if there is some way to use the productions to *derive* that string from the distinguished nonterminal by replacing (in any number of steps) a nonterminal symbol by any of the "or" clauses on the right-hand side of a production for that nonterminal symbol.

One natural way to describe the result of this derivation process is called a *parse tree*: a diagram of the complete grammatical structure of the string being parsed. For example, the following parse tree shows that the string (A*B+AC)D is in the language described by the above grammar.

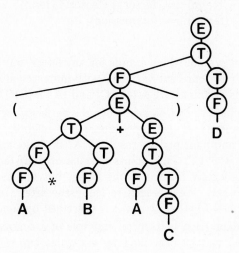

The circled internal nodes labeled E, F, and T represent ⟨*expression*⟩, ⟨*factor*⟩, and ⟨*term*⟩, respectively. Parse trees like this are sometimes used for English, to break down a "sentence" into "subject," "verb," "object," etc.

The main function of a parser is to accept strings which can be so derived and reject those that cannot, by attempting to construct a parse tree for any given string. That is, the parser can *recognize* whether a string is in the language described by the grammar by determining whether or not there exists a parse tree for the string. Top-down parsers do so by building the tree starting with the distinguished nonterminal at the top, working down towards the string to be recognized at the bottom; bottom-up parsers do this by starting with the string at the bottom, working backwards up towards the distinguished nonterminal at the top.

As we'll see, if the strings being recognized also have meanings implying further processing, then the parser can convert them into an internal representation which can facilitate such processing.

Another example of a context-free grammar may be found in the appendix of the *Pascal User Manual and Report*: it describes legal Pascal programs. The principles considered in this section for recognizing and using legal expressions apply directly to the complex job of compiling and executing Pascal programs. For example, the following grammar describes a very small subset of Pascal, arithmetic expressions involving addition and multiplication.

$$\langle expression \rangle ::= \langle term \rangle \mid \langle term \rangle + \langle expression \rangle$$
$$\langle term \rangle ::= \langle factor \rangle \mid \langle factor \rangle^* \langle term \rangle$$
$$\langle factor \rangle ::= (\langle expression \rangle) \mid v$$

Again, v is a special symbol which stands for any letter, but in this grammar the letters are likely to represent variables with numeric values. Examples of legal strings for this grammar are A+(B*C) and (A+B*C)*D*(A+(B+C)).

As we have defined things, some strings are perfectly legal both as arithmetic expressions and as regular expressions. For example, A*(B+C) might mean "add B to C and multiply the result by A" or "take any number of A's followed by either B or C." This points out the obvious fact that checking whether a string is legally formed is one thing, but understanding what it means is quite another. We'll return to this issue after we've seen how to parse a string to check whether or not it is described by some grammar.

Each regular expression is itself an example of a context-free grammar: any language which can be described by a regular expression can also be described by a context-free grammar. The converse is not true: for example, the concept of "balancing" parentheses can't be captured with regular expressions. Other types of grammars can describe languages which can't be described by context-free grammars. For example, *context-sensitive* grammars are the same as those above except that the left-hand sides of productions need not be single nonterminals. The differences between classes of languages and a hierarchy of grammars for describing them have been very carefully worked out and form a beautiful theory which lies at the heart of computer science.

Top-Down Parsing

One parsing method uses recursion to recognize strings from the language described exactly as specified by the grammar. Put simply, the grammar is such a complete specification of the language that it can be turned directly into a program!

Each production corresponds to a procedure with the name of the nonterminal on the left-hand side. Nonterminals on the right-hand side of the input correspond to (possibly recursive) procedure calls; terminals correspond to scanning the input string. For example, the following procedure is part of a top-down parser for our regular expression grammar:

```
procedure expression;
  begin
  term;
  if p[j]='+' then
    begin j:=j+1; expression end
  end;
```

An array p contains the regular expression being parsed, with an index j pointing to the character currently begin examined. To parse a given regular expression, we put it in $p[1..M]$, (with a sentinel character in $p[M+1]$ which is not used in the grammar) set j to 1, and call *expression*. If this results in j being set to $M+1$, then the regular expression is in the language described by the grammar. Otherwise, we'll see below how various error conditions are handled.

The first thing that *expression* does is call *term*, which has a slightly more complicated implementation:

```
procedure term;
  begin
  factor;
  if (p[j]='(') or letter(p[j]) then term;
  end;
```

A direct translation from the grammar would simply have *term* call *factor* and then *term*. This obviously won't work because it leaves no way to exit from *term*: this program would go into an infinite recursive loop if called. (Such loops have particularly unpleasant effects in many systems.) The implementation above gets around this by first checking the input to decide whether *term* should be called. The first thing that *term* does is call *factor*, which is the only one of the procedures that could detect a mismatch in the input. From the grammar, we know that when *factor* is called, the current input character must be either a "(" or an input letter (represented by v). This process of checking the next character (without incrementing j to decide what to do is called *lookahead*. For some grammars, this is not necessary; for others even more lookahead is required.

Now, the implementation of *factor* follows directly from the grammar. If the input character being scanned is not a "(" or an input letter, a procedure *error* is called to handle the error condition:

```
procedure factor;
  begin
  if p[j]='(' then
    begin
    j:=j+1;
    expression;
    if p[j]=')' then j:=j+1 else error
    end
  else if letter(p[j]) then j:=j+1 else error;
  if p[j]='*' then j:=j+1;
  end;
```

Another error condition occurs when a ")" is missing.

These procedures are obviously recursive; in fact they are so intertwined that they can't be compiled in Pascal without using the **forward** construct to get around the rule that a procedure can't be used without first being declared.

The parse tree for a given string gives the recursive call structure during parsing. The reader may wish to refer to the tree above and trace through the operation of the above three procedures when p contains (A*B+AC)D and *expression* is called with $j=1$. This makes the origin of the "top-down" name obvious. Such parsers are also often called *recursive descent* parsers because they move down the parse tree recursively.

The top-down approach won't work for all possible context-free grammars. For example, if we had the production $\langle expression \rangle ::= v \mid \langle expression \rangle + \langle term \rangle$ then we would have

```
procedure badexpression;
  begin
  if letter(p[j]) then j:=j+1 else
    begin
    badexpression;
    if p[j]<>'+' then error else
      begin j:=j+1; term end
    end
  end;
```

If this procedure were called with $p[j]$ a nonletter (as in our example, for $j=1$) then it would go into an infinite recursive loop. Avoiding such loops is a principal difficulty in the implementation of recursive descent parsers. For *term*, we used lookahead to avoid such a loop; in this case the proper way to get around the problem is to switch the grammar to say ⟨*term*⟩+⟨*expression*⟩. The occurrence of a nonterminal as the first thing on the right hand side of a replacement rule for itself is called *left recursion*. Actually, the problem is more subtle, because the left recursion can arise indirectly: for example if we were to have the productions ⟨*expression*⟩ ::= ⟨*term*⟩ and ⟨*term*⟩ ::= *v* | ⟨*expression*⟩ + ⟨*term*⟩. Recursive descent parsers won't work for such grammars: they have to be transformed to equivalent grammars without left recursion, or some other parsing method has to be used. In general, there is an intimate and very widely studied connection between parsers and the grammars they recognize. The choice of a parsing technique is often dictated by the characteristics of the grammar to be parsed.

Bottom-Up Parsing

Though there are several recursive calls in the programs above, it is an instructive exercise to remove the recursion systematically. Recall from Chapter 9 (where we removed the recursion from Quicksort) that each procedure call can be replaced by a stack push and each procedure return by a stack pop, mimicking what the Pascal system does to implement recursion. A reason for doing this is that many of the calls which seem recursive are not truly recursive. When a procedure call is the last action of a procedure, then a simple **goto** can be used. This turns *expression* and *term* into simple loops, which can be incorporated together and combined with *factor* to produce a single procedure with one true recursive call (the call to *expression* within *factor*).

This view leads directly to a quite simple way to check whether regular expressions are legal. Once all the procedure calls are removed, we see that each terminal symbol is simply scanned as it is encountered. The only real processing done is to check whether there is a right parenthesis to match each left parenthesis and whether each "+" is followed by either a letter or a "(". That is, checking whether a regular expression is legal is essentially equivalent to checking for balanced parentheses. This can be simply implemented by keeping a counter, initialized to 0, which is incremented when a left parenthesis is encountered, decremented when a right parenthesis is encountered. If the counter is zero when the end of the expression is reached, and each "+" of the expression is followed by either a letter or a "(", then the expression was legal.

Of course, there is more to parsing than simply checking whether the input string is legal: the main goal is to build the parse tree (even if in an implicit way, as in the top-down parser) for other processing. It turns out to be possible to do this with programs with the same essential structure as the parenthesis checker described in the previous paragraph. One type of parser which works in this way is the ·so-called *shift-reduce* parser. The idea is to maintain a pushdown stack which holds terminal and nonterminal symbols. Each step in the parse is either a *shift* step, in which the next input character is simply pushed onto the stack, or a *reduce* step, in which the top characters on the stack are matched to the right-hand side of some production in the grammar and "reduced to" (replaced by) the nonterminal on the left side of that production. Eventually all the input characters get shifted onto the stack, and eventually the stack gets reduced to a single nonterminal symbol.

The main difficulty in building a shift-reduce parser is deciding when to shift and when to reduce. This can be a complicated decision, depending on the grammar. Various types of shift-reduce parsers have been studied in great detail, an extensive literature has been developed on them, and they are quite often preferred over recursive descent parsers because they tend to be slightly more efficient and significantly more flexible. Certainly we don't have space here to do justice to this field, and we'll forgo even the details of an implementation for our example.

Compilers

A *compiler* may be thought of as a program which translates from one language to another. For example, a Pascal compiler translates programs from the Pascal language into the machine language of some particular computer. We'll illustrate one way that this might be done by continuing with our regular-expression pattern-matching example, where we wish to translate from the language of regular expressions to a "language" for pattern-matching machines, the *ch*, *next1*, and *next2* arrays of the *match* program of the previous chapter.

Essentially, the translation process is "one-to-one": for each character in the pattern (with the exception of parentheses) we want to produce a state for the pattern-matching machine (an entry in each of the arrays). The trick is to keep track of the information necessary to fill in the *next1* and *next2* arrays. To do so, we'll convert each of the procedures in our recursive descent parser into functions which create pattern-matching machines. Each function will add new states as necessary onto the end of the *ch*, *next1*, and *next2* arrays, and return the index of the initial state of the machine created (the final state will always be the last entry in the arrays).

For example, the function given below for the ⟨*expression*⟩ production creates the "or" states for the pattern matching machine.

```
function expression: integer;
    var t1, t2: integer;
    begin
    t1:=term; expression:=t1;
    if p[j]='+' then
        begin
        j:=j+1; state:=state+1;
        t2:=state; expression:=t2; state:=state+1;
        setstate(t2, ' ', expression, t1);
        setstate(t2-1, ' ', state, state);
        end;
    end;
```

This function uses a procedure *setstate* which simply sets the *ch*, *next1*, and *next2* array entries indexed by the first argument to the values given in the second, third, and fourth arguments, respectively. The index *state* keeps track of the "current" state in the machine being built. Each time a new state is created, *state* is simply incremented. Thus, the state indices for the machine corresponding to a particular procedure call range between the value of *state* on entry and the value of *state* on exit. The final state index is the value of *state* on exit. (We don't actually "create" the final state by incrementing *state* before exiting, since this makes it easy to "merge" the final state with later initial states, as we'll see below.)

With this convention, it is easy to check (beware of the recursive call!) that the above program implements the rule for composing two machines with the "or" operation as diagrammed in the previous chapter. First the machine for the first part of the expression is built (recursively), then two new null states are added and the second part of the expression built. The first null state (with index $t2-1$) is the final state of the machine of the first part of the expression which is made into a "no-op" state to skip to the final state for the machine for the second part of the expression, as required. The second null state (with index $t2$) is the initial state, so its index is the return value for *expression* and its *next1* and *next2* entries are made to point to the initial states of the two expressions. Note carefully that these are constructed in the opposite order than one might expect, because the value of *state* for the no-op state is not known until the recursive call to *expression* has been made.

The function for ⟨*term*⟩ first builds the machine for a ⟨*factor*⟩ then, if necessary, merges the final state of that machine with the initial state of the machine for another ⟨*term*⟩. This is easier done than said, since *state* is the final state index of the call to *factor*. A call to *term* without incrementing *state* does the trick:

```
function term;
  var t: integer;
  begin
  term:=factor;
  if (p[j]='(') or letter(p[j]) then t:=term
  end;
```

(We have no use for the initial state index returned by the second call to *term*, but Pascal requires us to put it somewhere, so we throw it away in a temporary variable *t*.)

The function for ⟨factor⟩ uses similar techniques to handle its three cases: a parenthesis calls for a recursive call on *expression*; a *v* calls for simple concatenation of a new state; and a * calls for operations similar to those in *expression*, according to the closure diagram from the previous section:

```
function factor;
  var t1, t2: integer;
  begin
  t1:=state;
  if p[j]='(' then
    begin
    j:=j+1; t2:=expression;
    if p[j]=')' then j:=j+1 else error
    end
  else if letter(p[j]) then
    begin
    setstate(state, p[j], state+1, 0);
    t2:=state; j:=j+1; state:=state+1
    end
  else error;
  if p[j]<>'*' then factor:=t2 else
    begin
    setstate(state, ' ', state+1, t2);
    factor:=state; next1[t1-1]:=state;
    j:=j+1; state:=state+1;
    end;
  end;
```

The reader may find it instructive to trace through the construction of the machine for the pattern (A*B+AC)D given in the previous chapter.

The final step in the development of a general regular expression pattern matching algorithm is to put these procedures together with the *match* procedure, as follows:

```
j:=1; state:=1;
next1[0]:=expression;
setstate(state, ' ', 0, 0);
for i:=1 to N−1 do
    if match(i)>=i then writeln(i);
```

This program will print out all character positions in a text string $a[1...N]$ where a pattern $p[1...M]$ leads to a match.

Compiler-Compilers

The program for general regular expression pattern matching that we have developed in this and the previous chapter is efficient and quite useful. A version of this program with a few added capabilities (for handling "don't-care" characters and other amenities) is likely to be among the most heavily used utilities on many computer systems.

It is interesting (some might say confusing) to reflect on this algorithm from a more philosophical point of view. In this chapter, we have considered parsers for unraveling the structure of regular expressions, based on a formal description of regular expressions using a context-free grammar. Put another way, we used the context-free grammar to specify a particular "pattern": sequences of characters with legally balanced parentheses. The parser then checks to see if the pattern occurs in the input (but only considers a match legal if it covers the entire input string). Thus parsers, which check that an input string is in the set of strings defined by some context-free grammar, and pattern matchers, which check that an input string is in the set of strings defined by some regular expression, are essentially performing the same function! The principal difference is that context-free grammars are capable of describing a much wider class of strings. For example, the set of all regular expressions can't be described with regular expressions.

Another difference in the way we've implemented the programs is that the context-free grammar is "built in" to the parser, while the *match* procedure is "table-driven": the same program works for all regular expressions, once they have been translated into the proper format. It turns out to be possible to build parsers which are table-driven in the same way, so that the same program can be used to parse all languages which can be described by context-free grammars. A *parser generator* is a program which takes a grammar as input and produces a parser for the language described by that grammar as output. This can be carried one step further: it is possible to build compilers which are table-driven in terms of both the input and the output languages. A *compiler-compiler* is a program which takes two grammars (and some formal specification of the relationships between them) as input and produces a compiler which translates strings from one language to the other as output.

Parser generators and compiler-compilers are available for general use in many computing environments, and are quite useful tools which can be used to produce efficient and reliable parsers and compilers with a relatively small amount of effort. On the other hand, top-down recursive descent parsers of the type considered here are quite serviceable for simple grammars which arise in many applications. Thus, as with many of the algorithms we have considered, we have a straightforward method which can be used for applications where a great deal of implementation effort might not be justified, and several advanced methods which can lead to significant performance improvements for large-scale applications. Of course, in this case, this is significantly understating the point: we've only scratched the surface of this extensively researched field.

Exercises

1. How does the recursive descent parser find an error in a regular expression such as (A+B)*BC+ which is incomplete?

2. Give the parse tree for the regular expression ((A+B)+(C+D)*)*.

3. Extend the arithmetic expression grammar to include exponentiation, **div** and **mod**.

4. Give a context-free grammar to describe all strings with no more than two consecutive 1's.

5. How many procedure calls are used by the recursive descent parser to recognize a regular expression in terms of the number of concatenation, *or*, and closure operations and the number of parentheses?

6. Give the *ch*, *next1* and *next2* arrays that result from building the pattern matching machine for the pattern ((A+B)+(C+D)*)*.

7. Modify the regular expression grammar to handle the "not" function and "don't-care" characters.

8. Build a general regular expression pattern matcher based on the improved grammar in your answer to the previous question.

9. Remove the recursion from the recursive descent compiler, and simplify the resulting code as much as possible. Compare the running time of the nonrecursive and recursive methods.

10. Write a compiler for simple arithmetic expressions described by the grammar in the text. It should produce a list of "instructions" for a machine capable of three operations: *push* the value of a variable onto a stack; *add* the top two values on the stack, removing them from the stack, then putting the result there; and *multiply* the top two values on the stack, in the same way.

22. File Compression

For the most part, the algorithms that we have studied have been designed primarily to use as little *time* as possible and only secondarily to conserve *space*. In this section, we'll examine some algorithms with the opposite orientation: methods designed primarily to reduce space consumption without using up too much time. Ironically, the techniques that we'll examine to save space are "coding" methods from information theory which were developed to minimize the amount of information necessary in communications systems and therefore originally intended to save time (not space).

In general, most files stored on computer systems have a great deal of redundancy. The methods we will examine save space by taking advantage of the fact that most files have a relatively low "information content." File compression techniques are often used for text files (in which certain characters appear much more often than others), "raster" files for encoding pictures (which can have large homogeneous areas), and files for the digital representation of sound and other analog signals (which can have large repeated patterns).

We'll look at an elementary algorithm for the problem (which is still quite useful) and an advanced "optimal" method. The amount of space saved by these methods will vary depending on characteristics of the file. Savings of 20% to 50% are typical for text files, and savings of 50% to 90% might be achieved for binary files. For some types of files, for example files consisting of random bits, little can be gained. In fact, it is interesting to note that any general-purpose compression method must make some files longer (otherwise we could continually apply the method to produce an arbitrarily small file).

On one hand, one might argue that file compression techniques are less important than they once were because the cost of computer storage devices has dropped dramatically and far more storage is available to the typical user than in the past. On the other hand, it can be argued that file compression techniques are more important than ever because, since so much storage is in use, the savings they make possible are greater. Compression techniques are also appropriate for storage devices which allow extremely high-speed access and are by nature relatively expensive (and therefore small).

Run-Length Encoding

The simplest type of redundancy in a file is long runs of repeated characters. For example, consider the following string:

AAAABBBAABBBBBCCCCCCCCCDABCBAAABBBBCCCD

This string can be encoded more compactly by replacing each repeated string of characters by a single instance of the repeated character along with a count of the number of times it was repeated. We would like to say that this string consists of 4 A's followed by 3 B's followed by 2 A's followed by 5 B's, etc. Compressing a string in this way is called *run-length encoding*. There are several ways to proceed with this idea, depending on characteristics of the application. (Do the runs tend to be relatively long? How many bits are used to encode the characters being encoded?) We'll look at one particular method, then discuss other options.

If we know that our string contains just letters, then we can encode counts simply by interspersing digits with the letters, thus our string might be encoded as follows:

4A3BAA5B8CDABCB3A4B3CD

Here "4A" means "four A's," and so forth. Note that is is not worthwhile to encode runs of length one or two, since two characters are needed for the encoding.

For binary files (containing solely 0's and 1's), a refined version of this method is typically used to yield dramatic savings. The idea is simply to store the run lengths, taking advantage of the fact that the runs alternate between 0 and 1 to avoid storing the 0's and 1's themselves. (This assumes that there are few short runs, but no run-length encoding method will work very well unless most of the runs are long.) For example, at the left in the figure below is a "raster" representation of the letter "q" lying on its side, which is representative of the type of information that might have to be processed by a text formatting system (such as the one used to print this book); at the right is a list of numbers which might be used to store the letter in a compressed form.

```
0000000000000000000000000000011111111111111000000000        28 14  9
0000000000000000000000000000111111111111111110000000        26 18  7
0000000000000000000000001111111111111111111111110000        23 24  4
0000000000000000000000011111111111111111111111111000        22 26  3
0000000000000000000001111111111111111111111111111110        20 30  1
0000000000000000000111111100000000000000000001111111        19  7 18 7
0000000000000000000111110000000000000000000000011111        19  5 22 5
0000000000000000000111000000000000000000000000000111        19  3 26 3
0000000000000000000111000000000000000000000000000111        19  3 26 3
0000000000000000000111000000000000000000000000000111        19  3 26 3
0000000000000000000111000000000000000000000000000111        19  3 26 3
0000000000000000000111000000000000000000000000001110        20  4 23 3 1
0000000000000000000011100000000000000000000000111000        22  3 20 3 3
0111111111111111111111111111111111111111111111111111         1 50
0111111111111111111111111111111111111111111111111111         1 50
0111111111111111111111111111111111111111111111111111         1 50
0111111111111111111111111111111111111111111111111111         1 50
0111111111111111111111111111111111111111111111111111         1 50
0110000000000000000000000000000000000000000000000011         1  2 46 2
```

That is, the first line consists of 28 0's followed by 14 1's followed by 9 more
0's, etc. The 63 counts in this table plus the number of bits per line (51)
contain sufficient information to reconstruct the bit array (in particular, note
that no "end of line" indicator is needed). If six bits are used to represent each
count, then the entire file is represented with 384 bits, a substantial savings
over the 975 bits required to store it explicitly.

Run-length encoding requires a separate representation for the file to be
encoded and the encoded version of the file, so that it can't work for all files.
This can be quite inconvenient: for example, the character file compression
method suggested above won't work for character strings that contain digits.
If other characters are used to encode the counts, it won't work for strings
that contain those characters. To illustrate a way to encode any string from
a fixed alphabet of characters using only characters from that alphabet, we'll
assume that we only have the 26 letters of the alphabet (and spaces) to work
with.

How can we make some letters represent digits and others represent
parts of the string to be encoded? One solution is to use some character
which is likely to appear rarely in the text as a so-called *escape character*.
Each appearance of that character signals that the next two letters form a
(count,character) pair, with counts represented by having the ith letter of
the alphabet represent the number i. Thus our example string would be
represented as follows with Q as the escape character:

QDABBBAAQEBQHCDABCBAAAQDBCCCD

The combination of the escape character, the count, and the one copy of the repeated character is called an *escape sequence*. Note that it's not worthwhile to encode runs less than four characters long since at least three characters are required to encode any run.

But what if the escape character itself happens to occur in the input? We can't afford to simply ignore this possibility, because it might be difficult to ensure that any particular character can't occur. (For example, someone might try to encode a string that has already been encoded.) One solution to this problem is to use an escape sequence with a count of zero to represent the escape character. Thus, in our example, the space character could represent zero, and the escape sequence "Q⟨space⟩" would be used to represent any occurrence of Q in the input. It is interesting to note that files which contain Q are the only files which are made longer by this compression method. If a file which has already been compressed is compressed again, it grows by at least the number of characters equal to the number of escape sequences used.

Very long runs can be encoded with multiple escape sequences. For example, a run of 51 A's would be encoded as QZAQYA using the conventions above. If many very long runs are expected, it would be worthwhile to reserve more than one character to encode the counts.

In practice, it is advisable to make both the compression and expansion programs somewhat sensitive to errors. This can be done by including a small amount of redundancy in the compressed file so that the expansion program can be tolerant of an accidental minor change to the file between compression and expansion. For example, it probably is worthwhile to put "end-of-line" characters in the compressed version of the letter "q" above, so that the expansion program can resynchronize itself in case of an error.

Run-length encoding is not particularly effective for text files because the only character likely to be repeated is the blank, and there are simpler ways to encode repeated blanks. (It was used to great advantage in the past to compress text files created by reading in punched-card decks, which necessarily contained many blanks.) In modern systems, repeated strings of blanks are never entered, never stored: repeated strings of blanks at the beginning of lines are encoded as "tabs," blanks at the ends of lines are obviated by the use of "end-of-line" indicators. A run-length encoding implementation like the one above (but modified to handle all representable characters) saves only about 4% when used on the text file for this chapter (and this savings all comes from the letter "q" example!).

Variable-Length Encoding

In this section we'll examine a file compression technique called *Huffman encoding* which can save a substantial amount of space on text files (and many other kinds of files). The idea is to abandon the way that text files are usually stored: instead of using the usual seven or eight bits for each character, Huffman's method uses only a few bits for characters which are used often, more bits for those which are rarely used.

It will be convenient to examine how the code is used before considering how it is created. Suppose that we wish to encode the string "A SIMPLE STRING TO BE ENCODED USING A MINIMAL NUMBER OF BITS." Encoding it in our standard compact binary code with the five-bit binary representation of i representing the ith letter of the alphabet (0 for blank) gives the following bit sequence:

$$00001000001001101001011011000001100001010000000$$
$$10011101001001001001011100011100000101000111$$
$$00000000100010100000000101011100001101111001000$$
$$00101001000000001010110011010010111000111000000$$
$$00001000000110101001011100100101101000001011000$$
$$00000011101010101101000100010110010000000011111$$
$$0011000000000100100110100100110$$

To "decode" this message, simply read off five bits at a time and convert according to the binary encoding defined above. In this standard code, the C, which appears only once, requires the same number of bits as the I, which appears six times. The Huffman code achieves economy in space by encoding frequently used characters with as few bits as possible so that the total number of bits used for the message is minimized.

The first step is to count the frequency of each character within the message to be encoded. The following code fills an array $count[0..26]$ with the frequency counts for a message in a character array $a[1..M]$. (This program uses the *index* procedure described in Chapter 19 to keep the frequency count for the ith letter of the alphabet in $count[i]$, with $count[0]$ used for blanks.)

```
for i:=0 to 26 do count[i]:=0;
for i:=1 to M do
   count[index(a[i])]:=count[index(a[i])]+1;
```

For our example string, the count table produced is

0	1	2	3	4	5	6	7	8	9	10	11	12	13	14	15	16	17	18	19	20	21	22	23	24	25	26
11	3	3	1	2	5	1	2	0	6	0	0	2	4	5	3	1	0	2	4	3	2	0	0	0	0	0

which indicates that there are eleven blanks, three A's, three B's, etc.

The next step is to build a "coding tree" from the bottom up according to the frequencies. First we create a tree node for each nonzero frequency from the table above:

11	3	3	1	2	5	1	2	6	2	4	5	3	1	2	4	3	2
A	B	C	D	E	F	G	I	L	M	N	O	P	R	S	T	U

Now we pick the two nodes with the smallest frequencies and create a new node with those two nodes as sons and with frequency value the sum of the values of the sons:

(It doesn't matter which nodes are used if there are more than two with the smallest frequency.) Continuing in this way, we build up larger and larger subtrees. The forest of trees after all nodes with frequency 2 have been put in is as follows:

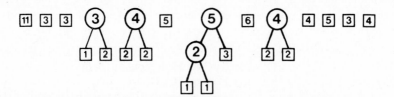

Next, the nodes with frequency 3 are put together, creating two new nodes of frequency 6, etc. Ultimately, all the nodes are combined together into a single tree:

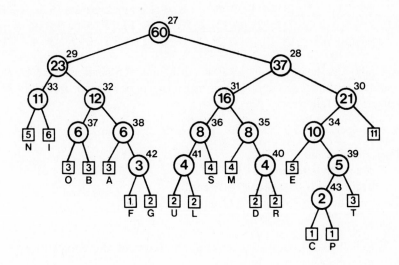

Note that nodes with low frequencies end up far down in the tree and nodes with high frequencies end up near the root of the tree. The numbers labeling the external (square) nodes in this tree are the frequency counts, while the number labeling each internal (round) node is the sum of the labels of its two sons. The small number above each node in this tree is the index into the *count* array where the label is stored, for reference when examining the program which constructs the tree below. (The labels for the internal nodes will be stored in *count*[27..51] in an order determined by the dynamics of the construction.) Thus, for example, the 5 in the leftmost external node (the frequency count for N) is stored in *count*[14], the 6 in the next external node (the frequency count for I) is stored in *count*[9], and the 11 in the father of these two is stored in *count*[33], etc.

It turns out that this structural description of the frequencies in the form of a tree is exactly what is needed to create an efficient encoding. Before looking at this encoding, let's look at the code for constructing the tree. The general process involves removing the smallest from a set of unordered elements, so we'll use the *pqdownheap* procedure from Chapter 11 to build and maintain an indirect heap on the frequency values. Since we're interested in small values first, we'll assume that the sense of the inequalities in *pqdownheap* has been reversed. One advantage of using indirection is that it is easy to ignore zero frequency counts. The following table shows the heap constructed for our example:

k		1	2	3	4	5	6	7	8	9	10	11	12	13	14	15	16	17	18
$heap[k]$		3	7	16	21	12	15	6	20	9	4	13	14	5	2	18	19	1	0
$count[heap[k]]$	1	2	1	2	2	3	1	3	6	2	4	5	5	3	2	4	3	11

Specifically, this heap is built by first initializing the *heap* array to point to the non-zero frequency counts, then using the *pqdownheap* procedure from Chapter 11, as follows:

```
N:=0;
for i:=0 to 26 do
    if count[i]<>0 then
        begin N:=N+1; heap[N]:=i end;
for k:=N downto 1 do pqdownheap(k);
```

As mentioned above, this assumes that the sense of the inequalities in the *pqdownheap* code has been reversed.

Now, the use of this procedure to construct the tree as above is straightforward: we take the two smallest elements off the heap, add them and put the result back into the heap. At each step we create one new count, and decrease the size of the heap by one. This process creates $N-1$ new counts, one for each of the internal nodes of the tree being created, as in the following code:

```
repeat
    t:=heap[1]; heap[1]:=heap[N]; N:=N−1;
    pqdownheap(1);
    count[26+N]:=count[heap[1]]+count[t];
    dad[t]:=26+N; dad[heap[1]]:=−26−N;
    heap[1]:=26+N; pqdownheap(1);
until N=1;
dad[26+N]:=0;
```

The first two lines of this loop are actually *pqremove*; the size of the heap is decreased by one. Then a new internal node is "created" with index $26+N$ and given a value equal to the sum of the value at the root and value just removed. Then this node is put at the root, which raises its priority, necessitating another call on *pqdownheap* to restore order in the heap. The tree itself is represented with an array of "father" links: $dad[t]$ is the index of the father of the node whose weight is in $count[t]$. The sign of $dad[t]$ indicates whether the node is a left or right son of its father. For example, in the tree above we might have $dad[0]=−30$, $count[30]=21$, $dad[30]=−28$, and $count[28]=37$

(indicating that the node of weight 21 has index 30 and its father has index 28 and weight 37).

The Huffman code is derived from this coding tree simply by replacing the frequencies at the bottom nodes with the associated letters and then viewing the tree as a radix search trie:

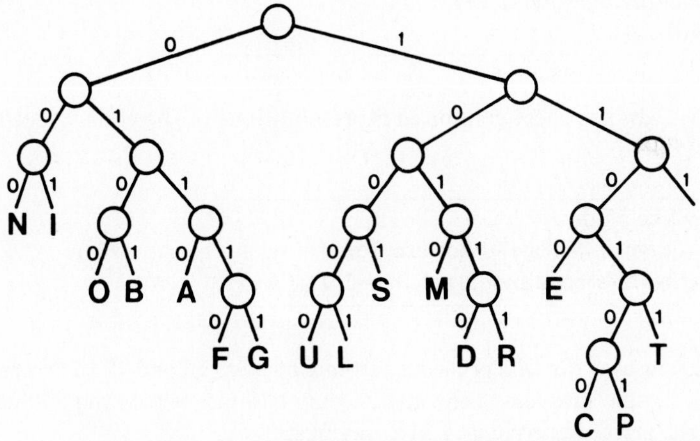

Now the code can be read directly from this tree. The code for N is 000, the code for I is 001, the code for C is 110100, etc. The following program fragment reconstructs this information from the representation of the coding tree computed during the sifting process. The code is represented by two arrays: *code*[k] gives the binary representation of the *k*th letter and *len*[k] gives the number of bits from *code*[k] to use in the code. For example, I is the 9th letter and has code 001, so *code*[9]=1 and *len*[9]=3.

```
for k:=0 to 26 do
  if count[k]=0 then
    begin code[k]:=0; len[k]:=0 end
  else
    begin
    i:=0; j:=1; t:=dad[k]; x:=0;
    repeat
      if t<0 then begin x:=x+j; t:=−t end;
      t:=dad[t]; j:=j+j; i:=i+1
    until t=0;
    code[k]:=x; len[k]:=i;
    end;
```

Finally, we can use these computed representations of the code to encode the message:

```
for j:=1 to M do
  for i:=len[index(a[j])] downto 1 do
    write(bits(code[index(a[j])], i−1, 1):1);
```

This program uses the *bits* procedure from Chapters 10 and 17 to access single bits. Our sample message is encoded in only 236 bits versus the 300 used for the straightforward encoding, a 21% savings:

```
0110111100100110101101011000111001111100111011
1011100100001111111110110100111010111100111110
0000110100010010110100101101111000010010010010
0011111110110111101000100000110100110100011110
0001000010100101110010111111101000111011101010
01110111001
```

An interesting feature of the Huffman code that the reader undoubtedly has noticed is that delimiters between characters are not stored, even though different characters may be coded with different numbers of bits. How can we determine when one character stops and the next begins to decode the message? The answer is to use the radix search trie representation of the code. Starting at the root, proceed down the tree according to the bits in the message: each time an external node is encountered, output the character at that node and restart at the root. But the tree is built at the time we *encode* the message: this means that we need to save the tree along with the message in order to decode it. Fortunately, this does not present any real difficulty. It is actually necessary only to store the *code* array, because the radix search trie which results from inserting the entries from that array into an initially empty tree is the decoding tree.

Thus, the storage savings quoted above is not entirely accurate, because the message can't be decoded without the trie and we must take into account the cost of storing the trie (i.e., the *code* array) along with the message. Huffman encoding is therefore only effective for long files where the savings in the message is enough to offset the cost, or in situations where the coding trie can be precomputed and used for a large number of messages. For example, a trie based on the frequencies of occurrence of letters in the English language could be used for text documents. For that matter, a trie based on the frequency of occurrence of characters in Pascal programs could be used for encoding programs (for example, ";" is likely to be near the top of such a trie). A Huffman encoding algorithm saves about 23% when run on the text for this chapter.

As before, for truly random files, even this clever encoding scheme won't work because each character will occur approximately the same number of times, which will lead to a fully balanced coding tree and an equal number of bits per letter in the code.

Exercises

1. Implement compression and expansion procedures for the run-length encoding method for a fixed alphabet described in the text, using Q as the escape character.

2. Could "QQ" occur somewhere in a file compressed using the method described in the text? Could "QQQ" occur?

3. Implement compression and expansion procedures for the binary file encoding method described in the text.

4. The letter "q" given in the text can be processed as a sequence of five-bit characters. Discuss the pros and cons of doing so in order to use a character-based run-length encoding method.

5. Draw a Huffman coding tree for the string "ABRACADABRA." How many bits does the encoded message require?

6. What is the Huffman code for a binary file? Give an example showing the maximum number of bits that could be used in a Huffman code for a N-character ternary (three-valued) file.

7. Suppose that the frequencies of the occurrence of all the characters to be encoded are different. Is the Huffman encoding tree unique?

8. Huffman coding could be extended in a straightforward way to encode in two-bit characters (using 4-way trees). What would be the main advantage and the main disadvantage of doing so?

9. What would be the result of breaking up a Huffman-encoded string into five-bit characters and Huffman encoding that string?

10. Implement a procedure to *decode* a Huffman-encoded string, given the *code* and *len* arrays.

23. Cryptology

In the previous chapter we looked at methods for encoding strings of characters to save space. Of course, there is another very important reason to encode strings of characters: to keep them secret.

Cryptology, the study of systems for secret communications, consists of two competing fields of study: *cryptography*, the design of secret communications systems, and *cryptanalysis*, the study of ways to compromise secret communications systems. The main application of cryptology has been in military and diplomatic communications systems, but other significant applications are becoming apparent. Two principal examples are computer file systems (where each user would prefer to keep his files private) and "electronic funds transfer" systems (where very large amounts of money are involved). A computer user wants to keep his computer files just as private as papers in his file cabinet, and a bank wants electronic funds transfer to be just as secure as funds transfer by armored car.

Except for military applications, we assume that cryptographers are "good guys" and cryptanalysts are "bad guys": our goal is to protect our computer files and our bank accounts from criminals. If this point of view seems somewhat unfriendly, it must be noted (without being over-philosophical) that by using cryptography one is assuming the existence of unfriendliness! Of course, even "good guys" must know something about cryptanalysis, since the very best way to be sure that a system is secure is to try to compromise it yourself. (Also, there are several documented instances of wars being brought to an end, and many lives saved, through successes in cryptanalysis.)

Cryptology has many close connections with computer science and algorithms, especially the arithmetic and string-processing algorithms that we have studied. Indeed, the art (science?) of cryptology has an intimate relationship with computers and computer science that is only beginning to be fully understood. Like algorithms, cryptosystems have been around far longer than computers. Secrecy system design and algorithm design have a common heritage, and the same people are attracted to both.

It is not entirely clear which branch of cryptology has been affected most by the availability of computers. The cryptographer now has available a much more powerful encryption machine than before, but this also gives him more room to make a mistake. The cryptanalyst has much more powerful tools for breaking codes, but the codes to be broken are more complicated than ever before. Cryptanalysis can place an incredible strain on computational resources; not only was it among the first applications areas for computers, but it still remains a principal applications area for modern supercomputers.

More recently, the widespread use of computers has led to the emergence of a variety of important new applications for cryptology, as mentioned above. New cryptographic methods have recently been developed appropriate for such applications, and these have led to the discovery of a fundamental relationship between cryptology and an important area of theoretical computer science that we'll examine briefly in Chapter 40.

In this chapter, we'll examine some of the basic characteristics of cryptographic algorithms because of the importance of cryptography in modern computer systems and because of close relationships with many of the algorithms we have studied. We'll refrain from delving into detailed implementations: cryptography is certainly a field that should be left to experts. While it's not difficult to "keep people honest" by encrypting things with a simple cryptographic algorithm, it is dangerous to rely upon a method implemented by a non-expert.

Rules of the Game

All the elements that go into providing a means for secure communications between two individuals together are called a *cryptosystem*. The canonical structure of a typical cryptosystem is diagramed below:

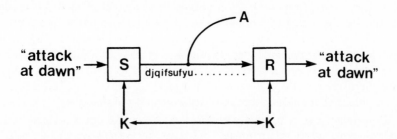

The *sender* (S) wishes to send a message (called the *plaintext*) to the *receiver* (R). To do so, he transforms the plaintext into a secret form suitable for transmission (called the *ciphertext*) using a cryptographic algorithm (the *encryption method*) and some *key* (K) parameters. To read the message, the receiver must have a matching cryptographic algorithm (the *decryption method*) and the same key parameters, which he can use to transform the ciphertext back into the plaintext, the message. It is usually assumed that the ciphertext is sent over insecure communications lines and is available to the cryptanalyst (A). It also is usually assumed that the encryption and decryption methods are known to the cryptanalyst: his aim is to recover the plaintext from the ciphertext without knowing the key parameters. Note that the whole system depends on some separate prior method of communication between the sender and receiver to agree on the key parameters. As a rule, the more key parameters, the more secure the cryptosystem is but the more inconvenient it is to use. This situation is akin to that for more conventional security systems: a combination safe is more secure with more numbers on the combination lock, but it is harder to remember the combination. The parallel with conventional systems also serves as a reminder that any security system is only as secure as the trustworthiness of the people that have the key.

It is important to remember that economic questions play a central role in cryptosystems. There is an economic motivation to build simple encryption and decryption devices (since many may need to be provided and complicated devices cost more). Also, there is an economic motivation to reduce the amount of key information that must be distributed (since a very secure and expensive method of communications must be used). Balanced against the cost of implementing cryptographic algorithms and distributing key information is the amount of money the cryptanalyst would be willing to pay to break the system. For most applications, it is the cryptographer's aim to develop a low-cost system with the property that it would cost the cryptanalyst much more to read messages than he would be willing to pay. For a few applications, a "provably secure" cryptosystem may be required: one for which it can be ensured that the cryptanalyst can never read messages no matter what he is willing to spend. (The very high stakes in some applications of cryptology naturally imply that very large amounts of money are used for cryptanalysis.) In algorithm design, we try to keep track of costs to help us choose the best algorithms; in cryptology, costs play a central role in the design process.

Simple Methods

Among the simplest (and among the oldest) methods for encryption is the *Caesar cipher*: if a letter in the plaintext is the Nth letter in the alphabet, replace it by the $(N + K)$th letter in the alphabet, where K is some fixed integer (Caesar used $K = 3$). For example, the table below shows how a message is encrypted using this method with $K = 1$:

> Plaintext: ATTACK AT DAWN
> Ciphertext: BUUBDLABUAEBXO

This method is weak because the cryptanalyst has only to guess the value of K: by trying each of the 26 choices, he can be sure that he will read the message.

A far better method is to use a general table to define the substitution to be made: for each letter in the plaintext, the table tells which letter to put in the ciphertext. For example, if the table gives the correspondence

> ABCDEFGHIJKLMNOPQRSTUVWXYZ
> THE QUICKBROWNFXJMPDVRLAZYG

then the message is encrypted as follows:

> Plaintext: ATTACK AT DAWN
> Ciphertext: HVVH OTHVTQHAF

This is much more powerful than the simple Caesar cipher because the cryptanalyst would have to try many more (about $27! > 10^{28}$) tables to be sure of reading the message. However, "simple substitution" ciphers like this are easy to break because of letter frequencies inherent in the language. For example, since E is the most frequent letter in English text, the cryptanalyst could get good start on reading the message by looking for the most frequent letter in the ciphertext and assuming that it is to be replaced by E. While this might not be the right choice, he certainly is better off than if he had to try all 26 letters. He can do even better by looking at two-letter combinations ("digrams"): certain digrams (such as QJ) never occur in English text while others (such as ER) are very common. By examining frequencies of letters and combinations of letters, a cryptanalyst can very easily break a simple substitution cipher.

One way to make this type of attack more difficult is to use more than one table. A simple example of this is an extension of the Caesar cipher called the *Vigenere cipher*: a small repeated key is used to determine the value of K for each letter. At each step, the key letter index is added to the plaintext letter index to determine the ciphertext letter index. Our sample plaintext, with the key ABC, is encrypted as follows:

> Key: ABCABCABCABCAB
> Plaintext: ATTACK AT DAWN
> Ciphertext: BVWBENACWAFDXP

For example, the last letter of the ciphertext is P, the 16th letter of the alphabet, because the corresponding plaintext letter is N (the 14th letter), and the corresponding key letter is B (the 2nd letter).

The Vigenere cipher can obviously be made more complicated by using different general tables for each letter of the plaintext (rather than simple offsets). Also, it is obvious that the longer the key, the better. In fact, if the key is as long as the plaintext, we have the *Vernam cipher*, more commonly called the *one-time pad*. This is the only provably secure cryptosystem known, and it is reportedly used for the Washington-Moscow hotline and other vital applications. Since each key letter is used only once, the cryptanalyst can do no better than try every possible key letter for every message position, an obviously hopeless situation since this is as difficult as trying all possible messages. However, using each key letter only once obviously leads to a severe key distribution problem, and the one-time pad is only useful for relatively short messages which are to be sent infrequently.

If the message and key are encoded in binary, a more common scheme for position-by-position encryption is to use the "exclusive-or" function: to encrypt the plaintext, "exclusive-or" it (bit by bit) with the key. An attractive feature of this method is that decryption is the same operation as encryption: the ciphertext is the exclusive-or of the plaintext and the key, but doing another exclusive-or of the ciphertext and the key returns the plaintext. Notice that the exclusive-or of the ciphertext and the plaintext is the key. This seems surprising at first, but actually many cryptographic systems have the property that the cryptanalyst can discover the key if he knows the plaintext.

Encryption/Decryption Machines

Many cryptographic applications (for example, voice systems for military communications) involve the transmission of large amounts of data, and this makes the one-time pad infeasible. What is needed is an approximation to the one-time pad in which a large amount of "pseudo-key" can be generated from a small amount of true key to be distributed.

The usual setup in such situations is as follows: an encryption machine is fed some *cryptovariables* (true key) by the sender, which it uses to generate a long stream of key bits (pseudo-key). The exclusive-or of these bits and the plaintext forms the ciphertext. The receiver, having a similar machine and the same cryptovariables, uses them to generate the same key stream to exclusive-or against the ciphertext and to retrieve the plaintext.

Key generation in this context is obviously very much like random number generation, and our random number generation methods are appropriate for key generation (the cryptovariables are the initial seeds of the random number generator). In fact, the linear feedback shift registers that we discussed in Chapter 3 were first developed for use in encryption/decryption machines such as described here. However, key generators have to be somewhat more complicated than random number generators, because there are easy ways to attack simple linear feedback shift registers. The problem is that it might be easy for the cryptanalyst to get some plaintext (for example, silence in a voice system), and therefore some key. If the cryptanalyst can get enough key that he has the entire contents of the shift register, then he can get all the key from that point on.

Cryptographers have several ways to avoid such problems. One way is to make the feedback function itself a cryptovariable. It is usually assumed that the cryptanalyst knows everything about the structure of the machine (maybe he stole one) except the cryptovariables, but if some of the cryptovariables are used to "configure" the machine, he may have difficulty finding their values. Another method commonly used to confuse the cryptanalyst is the *product cipher*, where two different machines are combined to produce a complicated key stream (or to drive each other). Another method is *nonlinear substitution*; here the translation between plaintext and ciphertext is done in large chunks, not bit-by-bit. The general problem with such complex methods is that they can be too complicated for even the cryptographer to understand and that there always is the possibility that things may degenerate badly for some choices of the cryptovariables.

Public-Key Cryptosystems

In commercial applications such as electronic funds transfer and (real) computer mail, the key distribution problem is even more onerous than in the traditional applications of cryptography. The prospect of providing long keys (which must be changed often) to every citizen, while still maintaining both security and cost-effectiveness, certainly inhibits the development of such systems. Methods have recently been developed, however, which promise to eliminate the key distribution problem completely. Such systems, called *public-key cryptosystems*, are likely to come into widespread use in the near future. One of the most prominent of these systems is based on some of the arithmetic algorithms that we have been studying, so we will take a close look at how it works.

The idea in public-key cryptosystems is to use a "phone book" of encryption keys. Everyone's encryption key (denoted by P) is public knowledge: a person's key could be listed, for example, next to his number in the telephone book. Everyone also has a secret key used for decryption; this secret key (denoted by S) is not known to anyone else. To transmit a message M, the sender looks up the receiver's public key, uses it to encrypt the message, and then transmits the message. We'll denote the encrypted message (ciphertext) by C=P(M). The receiver uses his private decryption key to decrypt and read the message. For this system to work we must have at least the following properties:

(i)   S(P(M))=M for every message M.
(ii) All (S,P) pairs are distinct.
(iii) Deriving S from P is as hard as reading M.
(iv) Both S and P are easy to compute.

The first of these is a fundamental cryptographic property, the second two provide the security, and the fourth makes the system feasible for use.

This general scheme was outlined by W. Diffie and M. Hellman in 1976, but they had no method which satisfied all of these properties. Such a method was discovered soon afterwards by R. Rivest, A. Shamir, and L. Adleman. Their scheme, which has come to be known as the *RSA public-key cryptosystem*, is based on arithmetic algorithms performed on very large integers. The encryption key P is the integer pair (N,p) and the decryption key S is the integer pair (N,s), where s is kept secret. These numbers are intended to be very large (typically, N might be 200 digits and p and s might be 100 digits). The encryption and decryption methods are then simple: first the message is broken up into numbers less than N (for example, by taking $\lg N$ bits at a time from the binary string corresponding to the character encoding of the message). Then these numbers are independently raised to a power modulo N: to *encrypt* a (piece of a) message M, compute $C = P(M) = M^p \bmod N$, and to *decrypt* a ciphertext C, compute $M = S(C) = C^s \bmod N$. This computation can be quickly and easily performed by modifying the elementary exponentiation algorithm that we studied in Chapter 4 to take the remainder when divided by N after each multiplication. (No more than $2 \log N$ such operations are required for each piece of the message, so the total number of operations (on 100 digit numbers!) required is linear in the number of bits in the message.)

Property (iv) above is therefore satisfied, and property (ii) can be easily enforced. We still must make sure that the cryptovariables N, p, and s can be chosen so as to satisfy properties (i) and (iii). To be convinced of these requires an exposition of number theory which is beyond the scope of this book, but we can outline the main ideas. First, it is necessary to generate three large (approximately 100-digit) "random" prime numbers: the largest will be s and we'll call the other two x and y. Then N is chosen to be the product of x and y, and p is chosen so that $ps \bmod (x-1)(y-1) = 1$. It is possible to prove that, with N, p, and s chosen in this way, we have $M^{ps} \bmod N = M$ for all messages M.

More specifically, each large prime can be generated by generating a large random number, then testing successive numbers starting at that point until a prime is found. One simple method performs a calculation on a random number that, with probability 1/2, will "prove" that the number to be tested is not prime. (A number which is not prime will survive 20 applications of this test less than one time out of a million, 30 applications less than 1 time out of a billion.) The last step is to compute p: it turns out that a variant of Euclid's algorithm (see Chapter 1) is just what is needed.

Furthermore, s seems to be difficult to compute from knowledge of p (and N), though no one has been able to *prove* that to be the case. Apparently, finding p from s requires knowledge of x and y, and apparently it is necessary to factor N to calculate x and y. But factoring N is thought to be very difficult: the best factoring algorithms known would take millions of years to factor a 200-digit number, using current technology.

An attractive feature of the RSA system is that the complicated computations involving N, p, and s are performed only once for each user who subscribes to the system, which the much more frequent operations of encryption and decryption involve only breaking up the message and applying the simple exponentiation procedure. This computational simplicity, combined with all the convenience features provided by public-key cryptosystems, make this system quite attractive for secure communications, especially on computer systems and networks.

The RSA method has its drawbacks: the exponentiation procedure is actually expensive by cryptographic standards, and, worse, there is the lingering possibility that it might be possible to read messages encrypted using the method. This is true with many cryptosystems: a cryptographic method must withstand serious cryptanalytic attacks before it can be used with confidence.

Several other methods have been suggested for implementing public-key cryptosystems. Some of the most interesting are linked to an important class of problems which are generally thought to be very hard (though this is not known for sure), which we'll discuss in Chapter 40. These cryptosystems have the interesting property that a successful attack could provide insight on how to solve some well-known difficult unsolved problems (as with factoring for the RSA method). This link between cryptology and fundamental topics in computer science research, along with the potential for widespread use of public-key cryptography, have made this an active area of current research.

Exercises

1. Decrypt the following message, which was encrypted with a Vigenere cipher using the pattern CAB (repeated as necessary) for the key (on a 27-letter alphabet, with blank preceding A): DOBHBUAASXFZWJQQ

2. What table should be used to *decrypt* messages that have been encrypted using the table substitution method?

3. Suppose that a Vigenere cipher with a two-character key is used to encrypt a relatively long message. Write a program to infer the key, based on the assumption that the frequency of occurrence of each character in odd positions should be roughly equal to the frequency of occurrence of each character in the even positions.

4. Write matching encryption and decryption procedures which use the "exclusive or" operation between a binary version of the message with a binary stream from one of the linear congruential random number generators of Chapter 3.

5. Write a program to "break" the method given in the previous exercise, assuming that the first 10 characters of the message are known to be blanks.

6. Could one encrypt plaintext by "and"ing it (bit by bit) with the key? Explain why or why not.

7. True or false: Public-key cryptography makes it convenient to send the same message to several different users. Discuss your answer.

8. What is P(S(M)) for the RSA method for public-key cryptography?

9. RSA encoding might involve computing M^n, where M might be a k digit number, represented in an array of k integers, say. About how many operations would be required for this computation?

10. Implement encryption/decryption procedures for the RSA method (assume that s, p and N are all given and represented in arrays of integers of size 25).

304

SOURCES for String Processing

The best references for further information on many of the algorithms in this section are the original sources. Knuth, Morris, and Pratt's 1977 paper and Boyer and Moore's 1977 paper form the basis for much of the material from Chapter 19. The 1968 paper by Thompson is the basis for the regular-expression pattern matcher of Chapters 20–21. Huffman's 1952 paper, though it predates many of the algorithmic considerations here, still makes interesting reading. Rivest, Shamir, and Adleman describe fully the implementation and applications of their public-key cryptosystem in their 1978 paper.

The book by Standish is a good general reference for many of the topics covered in these chapters, especially Chapters 19, 22, and 23. Parsing and compiling are viewed by many to be the heart of computer science, and there are a large number of standard references available, for example the book by Aho and Ullman. An extensive amount of background information on cryptography may be found in the book by Kahn.

A. V. Aho and J. D. Ullman, *Principles of Compiler Design*, Addison-Wesley, Reading, MA, 1977.

R. S. Boyer and J. S. Moore, "A fast string searching algorithm," *Communications of the ACM*, **20**, 10 (October, 1977).

D. A. Huffman, "A method for the construction of minimum-redundancy codes," *Proceedings of the IRE*, **40** (1952).

D. Kahn, *The Codebreakers*, Macmillan, New York, 1967.

D. E. Knuth, J. H. Morris, and V. R. Pratt, "Fast pattern matching in strings," *SIAM Journal on Computing*, **6**, 2 (June, 1977).

R. L. Rivest, A. Shamir and L. Adleman, "A method for obtaining digital signatures and public-key cryptosystems," *Communications of the ACM*, **21**, 2 (February, 1978).

T. A. Standish, *Data Structure Techniques*, Addison-Wesley, Reading, MA, 1980.

K. Thompson, "Regular expression search algorithm," *Communications of the ACM*, **11**, 6 (June, 1968).

GEOMETRIC ALGORITHMS

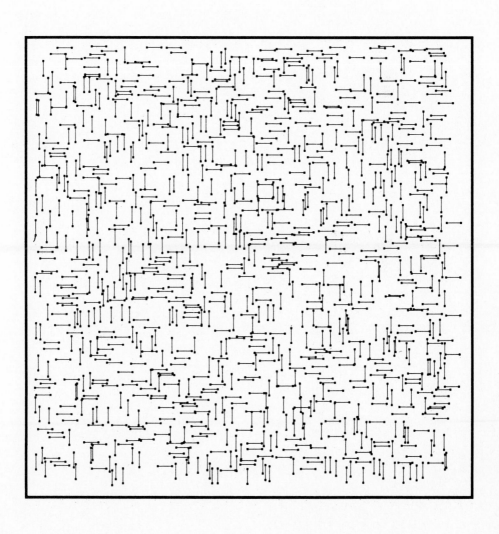

24. Elementary Geometric Methods

☐ Computers are being used more and more to solve large-scale problems which are inherently geometric. Geometric objects such as points, lines and polygons are the basis of a broad variety of important applications and give rise to an interesting set of problems and algorithms.

Geometric algorithms are important in design and analysis systems for physical objects ranging from buildings and automobiles to very large-scale integrated circuits. A designer working with a physical object has a geometric intuition which is difficult to support in a computer representation. Many other applications directly involve processing geometric data. For example, a political "gerrymandering" scheme to divide a district up into areas which have equal population (and which satisfy other criteria such as putting most of the members of the other party in one area) is a sophisticated geometric algorithm. Other applications abound in mathematics and statistics, where many types of problems arise which can be naturally set in a geometric representation.

Most of the algorithms that we've studied have involved text and numbers, which are represented and processed naturally in most programming environments. Indeed, the primitive operations required are implemented in the hardware of most computer systems. For geometric problems, we'll see that the situation is different: even the most elementary operations on points and lines can be computationally challenging.

Geometric problems are easy to visualize, but that can be a liability. Many problems which can be solved instantly by a person looking at a piece of paper (example: is a given point inside a given polygon?) require non-trivial computer programs. For more complicated problems, as in many other applications, the method of solution appropriate for implementation on a computer might be quite different from the method of solution appropriate for a person.

One might suspect that geometric algorithms would have a long history because of the constructive nature of ancient geometry and because useful applications are so widespread, but actually much of the work in the field has been quite recent. Of course, it is often the case that the work of ancient mathematicians has useful application in the development of algorithms for modern computers. The field of geometric algorithms is interesting to study because there is strong historical context, because new fundamental algorithms are still being developed, and because many important large-scale applications require these algorithms.

Points, Lines, and Polygons

Most of the programs that we'll study will operate on simple geometric objects defined in a two-dimensional space. (But we will consider a few algorithms which work in higher dimensions.) The fundamental object is a *point*, which we'll consider to be a pair of integers —the "coordinates" of the point in the usual Cartesian system. Considering only points with integer coordinates leads to slightly simpler and more efficient algorithms, and is not as severe a restriction as it might seem. A *line* is a pair of points, which we assume are connected together by a straight line segment. A *polygon* is a list of points: we assume that successive points are connected by lines and that the first point is connected to the last to make a closed figure.

To work with these geometric objects, we need to be decide how to represent them. Most of our programs will use the obvious representations

```
type point = record x, y: integer end;
     line = record p1, p2: point end;
```

Note that points are restricted to have integer coordinates. A *real* representation could also be used. However, it is important to note that restricting the algorithms to process only integers can be a very significant timesaver in many computing environments, because integer calculations are typically much more efficient that "floating-point" calculations. Thus, when we can get by with dealing only with integers without introducing much extra complication, we will do so.

More complicated geometric objects will be represented in terms of these basic components. For example, polygons will be represented as arrays of *points*. Note that using arrays of *lines* would result in each point on the polygon being included twice (though that still might be the natural representation for some algorithms). Also, it is useful for some applications to include extra information associated with each point or line. This can clearly be handled by adding an *info* field to the records.

We'll use the following set of sixteen points to illustrate the operation of several geometric algorithms:

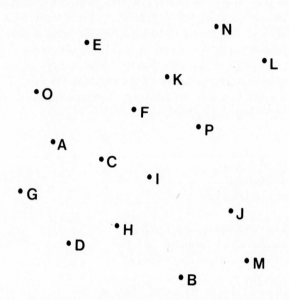

The points are labeled with single letters for reference in explaining the examples. The programs usually have no reason to refer to points by "name"; they are simply stored in an array and are referred to by index. Of course, the order in which the points appear in the array may be important in some of the programs: indeed, it is the goal of some geometric algorithms to "sort" the points into some particular order. The labels that we use are assigned in the order in which the points are assumed to appear in the input. These points have the following integer coordinates:

$$
\begin{array}{ccccccccccccccccc}
 & A & B & C & D & E & F & G & H & I & J & K & L & M & N & O & P \\
x\colon & 3 & 11 & 6 & 4 & 5 & 8 & 1 & 7 & 9 & 14 & 10 & 16 & 15 & 13 & 2 & 12 \\
y\colon & 9 & 1 & 8 & 3 & 15 & 11 & 6 & 4 & 7 & 5 & 13 & 14 & 2 & 16 & 12 & 10 \\
\end{array}
$$

A typical program will maintain an **array** $p[1..N]$ of points and simply read in N pairs of integers, assigning the first pair to the x and y coordinates of $p[1]$, the second pair to the x and y coordinates of $p[2]$, etc. When p is representing a polygon, it is sometimes convenient to maintain "sentinel" values $p[0]=p[N]$ and $p[N+1]=p[1]$.

At some point, we usually want to "draw" our geometric objects. Even if this is not inherent in the application, it is certainly more pleasant to work with pictures than numbers when developing or debugging a new implementation. It is clear that the fundamental operation here is drawing a line. Many types of hardware devices have this capability, but it is reasonable to consider drawing lines using just characters, so that we can print approximations to picture of our objects directly as output to Pascal programs (for example). Our restriction to integer point coordinates helps us here: given a line, we simply want a collection of points that approximates the line. The following recursive program draws a line by drawing the endpoints, then splitting the line in half and drawing the two halves.

```
procedure draw(l: line);
  var dx, dy: integer;
      t: point; l1, l2: line;
  begin
  dot(l.p1.x, l.p1.y); dot(l.p2.x, l.p2.y);
  dx:=l.p2.x−l.p1.x; dy:=l.p2.y−l.p1.y;
  if (abs(dx)>1) or (abs(dy)>1) then
    begin
    t.x:=l.p1.x+dx div 2; t.y:=l.p1.y+dy div 2;
    l1.p1:=l.p1; l1.p2:=t; draw(l1);
    l2.p1:=t; l2.p2:=l.p2; draw(l2);
    end;
  end;
```

The procedure *dot* is assumed to "draw" a single point. One way to implement this is to maintain a two-dimensional array of characters with one character per point allowed, initialized to ".". The *dot* simply corresponds to storing a different character (say "*") in the array position corresponding to the referenced point. The picture is "drawn" by printing out the whole array.

For example, when the above procedure is used with such a *dot* procedure with a 33×33 picture array to "draw" the lines connecting our sample points BG, CD, EO, IL, and PK at a resolution of two characters per unit of measure, we get the following picture:

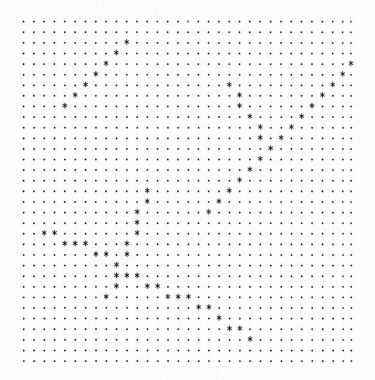

Algorithms for converting geometric objects to points in this manner are called *scan-conversion* algorithms. This example illustrates that it is easy to draw nice-looking diagonal lines like EO and IL, but that it is somewhat harder to make lines of arbitrary slope look nice using a coarse matrix of characters. The recursive method given above has the disadvantages that it is not particularly efficient (some points get plotted several times) and that it doesn't draw certain lines very well (for example lines which are nearly horizontal and nearly vertical). It has the advantages that it is relatively simple and that it handles all the cases of the various orientation of the endpoints of the line in a uniform way. Many sophisticated scan-conversion algorithms have been developed which are more efficient and more accurate than this recursive one.

If the array has a very large number of dots, then the ragged edges of the lines aren't discernible, but the same types of algorithms are appropriate. However, the very high resolution necessary to make high-quality lines can require very large amounts of memory (or computer time), so sophisticated algorithms are called for, or other technologies might be appropriate. For example, the text of this book was printed on a device capable of printing millions of dots per square inch, but most of the lines in the figures were drawn with pen and ink.

Line Intersection

As our first example of an elementary geometric problem, we'll consider the problem of determining whether or not two given line segments intersect. The following diagram illustrates some of the situations that can arise.

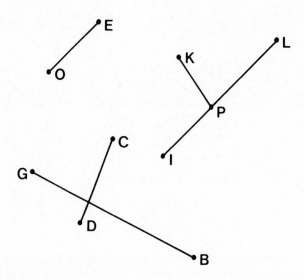

When line segments actually intersect, as do CD and BG, the situation is quite straightforward except that the endpoint of one line might fall on the other, as in the case of KP and IL. When they don't intersect, then we extend the line segments and consider the position of the intersection point of these extended lines. This point could fall on one of the segments, as in the case of IL and BG, or it could lie on neither of the segments, as in the case of CD and OE, or the lines could be parallel, as in the case of OE and IL. The straightforward way to solve this problem is to find the intersection point of the lines defined by the line segments, then check whether this intersection point falls between the endpoints of both of the segments. Another easy method uses a tool that we'll find useful later, so we'll consider it in more detail: Given a line and two points, we're often interested in whether the two points fall on the same side of the line or not. This function is straightforward to compute from the equations for the lines as follows:

```
function same(l: line; p1, p2: point): integer;
var dx, dy, dx1, dx2, dy1, dy2: integer;
begin
dx:=l.p2.x−l.p1.x;  dy:=l.p2.y−l.p1.y;
dx1:=p1.x−l.p1.x;  dy1:=p1.y−l.p1.y;
dx2:=p2.x−l.p2.x;  dy2:=p2.y−l.p2.y;
same:=(dx*dy1−dy*dx1)*(dx*dy2−dy*dx2)
end;
```

In terms of the variables in this program, it is easy to check that the quantity $(dx\,dy_1 - dy\,dx_1)$ is 0 if $p1$ is on the line, positive if $p1$ is on one side, and negative if it is on the other side. The same holds true for the other point, so the product of the quantities for the two points is positive if and only if the points fall on the same side of the line, negative if and only if the points fall on different sides of the line, and 0 if and only if one or both points fall on the line. We'll see that different algorithms need to treat points which fall on lines in different ways, so this three-way test is quite useful.

This immediately gives an implementation of the *intersect* function. If the endpoints of both line segments are on opposite sides of the other then they must intersect.

```
function intersect(l1, l2: line): boolean;
begin
intersect:=(same(l1, l2.p1, l2.p2)<=0)
               and (same(l2, l1.p1, l1.p2)<=0)
end;
```

Unfortunately, there is one case where this function returns the wrong answer: if the four line endpoints are collinear, it always will report intersection, even though the lines may be widely separated. Special cases of this type are the bane of geometric algorithms. The reader may gain some appreciation for the kind of complications such cases can lead to by finding a clean way to repair *intersect* and *same* to handle all cases.

If many lines are involved, the situation becomes much more complicated. Later on, we'll see a sophisticated algorithm for determining whether any pair in a set of N lines intersects.

Simple Closed Path

To get the flavor of problems dealing with sets of points, let's consider the problem of finding a path through a set of N given points which doesn't intersect itself, visits all the points, and returns to the point at which it started. Such a path is called a *simple closed path*. One can imagine many applications for this: the points might represent homes and the path the route that a mailman might take to get to each of the homes without crossing his path. Or we might simply want a reasonable way to draw the points using a mechanical plotter. This is an elementary problem because it asks only for any closed path connecting the points. The problem of finding the best such path, called the *traveling salesman problem*, is much, much more difficult. We'll look at this problem in some detail in the last few chapters of this book. In the next chapter, we'll consider a related but much easier problem: finding the shortest path that surrounds a set of N given points. In Chapter 31, we'll see how to find the best way to "connect" a set of points.

An easy way to solve the elementary problem at hand is the following: Pick one of the points to serve as an "anchor." Then compute the angle made by drawing a line from each of the points in the set to the anchor and then out the positive horizontal direction (this is part of the polar coordinate of each point with the anchor point as origin). Next, sort the points according to that angle. Finally, connect adjacent points. The result is a simple closed path connecting the points, as drawn below:

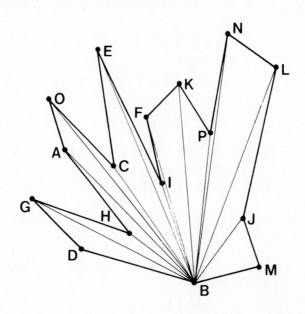

In this example, B is used as the anchor. If the points are visited in the order B M J L N P K F I E C O A H G D B then a simple closed polygon will be traced out.

If dx and dy are the delta x and y distances from some point to the anchor point, then the angle needed in this algorithm is $\tan^{-1} dy/dx$. Although the arctangent is a built-in function in Pascal (and some other programming environments), it is likely to be slow and it leads to at least two annoying extra conditions to compute: whether dx is zero, and which quadrant the point is in. Since the angle is used only for the sort in this algorithm, it makes sense to use a function that is much easier to compute but has the same ordering properties as the arctangent (so that when we sort, we get the same result). A good candidate for such a function is simply $dy/(dy + dx)$. Testing for exceptional conditions is still necessary, but simpler. The following program returns a number between 0 and 360 that is *not* the angle made by *p1* and *p2* with the horizontal but which has the same order properties as the true angle.

```
function theta(p1, p2: point): real;
  var dx, dy, ax, ay: integer;
      t: real;
  begin
  dx:=p2.x−p1.x; ax:=abs(dx);
  dy:=p2.y−p1.y; ay:=abs(dy);
  if (dx=0) and (dy=0) then t:=0
                        else t:=dy/(ax+ay);
  if dx<0 then t:= 2−t
          else if dy<0 then t:=4+t;
  theta:=t*90.0;
  end;
```

In some programming environments it may not be worthwhile to use such programs instead of standard trigonometric functions; in others it might lead to significant savings. (In some cases it might be worthwhile to change *theta* to have an integer value, to avoid using *real* numbers entirely.)

Inclusion in a Polygon

The next problem that we'll consider is a natural one: given a polygon represented as an array of points and another point, determine whether the point is inside or outside. A straightforward solution to this problem immediately suggests itself: draw a long line segment from the point in any direction (long enough so that its other endpoint is guaranteed to be outside the polygon) and count the number of lines from the polygon that it crosses. If the number is odd, the point must be inside; if it is even, the point is outside. This is easily seen by tracing what happens as we come in from the endpoint on the outside: after the first line we hit, we are inside, after the second we are outside, etc. If we proceed an even number of times, the point at which we end up (the original point) must be outside.

The situation is not quite so simple, because some intersections might occur right at the vertices of the input polygon. The drawing below shows some of the situations that need to be handled.

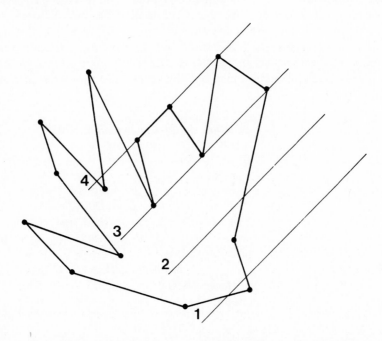

Lines 1 and 2 are straightforward; line 3 leaves the polygon at a vertex; and line 4 coincides with one of the edges of the polygon. The point where line 3 exits should count as 1 intersection with the polygon; all the other points where lines intersect vertices should count as 0 (or 2). The reader may be amused to try to find a simple test to distinguish these cases before reading further.

The need to handle cases where polygon vertices fall on the test lines forces us to do more than just count the line segments in the polygon which intersect the test line. Essentially, we want to travel around the polygon, incrementing an intersection counter whenever we go from one side of the test line to another. One way to implement this is to simply ignore points which fall on the test line, as in the following program:

```
function inside(t: point): boolean;
  var count, i, j: integer;
      lt, lp: line;
  begin
  count:=0; j:=0;
  p[0]:=p[N]; p[N+1]:=p[1];
  lt.p1:=t; lt.p2:=t; lt.p2.x:=maxint;
  for i:=1 to N do
    begin
    lp.p1:=p[i]; lp.p2:=p[i];
    if not intersect(lp, lt) then
      begin
      lp.p2:=p[j]; j:=i;
      if intersect(lp, lt) then count:=count+1;
      end;
    end;
  inside:=((count mod 2)=1);
  end;
```

This program uses a horizontal test line for ease of calculation (imagine the above diagram as rotated 45 degrees). The variable j is maintained as the index of the last point on the polygon known not to lie on the test line. The program assumes that $p[1]$ is the point with the smallest x coordinate among all the points with the smallest y coordinate, so that if $p[1]$ is on the test line, then $p[0]$ cannot be. For example, this choice might be used for $p[1]$ as the "anchor" for the procedure suggested above for computing a simple closed polygon. The same polygon can be represented by N different p arrays, but as this illustrates it is sometimes convenient to fix a standard rule for $p[1]$. If the next point on the polygon which is not on the test line is on the same side of the test line as the jth point, then we need not increment the intersection counter (*count*); otherwise we have an intersection. The reader may wish to check that this algorithm works properly for lines like lines 3 and 4 in the diagram above.

If the polygon has only three or four sides, as is true in many applications, then such a complex program is not called for: a simpler procedure based on calls to *same* will be adequate.

Perspective·

From the few examples given, it should be clear that it is easy to underestimate the difficulty of solving a particular geometric problem with a computer. There are many other elementary geometric computations that we have not treated at all. For example, a program to compute the area of a polygon makes an interesting exercise. However, the problems that we have studied have provided some basic tools that we will find useful in later sections for solving the more difficult problems.

Some of the algorithms that we'll study involve building geometric structures from a given set of points. The "simple closed polygon" is an elementary example of this. We will need to decide upon appropriate representations for such structures, develop algorithms to build them, and investigate their use for particular applications areas. As usual, these considerations are intertwined. For example, the algorithm used in the *inside* procedure in this chapter depends in an essential way on the representation of the simple closed polygon as an ordered set of points (rather than as an unordered set of lines).

Many of the algorithms that we'll study involve *geometric search*: we want to know which points from a given set are close to a given point, or which points fall in a given rectangle, or which points are closest to each other. Many of the algorithms appropriate for such search problems are closely related to the search algorithms that we studied in Chapters 14–17. The parallels will be quite evident.

Few geometric algorithms have been analyzed to the point where precise statements can be made about their relative performance characteristics. As we've already seen, the running time of a geometric algorithm can depend on many things. The distribution of the points themselves, the order in which they appear in the input, and whether trigonometric functions are needed or used can all significantly affect the running time of geometric algorithms. As usual in such situations, we do have empirical evidence which suggests good algorithms for particular applications. Also, many of the algorithms are designed to perform well in the worst case, no matter what the input is.

Exercises

1. List the points plotted by *draw* when plotting a line from $(0,0)$ to $(1,21)$.

2. Give a quick algorithm for determining whether two line segments are parallel, without using any divisions.

3. Given an array of *lines*, how would you test to see if they form a simple closed polygon?

4. Draw the simple closed polygons that result from using A, C, and D as "anchors" in the method described in the text.

5. Suppose that we use an arbitrary point for the "anchor" in the method for computing a simple closed polygon described in the text. Give conditions which such a point must satisfy for the method to work.

6. What does the *intersect* function return when called with two copies of the same line segment?

7. Write a program like *draw* to "fill in" an arbitrary triangle. (Your program should call *dot* for all the points inside the triangle.)

8. Does *inside* call a vertex of the polygon inside or outside?

9. What is the maximum value achievable by *count* when *inside* is executed on a polygon with N vertices? Give an example supporting your answer.

10. Write an efficient program for determining if a given point is inside a given quadrilateral.

25. Finding the Convex Hull

Often, when we have a large number of points to process, we're interested in the boundaries of the point set. When looking at a diagram of a set of points plotted in the plane, a human has little trouble distinguishing those on the "inside" of the point set from those which lie on the edge. This distinction is a fundamental property of point sets; in this chapter we'll see how it can be precisely characterized by looking at algorithms for separating out the "boundary" points of a point set.

The mathematical notion of the natural boundary of a point set depends on a geometric property called *convexity*. This is a simple concept that the reader may have encountered before: a *convex polygon* is a polygon with the property that any line connecting any two points inside the polygon must itself lie inside the polygon. For example, the "simple closed polygon" that we computed in the previous chapter is decidedly nonconvex, but any triangle or rectangle is convex.

Now, the mathematical name for the natural boundary of a point set is the *convex hull*. The convex hull of a set of points in the plane is defined to be the smallest convex polygon containing them all. Equivalently, the convex hull is the shortest path which surrounds the points. An obvious property of the convex hull that is easy to prove is that the vertices of the convex polygon defining the hull are points from the original point set. Given N points, some of them form a convex polygon within which all the others are contained. The problem is to find those points. Many algorithms have been developed to find the convex hull; in this chapter we'll examine a few representative ones.

For a large number N of points, the convex hull could contain as few as 3 points (if three points form a large triangle containing all the others) or as many as N points (if all the points fall on a convex polygon, then they all comprise their own convex hull). Some algorithms work well when there are many points on the convex hull; others work better when there are only a few.

Below is diagramed our sample set of points and their convex hull.

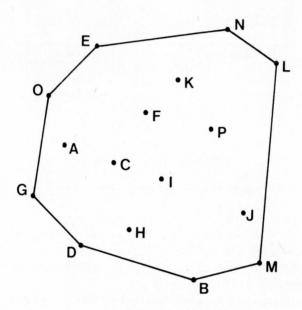

A fundamental property of the convex hull is that any line outside the hull, when moved in any direction towards the hull, hits the hull at one of its vertex points. (This is an alternate way to define the hull: it is the subset of points from the point set that could be hit by a line moving in at some angle from infinity.) In particular, it's easy to find a few points that are guaranteed to be on the hull by applying this rule for horizontal and vertical lines: the points with the smallest and largest x and y coordinates are all on the convex hull.

The convex hull naturally defines the "boundaries" of a point set and is a fundamental geometric computation. It plays an important role in many statistical computations, especially when generalized to higher dimensions.

Rules of the Game

The input to an algorithm for finding the convex hull will obviously be an array of points; we can use the *point* type defined in the previous chapter. The output will be a polygon, also represented as an array of points with the property that tracing through the points in the order in which they appear in the array traces the outline of the polygon. On reflection, this might appear to require an extra ordering condition on the computation of the convex hull (why not just return the points on the hull in any order?), but output in the ordered form is obviously more useful, and it has been shown that the unordered computation is no easier to do. For all of the algorithms that we consider, it is convenient to do the computation *in place*: the array used for the original point set is also used to hold the output. The algorithms simply rearrange the points in the original array so that the convex hull appears in the first M positions, in order.

From the description above, it may be clear that computing the convex hull is closely related to sorting. In fact, a convex hull algorithm can be used to sort, in the following way. Given N numbers to sort, turn them into points (in polar coordinates) by treating the numbers as angles (suitably normalized) with a fixed radius for each point. The convex hull of this point set is an N-gon containing all of the points. Now, since the output must be ordered in the order the points appear on this polygon, it can be used to find the sorted order of the original values (remember that the input was unordered). This is not a formal proof that computing the convex hull is no easier than sorting, because, for example, the cost of the trigonometric functions required to convert the original numbers to be sorted into points on the polygon must be considered. Comparing convex hull algorithms (which involve trigonometric operations) to sorting algorithms (which involve comparisons between keys) is a bit like comparing apples to oranges, but it has been shown that any convex hull algorithm must require about $N \log N$ operations, the same as sorting (even though the operations allowed are likely to be quite different). It is helpful to view finding the convex hull of a set of points as a kind of "two-dimensional sort" because frequent parallels to sorting algorithms arise in the study of algorithms for finding the convex hull.

In fact, the algorithms that we'll study show that finding the convex hull is no harder than sorting either: there are several algorithms that run in time proportional to $N \log N$ in the worst case. Many of the algorithms tend to use even less time on actual point sets, because their running time depends on the way that the points are distributed and on the number of points on the hull.

We'll look at three quite different methods for finding the convex hull of a set of points and then discuss their relative running times.

Package Wrapping

The most natural convex hull algorithm, which parallels the method a human would use to draw the convex hull of a set of points, is a systematic way to "wrap up" the set of points. Starting with some point guaranteed to be on the convex hull (say the one with the smallest y coordinate), take a horizontal ray in the positive direction and "sweep" it upward until hitting another point; this point must be on the hull. Then anchor at that point and continue "sweeping" until hitting another point, etc., until the "package" is fully "wrapped" (the beginning point is included again). The following diagram shows how the hull is discovered in this way.

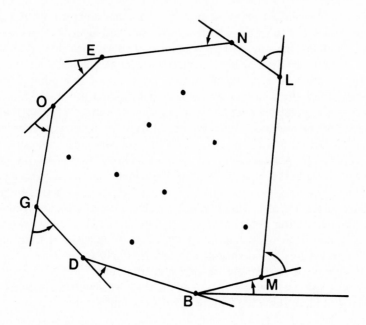

Of course, we don't actually sweep through all possible angles, we just do a standard find-the-minimum computation to find the point that would be hit next. This method is easily implemented by using the function *theta*(*p1, p2*: *point*) developed in the previous chapter, which can be thought of as returning the angle between *p1, p2* and the horizontal (though it actually returns a more easily computed number with the same ordering properties). The following program finds the convex hull of an array *p*[*1..N*] of points, represented as described in the previous chapter (the array position *p*[*N+1*] is also used, to hold a sentinel):

```
function wrap: integer;
  var i, min, M: integer;
      minangle, v: real;
      t: point;
  begin
  min:=1;
  for i:=2 to N do
    if p[i].y<p[min].y then min:=i;
  M:=0; p[N+1]:=p[min]; minangle:=0.0;
  repeat
    M:=M+1; t:=p[M]; p[M]:=p[min]; p[min]:=t;
    min:=N+1; v:=minangle; minangle:=360.0;
    for i:=M+1 to N+1 do
      if theta(p[M],p[i])>v then
        if theta(p[M],p[i])<minangle then
          begin min:=i; minangle:=theta(p[M],p[min]) end;
  until min=N+1;
  wrap:=M;
  end;
```

First, the point with the lowest y coordinate is found and copied into $p[N+1]$ in order to stop the loop, as described below. The variable M is maintained as the number of points so far included on the hull, and v is the current value of the "sweep" angle (the angle from the horizontal to the line between $p[M-1]$ and $p[M]$). The **repeat** loop puts the last point found into the hull by exchanging it with the Mth point, and uses the *theta* function from the previous chapter to compute the angle from the horizontal made by the line between that point and each of the points not yet included on the hull, searching for the one whose angle is smallest among those with angles bigger than v. The loop stops when the first point (actually the copy of the first point that was put into $p[N+1]$) is encountered again.

This program may or may not return points which fall on a convex hull edge. This happens when more than one point has the same *theta* value with $p[M]$ during the execution of the algorithm; the implementation above takes the first value. In an application where it is important to find points falling on convex hull edges, this could be achieved by changing *theta* to take the distance between the points given as its arguments into account and give the closer point a smaller value when two points have the same angle.

The following table traces the operation of this algorithm: the Mth line of the table gives the value of v and the contents of the p array after the Mth point has been added to the hull.

7.50	[B]	A	C	D	E	F	G	H	I	J	K	L	M	N	O	P
18.00	B	[M]	C	D	E	F	G	H	I	J	K	L	A	N	O	P
83.08	B	M	[L]	D	E	F	G	H	I	J	K	C	A	N	O	P
144.00	B	M	L	[N]	E	F	G	H	I	J	K	C	A	D	O	P
190.00	B	M	L	N	[E]	F	G	H	I	J	K	C	A	D	O	P
225.00	B	M	L	N	E	[O]	G	H	I	J	K	C	A	D	F	P
257.14	B	M	L	N	E	O	[G]	H	I	J	K	C	A	D	F	P
315.00	B	M	L	N	E	O	G	[D]	I	J	K	C	A	H	F	P

One attractive feature of this method is that it generalizes to three (or more) dimensions. The convex hull of a set of points in 3-space is a convex three-dimensional object with flat faces. It can be found by "sweeping" a plane until the hull is hit, then "folding" faces of the plane, anchoring on different lines on the boundary of the hull, until the "package" is "wrapped."

The program is quite similar to selection sorting, in that we successively choose the "best" of the points not yet chosen, using a brute-force search for the minimum. The major disadvantage of the method is that in the worst case, when all the points fall on the convex hull, the running time is proportional to N^2.

The Graham Scan

The next method that we'll examine, invented by R. L. Graham in 1972, is interesting because most of the computation involved is for sorting: the algorithm includes a sort followed by a relatively inexpensive (though not immediately obvious) computation. The algorithm starts with the construction of a simple closed polygon from the points using the method of the previous chapter: sort the points using as keys the *theta* function values corresponding to the angle from the horizontal made from the line connecting each point with an 'anchor' point $p[1]$ (with the lowest y coordinate) so that tracing $p[1], p[2], \ldots, p[N], p[1]$ gives a closed polygon. For our example set of points, we get the simple closed polygon of the previous section. Note that $p[N]$, $p[1]$, and $p[2]$ are consecutive points on the hull; we've essentially run the first iteration of the package wrapping procedure (in both directions).

Computation of the convex hull is completed by proceeding around, trying to place each point on the hull and eliminating previously placed points that couldn't possibly be on the hull. For our example, we consider the points in the order B M J L N P K F I E C O A H G D. The test for which points to eliminate is not difficult. After each point has been added, we assume that we have eliminated enough points so that what we have traced out so far could be part of the convex hull, based on the points so far seen. The algorithm is based on the fact that all of the points in the point set must be on the same side of each edge of the convex hull. Each time we consider a point, we eliminate from the hull any edge which violates this condition.

Specifically, the test for eliminating a point is the following: when we come to examine a new point $p[i]$, we eliminate $p[k]$ from the hull if the line between $p[k]$ and $p[k-1]$ goes between $p[i]$ and $p[1]$. If $p[i]$ and $p[1]$ are on the same side of the line, then $p[k]$ could still be on the hull, so we don't eliminate it. The following diagram shows the situation for our example when L is considered:

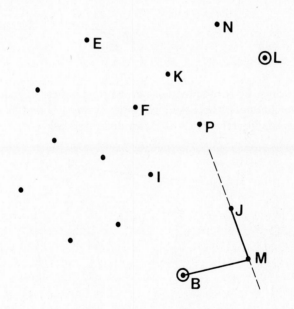

The extended line JM runs between L and B, so J couldn't be on the hull. Now L, N, and P are added to the hull, then P is eliminated when K is considered (because the extended line NP goes between B and K), then F and I are added, leaving the following situation when E is considered.

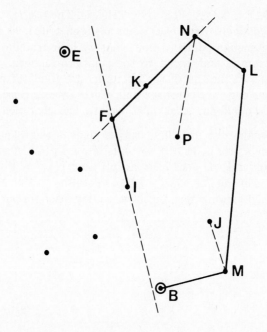

At this point, I must be eliminated because FI runs between E and B, then F and K must be eliminated because NKF runs between E and B. Continuing in this way, we finally arrive back at B, as illustrated below:

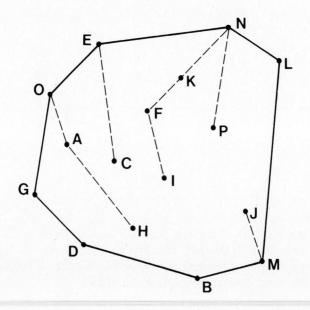

The dotted lines in the diagrams are all the edges that were included, then eliminated. The initial sort guarantees that each point is considered as a possible hull point in turn, because all points considered earlier have a smaller *theta* value. Each line that survives the "eliminations" has the property that every point is on the same side of it as $p[1]$, which implies that it must be on the hull.

Once the basic method is understood, the implementation is straightforward. First, the point with the minimum y value is exchanged with $p[1]$. Next, *shellsort* (or some other appropriate sorting routine) is used to rearrange the points, modified as necessary to compare two points using their *theta* values with $p[1]$. Finally, the scan described above is performed. The following program finds the convex hull of the point set $p[1..N]$ (no sentinel is needed):

```
function grahamscan: integer;
    var i, j, min, M: integer;
        l: line;  t: point;
    begin
    min:=1;
    for i:=2 to N do
        if p[i].y<p[min].y then min:=i;
    t:=p[1];  p[1]:=p[min];  p[min]:=t;
    shellsort;
    M:=2;
    for i:=4 to N do
        begin
        M:=M+2;
        repeat
            M:=M-1;
            l.p1:=p[M];  l.p2:=p[M-1];
        until same(l, p[1], p[i])>=0;
        t:=p[M+1];  p[M+1]:=p[i];  p[i]:=t;
        end;
    grahamscan:=M;
    end;
```

The loop maintains a partial hull in $p[1], \ldots, p[M]$, as described in the text above. For each new i value considered, M is decremented if necessary to eliminate points from the partial hull and then $p[i]$ is exchanged with $p[M+1]$ to (tentatively) add it to the partial hull. The following table shows the contents of the p array each time a new point is considered for our example:

1	2	3	4	5	6	7	8	9	10	11	12	13	14	15	16
B	M	J̲	L̲	N	P	K	F	I	E	C	O	A	H	G	D
B	M	L̲	J	N̲	P	K	F	I	E	C	O	A	H	G	D
B	M	L	N̲	J	P̲	K	F	I	E	C	O	A	H	G	D
B	M	L	N	P̲	J	K̲	F	I	E	C	O	A	H	G	D
B	M	L	N	K̲	J	P	F̲	I	E	C	O	A	H	G	D
B	M	L	N	K	F̲	P	J	I̲	E	C	O	A	H	G	D
B	M	L	N	K	F	I̲	J	P	E̲	C	O	A	H	G	D
B	M	L	N	E̲	F	I	J	P	K	C̲	O	A	H	G	D
B	M	L	N	E	C̲	I	J	P	K	F	O̲	A	H	G	D
B	M	L	N	E	O̲	I	J	P	K	F	C	A̲	H	G	D
B	M	L	N	E	O	A̲	J	P	K	F	C	I	H̲	G	D
B	M	L	N	E	O	A	H̲	P	K	F	C	I	J	G̲	D
B	M	L	N	E	O	G̲	H	P	K	F	C	I	J	A	D̲
B	M	L	N	E	O	G	D̲	P	K	F	C	I	J	A	H

This table depicts, for *i* from 4 to *N*, the solution when *p*[*i*] is first considered, with *p*[*M*] and *p*[*i*] boxed.

The program as given above could fail if there is more than one point with the lowest *y* coordinate, unless *theta* is modified to properly sort collinear points, as described above. (This is a subtle point which the reader may wish to check.) Alternatively, the *min* computation could be modified to find the point with the lowest *x* coordinate among all points with the lowest *y* coordinate, the canonical form described in Chapter 24.

One reason that this method is interesting to study is that it is a simple form of *backtracking*, the algorithm design technique of "try something, if it doesn't work then try something else" which we'll see in much more complicated forms in Chapter 39.

Hull Selection

Almost any convex hull method can be vastly improved by a method developed independently by W. F. Eddy and R. W. Floyd. The general idea is simple: pick four points known to be on the hull, then throw out everything inside the quadrilateral formed by those four points. This leaves many fewer points to be considered by, say, the Graham scan or the package wrapping technique. The method could be also applied recursively, though this is usually not worth the trouble.

The four points known to be on the hull should be chosen with an eye towards any information available about the input points. In the absence of any information at all, the simplest four points to use are those with the smallest and largest x and y coordinates. However, it might be better to adapt the choice of points to the distribution of the input. For example, if all x and y values within certain ranges are equally likely (a rectangular distribution), then choosing four points by scanning in from the corners might be better (find the four points with the largest and smallest sum and difference of the two coordinates). The diagram below shows that only A and J survive the application of this technique to our example set of points.

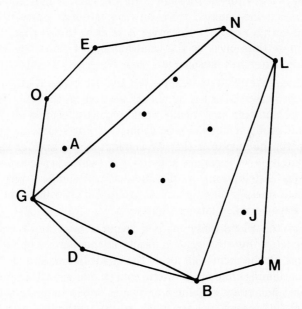

The recursive version of this technique is very similar to the Quicksort-like *select* procedure for selection that we discussed in Chapter 12. Like that procedure, it is vulnerable to an N^2 worst-case running time. For example, if all the original points are on the convex hull, then no points will get thrown out in the recursive step. Like *select*, the running time is linear on the average, as discussed further below.

Performance Issues

As mentioned in the previous chapter, geometric algorithms are somewhat harder to analyze than algorithms from some of the other areas we've studied because the input (and the output) is more difficult to characterize. It often doesn't make sense to speak of "random" point sets: for example, as N gets large, the convex hull of points drawn from a rectangular distribution is extremely likely to be very close to the rectangle defining the distribution. The algorithms that we've looked at depend on different properties of the point set distribution and are thus in practice incomparable, because to compare them analytically would require an understanding of very complicated interactions between little-understood properties of point sets. On the other hand, we can say some things about the performance of the algorithms that will help choosing one for a particular application.

The easiest of the three to analyze is the Graham scan. It requires time proportional to $N \log N$ for the sort and N for the scan. A moment's reflection is necessary to convince oneself that the scan is linear, since it does have a **repeat** "loop-within-a-loop." However, it is easy to see that no point is "eliminated" more than once, so the total number of times the code within that **repeat** loop is iterated must be less than N.

The "package-wrapping" technique, on the other hand, obviously takes about MN steps, where M is the number of vertices on the hull. To compare this with the Graham scan analytically would require a formula for M in terms of N, a difficult problem in stochastic geometry. For a circular distribution (and some others) the answer is that M is about $N^{1/3}$, and for values of N which are not large $N^{1/3}$ is comparable to $\log N$ (which is the expected value for a rectangular distribution), so this method will compete very favorably with the Graham scan for many practical problems. Of course, the N^2 worst case should always be taken into consideration.

Analysis of the Floyd-Eddy method requires even more sophisticated stochastic geometry, but the general result is the same as that given by intuition: almost all the points fall inside the quadrilateral and are discarded. This makes the running time of the whole convex hull algorithm proportional to N, since most points are examined only once (when they are thrown out). On the average, it doesn't matter much which method is used after one application of the Floyd-Eddy method, since so few points are likely to be left. However, to protect against the worst case (when all points are on the hull), it is prudent to use the Graham scan. This gives an algorithm which is almost sure to run in linear time in practice and is guaranteed to run in time proportional to $N \log N$.

Exercises

1. Suppose it is known in advance that the convex hull of a set of points is a triangle. Give an easy algorithm for finding the triangle. Answer the same question for a quadrilateral.

2. Give an efficient method for determining whether a point falls within a given convex polygon.

3. Implement a convex hull algorithm like insertion sort, using your method from the previous exercise.

4. Is it strictly necessary for the Graham scan to start with a point guaranteed to be on the hull? Explain why or why not.

5. Is it strictly necessary for the package-wrapping method to start with a point guaranteed to be on the hull? Explain why or why not.

6. Draw a set of points that makes the Graham scan for finding the convex hull particularly inefficient.

7. Does the Graham scan work for finding the convex hull of the points which make up the vertices of *any* simple polygon? Explain why or give a counterexample showing why not.

8. What four points should be used for the Floyd-Eddy method if the input is assumed to be randomly distributed within a circle (using random polar coordinates)?

9. Run the package-wrapping method for large points sets with both x and y equally likely to be between 0 and 1000. Use your curve fitting routine to find an approximate formula for the running time of your program for a point set of size N.

10. Use your curve-fitting routine to find an approximate formula for the number of points left after the Floyd-Eddy method is used on point sets with x and y equally likely to be between 0 and 1000.

26. Range Searching

Given a set of points in the plane, it is natural to ask which of those points fall within some specified area. "List all cities within 50 miles of Providence" is a question of this type which could reasonably be asked if a set of points corresponding to the cities of the U.S. were available. When the geometric shape is restricted to be a rectangle, the issue readily extends to non-geometric problems. For example, "list all those people between 21 and 25 with incomes between $60,000 and $100,000" asks which "points" from a file of data on people's names, ages, and incomes fall within a certain rectangle in the age-income plane.

Extension to more than two dimensions is immediate. If we want to list all stars within 50 light years of the sun, we have a three-dimensional problem, and if we want the rich young people of the second example in the paragraph above to be tall and female as well, we have a four-dimensional problem. In fact, the dimension can get very high for such problems.

In general, we assume that we have a set of *records* with certain *attributes* that take on values from some ordered set. (This is sometimes called a *database*, though more precise and complete definitions have been developed for this important term.) The problem of finding all records in a database which satisfy specified range restrictions on a specified set of attributes is called *range searching*. For practical applications, this is a difficult and important problem. In this chapter, we'll concentrate on the two-dimensional geometric problem in which records are points and attributes are their coordinates, then we'll discuss appropriate generalizations.

The methods that we'll look at are direct generalizations of methods that we have seen for searching on single keys (in one dimension). We presume that many queries will be made on the same set of points, so the problem splits into two parts: we need a *preprocessing* algorithm, which builds the given points into a structure supporting efficient range searching, and a *range-searching* algorithm, which uses the structure to return points falling within any given (multidimensional) range. This separation makes different methods difficult to compare, since the total cost depends not only on the distribution of the points involved but also on the number and nature of the queries.

The range-searching problem in one dimension is to return all points falling within a specified interval. This can be done by sorting the points for preprocessing and, then using binary search (to find all points in a given interval, do a binary search on the endpoints of the interval and return all the points that fall in between). Another solution is to build a binary search tree and then do a simple recursive traversal of the tree, returning points that are within the interval and ignoring parts of the tree that are outside the interval. For example, the binary search tree that is built using the x coordinates of our points from the previous chapter, when inserted in the given order, is the following:

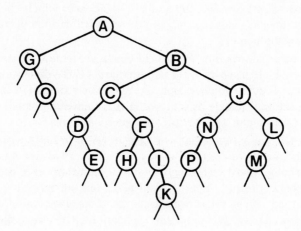

Now, the program required to find all the points in a given interval is a direct generalization of the *treeprint* procedure of Chapter 14. If the left endpoint of the interval falls to the left of the point at the root, we (recursively) search the left subtree, similarly for the right, checking each node we encounter to see whether its point falls within the interval:

```
type interval = record x1, x2: integer end;
procedure bstrange(t: link; int: interval);
   var tx1, tx2: boolean;
   begin
   if t<>z then
      begin
      tx1:=t↑.key>=int.x1;
      tx2:=t↑.key<=int.x2;
      if tx1 then bstrange(t↑.l, int);
      if tx1 and tx2 then write(name(t↑.id), ' ');
      if tx2 then bstrange(t↑.r, int);
      end
   end;
```

(This program could be made slightly more efficient by maintaining the interval *int* as a global variable rather than passing its unchanged values through the recursive calls.) For example, when called on the interval [5,9] for the example tree above, *range* prints out E C H F I. Note that the points returned do not necessarily need to be connected in the tree.

These methods require time proportional to about $N \log N$ for preprocessing, and time proportional to about $R + \log N$ for *range*, where R is the number of points actually falling in the range. (The reader may wish to check that this is true.) Our goal in this chapter will be to achieve these same running times for multidimensional range searching.

The parameter R can be quite significant: given the facility to make range queries, it is easy for a user to formulate queries which could require all or nearly all of the points. This type of query could reasonably occur in many applications, but sophisticated algorithms are not necessary if *all* queries are of this type. The algorithms that we consider are designed to be efficient for queries which are not expected to return a large number of points.

Elementary Methods

In two dimensions, our "range" is an area in the plane. For simplicity, we'll consider the problem of finding all points whose x coordinates fall within a given x-interval and whose y coordinates fall within a given y-interval: that is, we seek all points falling within a given rectangle. Thus, we'll assume a type *rectangle* which is a record of four integers, the horizontal and vertical interval endpoints. The basic operation that we'll use is to test whether a point falls within a given rectangle, so we'll assume a function *insiderect(p: point; rect: rectangle)* which checks this in the obvious way, returning *true* if *p* falls within *rect*. Our goal is to find all the points which fall within a given rectangle, using as few calls to *insiderect* as possible.

The simplest way to solve this problem is *sequential search*: scan through all the points, testing each to see if it falls within the specified range (by calling *insiderect* for each point). This method is in fact used in many database applications because it is easily improved by "batching" the range queries, testing for many different ones in the same scan through the points. In a very large database, where the data is on an external device and the time to read the data is by far the dominating cost factor, this can be a very reasonable method: collect as many queries as will fit in the internal memory and search for them all in one pass through the large external data file. If this type of batching is inconvenient or the database is somewhat smaller, however, there are much better methods available.

A simple first improvement to sequential search is to apply directly a known one-dimensional method along one or more of the dimensions to be searched. For example, suppose the following search rectangle is specified for our sample set of points:

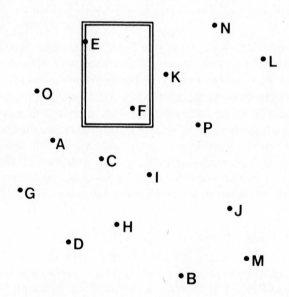

One way to proceed is to find the points whose *x* coordinates fall within the *x* range specified by the rectangle, then check the *y* coordinates of those points to determine whether or not they fall within the rectangle. Thus, points that could not be within the rectangle because their x coordinates are out of range are never examined. This technique is called *projection*; obviously we could also project on y. For our example, we would check E C H F and I for an x projection, as described above and we would check O E F K P N and L for a y projection.

If the points are uniformly distributed in a rectangular shaped region, then it's trivial to calculate the average number of points checked. The fraction of points we would expect to find in a given rectangle is simply the ratio of the area of that rectangle to the area of the full region; the fraction of points we would expect to check for an x projection is the ratio of the width of the rectangle to the width of the region, and similarly for a y projection. For our example, using a 4-by-6 rectangle in a 16-by-16 region means that we would expect to find 3/32 of the points in the rectangle, 1/4 of them in an x projection, and 3/8 of them in a y projection. Obviously, under such circumstances, it's best to project onto the axis corresponding to the narrower of the two rectangle dimensions. On the other hand, it's easy to construct situations in which the projection technique could fail miserably: for example if the point set forms an "L" shape and the search is for a range that encloses only the point at the corner of the "L," then projection on either axis would eliminate only half the points.

At first glance, it seems that the projection technique could be improved somehow to "intersect" the points that fall within the x range and the points that fall within the y range. Attempts to do this without examining either all the points in the x range or all the points in the y range in the worst case serve mainly to give one an appreciation for the more sophisticated methods that we are about to study.

Grid Method

A simple but effective technique for maintaining proximity relationships among points in the plane is to construct an artificial grid which divides the area to be searched into small squares and keep short lists of points that fall into each square. (This technique is reportedly used in archaeology, for example.) Then, when points that fall within a given rectangle are sought, only the lists corresponding to squares that intersect the rectangle have to be searched. In our example, only E, C, F, and K are examined, as sketched below.

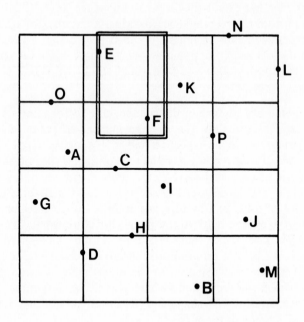

The main decision to be made in implementing this method is determining the size of the grid: if it is too coarse, each grid square will contain too many points, and if it is too fine, there will be too many grid squares to search (most of which will be empty). One way to strike a balance between these two is to choose the grid size so that the number of grid squares is a constant fraction of the total number of points. Then the number of points in each square is expected to be about equal to some small constant. For our example, using a 4 by 4 grid for a sixteen-point set means that each grid square is expected to contain one point.

Below is a straightforward implementation of a program to read in xy coordinates of a set of points, then build the grid structure containing those points. The variable *size* is used to control how big the grid squares are and thus determine the resolution of the grid. For simplicity, assume that the coordinates of all the points fall between 0 and some maximum value *max*. Then, to get a G-by-G grid, we set *size* to the value *max/G*, the width of the grid square. To find which grid square a point belongs to, we divide its coordinates by *size*, as in the following implementation:

```
program rangegrid(input, output);
const Gmax=20;
type   point = record x, y, info: integer end;
       link=↑node;
       node=record p: point; next: link end;
var grid: array [0..Gmax, 0..Gmax] of link;
    p: point;
    i, j, k, size, N: integer;
    z: link;
procedure insert(p: point);
  var t: link;
  begin
  new(t); t↑.p:=p;
  t↑.next:=grid[p.x div size, p.y div size];
  grid[p.x div size, p.y div size]:=t;
  end;
begin
new(z);
for i:=0 to Gmax do
  for j:=0 to Gmax do grid[i, j]:=z;
readln(N);
for k:=1 to N do
  begin
  readln(p.x, p.y);  p.info:=k;
  insert(p)
  end;
end.
```

This program uses our standard linked list representations, with dummy tail node z. The point type is extended to include a field *info* which contains the integer k for the kth point read in, for convenience in referencing the points. In keeping with the style of our examples, we'll assume a function $name(k)$ to return the kth letter of the alphabet: clearly a more general naming mechanism will be appropriate for actual applications.

As mentioned above, the setting of the variable *size* (which is omitted from the above program) depends on the number of points, the amount of memory available, and the range of coordinate values. Roughly, to get M points per grid square, *size* should be chosen to be the nearest integer to *max* divided by $\sqrt{N/M}$. This leads to about N/M grid squares. These estimates aren't accurate for small values of the parameters, but they are useful for most situations, and similar estimates can easily be formulated for specialized applications.

Now, most of the work for range searching is handled by simply indexing into the *grid* array, as follows:

```
procedure gridrange(rect: rectangle);
  var t: link;
      i, j: integer;
  begin
  for i:=(rect.x1 div size) to (rect.x2 div size) do
    for j:=(rect.y1 div size) to (rect.y2 div size) do
      begin
      t:=grid[i,j];
      while t<>z do
        begin
        if insiderect(t↑.p, rect) then write(name(t↑.p.info));
        t:=t↑.next
        end
      end
  end;
```

The running time of this program is proportional to the number of grid squares touched. Since we were careful to arrange things so that each grid square contains a constant number of points on the average, this is also proportional, on the average, to the number of points examined. If the number of points in the search rectangle is R, then the number of grid squares examined is proportional to R. The number of grid squares examined which do not fall completely inside the search rectangle is certainly less than a small constant times R, so the total running time (on the average) is linear in R, the number of points sought. For large R, the number of points examined which don't fall in the search rectangle gets quite small: all such points fall in a grid square which intersects the edge of the search rectangle, and the number of such squares is proportional to $\sqrt{R}$ for large R. Note that this argument falls apart if the grid squares are too small (too many empty grid squares inside the search rectangle) or too large (too many points in grid squares on the perimeter of the search rectangle) or if the search rectangle is thinner than the grid squares (it could intersect many grid squares, but have few points inside it).

The grid method works well if the points are well distributed over the assumed range but badly if they are clustered together. (For example, all the points could fall in one grid box, which would mean that all the grid machinery gained nothing.) The next method that we will examine makes this worst case very unlikely by subdividing the space in a nonuniform way, adapting to the point set at hand.

2D Trees

Two-dimensional trees are dynamic, adaptable data structures which are very similar to binary trees but divide up a geometric space in a manner convenient for use in range searching and other problems. The idea is to build binary search trees with points in the nodes, using the y and x coordinates of the points as keys in a strictly alternating sequence.

The same algorithm is used for inserting points into 2D trees as for normal binary search trees, except at the root we use the y coordinate (if the point to be inserted has a smaller y coordinate than the point at the root, go left; otherwise go right), then at the next level we use the x coordinate, then at the next level the y coordinate, etc., alternating until an external node is encountered. For example, the following 2D tree is built for our sample set of points:

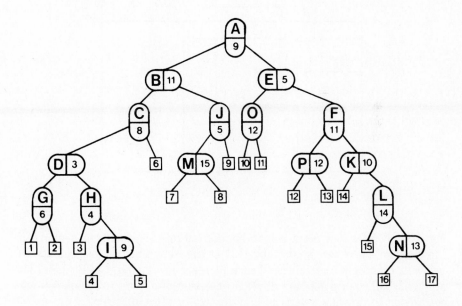

The particular coordinate used is given at each node along with the point name: nodes for which the y coordinate is used are drawn vertically, and those for which the x coordinates is used are drawn horizontally.

This technique corresponds to dividing up the plane in a simple way: all the points below the point at the root go in the left subtree, all those above in the right subtree, then all the points above the point at the root and to the left of the point in the right subtree go in the left subtree of the right subtree of the root, etc. Every external node of the tree corresponds to some rectangle in the plane. The diagram below shows the division of the plane corresponding to the above tree. Each numbered region corresponds to an external node in the tree; each point lies on a horizontal or vertical line segment which defines the division made in the tree at that point.

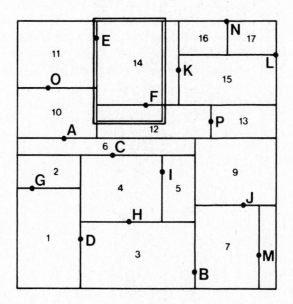

For example, if a new point was to be inserted into the tree from region 9 in the diagram above, we would move left at the root, since all such points are below A, then right at B, since all such points are to the right of B, then right at J, since all such points are above J. Insertion of a point in region 9 would correspond to drawing a vertical line through it in the diagram.

The code for the construction of 2D trees is a straightforward modification of standard binary tree search to switch between x and y coordinates at each level:

```
function twoDinsert(p: point; t: link): link;
  var f: link;
      d, td: boolean;
  begin
  d:=true;
  repeat
    if d then td:=p.x<t↑.p.x
        else  td:=p.y<t↑.p.y;
    f:=t;
    if td then t:=t↑.l else t:=t↑.r;
    d:= not d;
  until t=z;
  new(t); t↑.p:=p; t↑.l:=z; t↑.r:=z;
  if td then f↑.l:=t else f↑.r:=t;
  twoDinsert:=t
  end;
```

As usual, we use a header node *head* with an artificial point $(0,0)$ which is "less" than all the other points so that the tree hangs off the right link of *head*, and an artificial node z is used to represent all the external nodes. The call *twoDinsert(p, head)* will insert a new node containing p into the tree. A boolean variable d is toggled on the way down the tree to effect the alternating tests on x and y coordinates. Otherwise the procedure is identical to the standard procedure from Chapter 14. In fact, it turns out that for randomly distributed points, 2D trees have all the same performance characteristics of binary search trees. For example, the average time to build such a tree is proportional to $N \log N$, but there is an N^2 worst case.

To do range searching using 2D trees, we test the point at each node against the range along the dimension that is used to divide the plane of that node. For our example, we begin by going right at the root and right at node E, since our search rectangle is entirely above A and to the right of E. Then, at node F, we must go down both subtrees, since F falls in the x range defined by the rectangle (note carefully that this is *not* the same as F falling within the rectangle). Then the left subtrees of P and K are checked, corresponding to checking areas 12 and 14 of the plane, which overlap the search rectangle. This process is easily implemented with a straightforward generalization of the 1D *range* procedure that we examined at the beginning of this chapter:

```
procedure twoDrange(t: link; rect: rectangle; d: boolean);
  var t1, t2, tx1, tx2, ty1, ty2: boolean;
  begin
  if t<>z then
    begin
    tx1:=rect.x1<t↑.p.x;  tx2:=t↑.p.x<=rect.x2;
    ty1:=rect.y1<t↑.p.y;  ty2:=t↑.p.y<=rect.y2;
    if d then begin t1:=tx1;  t2:=tx2 end
        else  begin t1:=ty1;  t2:=ty2 end;
    if t1 then twoDrange(t↑.l, rect, not d);
    if insiderect(t↑.p,  rect) then write(name(t↑.p.info), ' ');
    if t2 then twoDrange(t↑.r, rect,  not d);
    end
  end;
```

This procedure goes down both subtrees only when the dividing line cuts the rectangle, which should happen infrequently for relatively small rectangles. Although the method hasn't been fully analyzed, its running time seems sure to be proportional to $R + \log N$ to retrieve R points from reasonable ranges in a region containing N points, which makes it very competitive with the grid method.

Multidimensional Range Searching

Both the grid method and 2D trees generalize directly to more than two dimensions: simple, straightforward extensions to the above algorithms immediately yield range-searching methods which work for more than two dimensions. However, the nature of multidimensional space dictates that some caution is called for and that the performance characteristics of the algorithms might be difficult to predict for a particular application.

To implement the grid method for k-dimensional searching, we simply make *grid* a k-dimensional array and use one index per dimension. The main problem is to pick a reasonable value for *size*. This problem becomes quite obvious when large k is considered: what type of grid should we use for 10-dimensional search? The problem is that even if we use only three divisions per dimension, we need 3^{10} grid squares, most of which will be empty, for reasonable values of N.

The generalization from 2D to kD trees is also straightforward: simply cycle through the dimensions (as we did for two dimensions by alternating between x and y) while going down the tree. As before, in a random situation, the resulting trees have the same characteristics as binary search trees. Also as before, there is a natural correspondence between the trees and a simple geometric process. In three dimensions, branching at each node corresponds to cutting the three-dimensional region of interest with a plane; in general we cut the k-dimensional region of interest with a $(k-1)$-dimensional hyperplane.

If k is very large, there is likely to be a significant amount of imbalance in the kD trees, again because practical point sets can't be large enough to take notice of randomness over a large number of dimensions. Typically, all points in a subtree will have the same value across several dimensions, which leads to several one-way branches in the trees. One way to help alleviate this problem is, rather than simply cycle through the dimensions, always to use the dimension that will divide up the point set in the best way. This technique can also be applied to 2D trees. It requires that extra information (which dimension should be discriminated upon) be stored in each node, but it does relieve imbalance, especially in high-dimensional trees.

In summary, though it is easy to see how to to generalize the programs for range searching that we have developed to handle multidimensional problems, such a step should not be taken lightly for a large application. Large databases with many attributes per record can be quite complicated objects indeed, and it is often necessary to have a good understanding of the characteristics of the database in order to develop an efficient range-searching method for a particular application. This is a quite important problem which is still being actively studied.

$\square$

Exercises

1. Write a nonrecursive version of the 1D *range* program given in the text.

2. Write a program to print out all points from a binary tree which do *not* fall in a specified interval.

3. Give the maximum and minimum number of grid squares that will be searched in the grid method as functions of the dimensions of the grid squares and the search rectangle.

4. Discuss the idea of avoiding the search of empty grid squares by using linked lists: each grid square could be linked to the next nonempty grid square in the same row and the next nonempty grid square in the same column. How would the use of such a scheme affect the grid square size to be used?

5. Draw the tree and the subdivision of the plane that results if we build a 2D tree for our sample points starting with a vertical dividing line. (That is, call *range* with a third argument of *false* rather than *true*.)

6. Give a set of points which leads to a worst-case 2D tree having no nodes with two sons; give the subdivision of the plane that results.

7. Describe how you would modify each of the methods to return all points that fall within a given circle.

8. Of all search rectangles with the same area, what shape is likely to make each of the methods perform the worst?

9. Which method should be preferred for range searching in the case that the points cluster together in large groups spaced far apart?

10. Draw the 3D tree that results when the points $(3, 1, 5)$ $(4, 8, 3)$ $(8, 3, 9)$ $(6, 2, 7)$ $(1, 6, 3)$ $(1, 3, 5)$ $(6, 4, 2)$ are inserted into an initially empty tree.

27. Geometric Intersection

A natural problem arising frequently in applications involving geometric data is: "Given a set of N objects, do any two intersect?" The "objects" involved may be lines, rectangles, circles, polygons, or other types of geometric objects. For example, in a system for designing and processing integrated circuits or printed circuit boards, it is important to know that no two wires intersect to make a short circuit. In an industrial application for designing layouts to be executed by a numerically controlled cutting tool, it is important to know that no two parts of the layout intersect. In computer graphics, the problem of determining which of a set of objects is obscured from a particular viewpoint can be formulated as a geometric intersection problem on the projections of the objects onto the viewing plane. And in operations research, the mathematical formulation of many important problems leads naturally to a geometric intersection problem.

The obvious solution to the intersection problem is to check each pair of objects to see if they intersect. Since there are about $\frac{1}{2}N^2$ pairs of objects, the running time of this algorithm is proportional to N^2. For many applications, this is not a problem because other factors limit the number of objects which can be processed. However, geometric applications systems have become much more ambitious, and it is not uncommon to have to process hundreds of thousands or even millions of objects. The brute-force N^2 algorithm is obviously inadequate for such applications. In this section, we'll study a general method for determining whether any two out of a set of N objects intersect in time proportional to $N \log N$, based on algorithms presented by M. Shamos and D. Hoey in a seminal paper in 1976.

First, we'll consider an algorithm for returning all intersecting pairs among a set of lines that are constrained to be horizontal or vertical. This makes the problem easier in one sense (horizontal and vertical lines are relatively simple geometric objects), more difficult in another sense (returning all intersecting pairs is more difficult than simply determining whether one such pair exists). The implementation that we'll develop applies binary search trees and the interval range-searching program of the previous chapter in a doubly recursive program.

Next, we'll examine the problem of determining whether any two of a set of N lines intersect, with no constraints on the lines. The same general strategy as used for the horizontal-vertical case can be applied. In fact, the same basic idea works for detecting intersections among many other types of geometric objects. However, for lines and other objects, the extension to return all intersecting pairs is somewhat more complicated than for the horizontal-vertical case.

Horizontal and Vertical Lines

To begin, we'll assume that all lines are either horizontal or vertical: the two points defining each line either have equal x coordinates or equal y coordinates, as in the following sample set of lines:

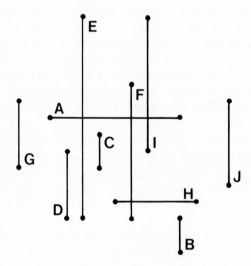

(This is sometimes called *Manhattan* geometry because, the Chrysler building notwithstanding, the Manhattan skyline is often sketched using only horizontal and vertical lines.) Constraining lines to be horizontal or vertical is certainly a severe restriction, but this is far from a "toy" problem. It is often the case that this restriction is imposed for some other reason for a particular application. For example, very large-scale integrated circuits are typically designed under this constraint.

The general plan of the algorithm to find an intersection in such a set of lines is to imagine a horizontal scan line sweeping from bottom to top in the diagram. Projected onto this scan line, vertical lines are points, and horizontal lines are intervals: as the scan line proceeds from bottom to top, points (representing vertical lines) appear and disappear, and horizontal lines are periodically encountered. An intersection is found when a horizontal line is encountered which represents an interval on the scan line that contains a point representing a vertical line. The point means that the vertical line intersects the scan line, and the horizontal line lies on the scan line, so the horizontal and vertical lines must intersect. In this way, the two-dimensional problem of finding an intersecting pair of lines is reduced to the one-dimensional range-searching problem of the previous chapter.

Of course, it is not necessary actually to "sweep" a horizontal line all the way up through the set of lines: since we only need to take action when endpoints of the lines are encountered, we can begin by sorting the lines according to their y coordinate, then processing the lines in that order. If the bottom endpoint of a vertical line is encountered, we add the x coordinate of that line to the tree; if the top endpoint of a vertical line is encountered, we delete that line from the tree; and if a horizontal line is encountered, we do an interval range search using its two x coordinates. As we'll see, some care is required to handle equal coordinates among line endpoints (the reader should now be accustomed to encountering such difficulties in geometric algorithms).

To trace through the operation of our algorithm on our set of sample points, we first must sort the line endpoints by their y coordinate:

$$B\ B\ D\ E\ F\ H\ J\ C\ G\ D\ I\ C\ A\ G\ J\ F\ E\ I$$

Each vertical line appears twice in this list, each horizontal line appears once. For the purposes of the line intersection algorithm, this sorted list can be thought of as a sequence of *insert* (vertical lines when the bottom endpoint is encountered), *delete* (vertical lines when the top endpoint is encountered), and *range* (for the endpoints of horizontal lines) commands. All of these "commands" are simply calls on the standard binary tree routines from Chapters 14 and 26, using x coordinates as keys.

For our example, we begin with the following sequence of binary search trees:

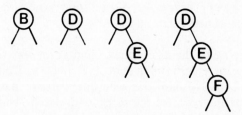

First B is inserted into an empty tree, then deleted. Then D, E, and F are inserted. At this point, H is encountered, and a range search for the interval defined by H is performed on the rightmost tree in the above diagram. This search discovers the intersection between H and F. Proceeding down the list above in order, we add J, C, then G to get the following sequence of trees:

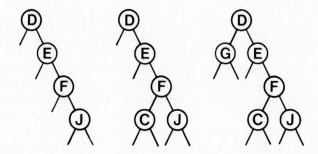

Next, the upper endpoint of D is encountered, so it is deleted; then I is added and C deleted, which gives the following sequence of trees:

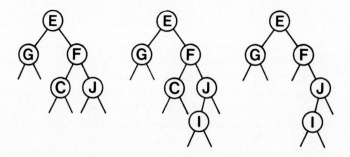

At this point A is encountered, and a range search for the interval defined by A is performed on the rightmost tree in the diagram above. This search discovers the intersections between A and E, F, and I. (Recall that although G and J are visited during this search, any points to the left of G or to the right of J would not be touched.) Finally, the upper endpoints of G, J, F, E, and I are encountered, so those points are successively deleted, leading back to the empty tree.

The first step in the implementation is to sort the line endpoints on their y coordinate. But since binary trees are going to be used to maintain the status of vertical lines with respect to the horizontal scan line, they may as well be used for the initial y sort! Specifically, we will use two "indirect" binary trees on the line set, one with header node hy and one with header node hx. The y tree will contain all the line endpoints, to be processed in order one at a time; the x tree will contain the lines that intersect the current horizontal scan line. We begin by initializing both hx and hy with 0 keys and pointers to a dummy external node z, as in *treeinitialize* in Chapter 14. Then the hy tree is constructed by inserting both y coordinates from vertical lines and the y coordinate of horizontal lines into the binary search tree with header node hy, as follows:

```
procedure buildytree;
  var N, k, x1, y1, x2, y2: integer;
  begin
  readln(N);
  for k:=1 to N do
    begin
    readln(x1, y1, x2, y2);
    lines[k].p1.x:=x1;  lines[k].p1.y:=y1;
    lines[k].p2.x:=x2;  lines[k].p2.y:=y2;
    bstinsert(k, y1, hy);
    if y2<>y1 then bstinsert(k, y2, hy);
    end;
  end;
```

This program reads in groups of four numbers which specify lines, and puts them into the *lines* array and the binary search tree on the y coordinate. The standard *bstinsert* routine from Chapter 14 is used, with the y coordinates as keys, and indices into the array of lines as the *info* field. For our example set of lines, the following tree is constructed:

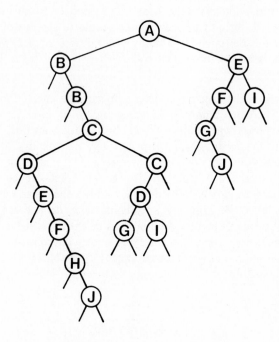

Now, the sort on y is effected by a recursive program with the same recursive structure as the *treeprint* routine for binary search trees in Chapter 14. We visit the nodes in increasing y order by visiting all the nodes in the left subtree of the *hy* tree, then visiting the root, then visiting all the nodes in the right subtree of the *hy* tree. At the same time, we maintain a separate tree (rooted at *hx*) as described above, to simulate the operation of passing a horizontal scan line through. The code at the point where each node is "visited" is rather straightforward from the description above. First, the coordinates of the endpoint of the corresponding line are fetched from the *lines* array, indexed by the *info* field of the node. Then the *key* field in the node is compared against these to determine whether this node corresponds to the upper or the lower endpoint of the line: if it is the lower endpoint, it is inserted into the *hx* tree, and if it is the upper endpoint, it is deleted from the *hx* tree and a range search is performed. The implementation differs slightly from this description in that horizontal lines are actually inserted into the *hx* tree, then immediately deleted, and a range search for a one-point interval is performed for vertical lines. This makes the code properly handle the case of overlapping vertical lines, which are considered to "intersect."

```
procedure scan(next: link);
  var t, x1, x2, y1, y2: integer;
    int: interval;
  begin
  if next<>z then
    begin
    scan(next↑.l);
    x1:=lines[next↑.info].p1.x;  y1:=lines[next↑.info].p1.y;
    x2:=lines[next↑.info].p2.x;  y2:=lines[next↑.info].p2.y;
    if x2<x1 then begin t:=x2;  x2:=x1;  x1:=t end;
    if y2<y1 then begin t:=y2;  y2:=y1;  y1:=t end;
    if next↑.key=y1 then bstinsert(next↑.info, x1, hx);
    if next↑.key=y2 then
      begin
      bstdelete(next↑.info, x1, hx);
      int.x1:=x1;  int.x2:=x2;
      write(name(next↑.info), ': ');
      bstrange(hx↑.r, int);
      writeln;
      end;
    scan(next↑.r)
    end
  end;
```

The running time of this program depends on the number of intersections that are found as well as the number of lines. The tree manipulation operations take time proportional to $\log N$ on the average (if balanced trees were used, a $\log N$ worst case could be guaranteed), but the time spent in *bstrange* also depends on the total number of intersections it returns, so the total running time is proportional to $N \log N + I$, where I is the number of intersecting pairs. In general, the number of intersections could be quite large. For example, if we have $N/2$ horizontal lines and $N/2$ vertical lines arranged in a crosshatch pattern, then the number of intersections is proportional to N^2. As with range searching, if it is known in advance that the number of intersections will be very large, then some brute-force approach should be used. Typical applications involve a "needle-in-haystack" sort of situation where a large set of lines is to be checked for a few possible intersections.

This approach of intermixed application of recursive procedures operating on the x and y coordinates is quite important in geometric algorithms. Another example of this is the 2D tree algorithm of the previous chapter, and we'll see yet another example in the next chapter.

General Line Intersection

When lines of arbitrary slope are allowed, the situation can become more complicated, as illustrated by the following example.

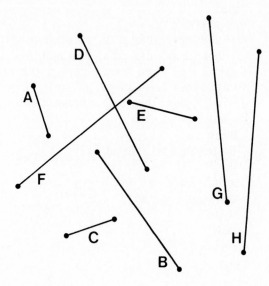

First, the various line orientations possible make it necessary to test explicitly whether certain pairs of lines intersect: we can't get by with a simple interval range test. Second, the ordering relationship between lines for the binary tree is more complicated than before, as it depends on the current y range of interest. Third, any intersections which do occur add new "interesting" y values which are likely to be different from the set of y values that we get from the line endpoints.

It turns out that these problems can be handled in an algorithm with the same basic structure as given above. To simplify the discussion, we'll consider an algorithm for detecting whether or not there exists an intersecting pair in a set of N lines, and then we'll discuss how it can be extended to return all intersections.

As before, we first sort on y to divide the space into strips within which no line endpoints appear. Just as before, we proceed through the sorted list of points, adding each line to a binary search tree when its bottom point is encountered and deleting it when its top point is encountered. Just as before, the binary tree gives the order in which the lines appear in the horizontal "strip" between two consecutive y values. For example, in the strip between the bottom endpoint of D and the top endpoint of B in the diagram above, the lines should appear in the order F B D G H. We assume that there are no intersections within the current horizontal strip of interest: our goal is to maintain this tree structure and use it to help find the first intersection.

To build the tree, we can't simply use x coordinates from line endpoints as keys (doing this would put B and D in the wrong order in the example above, for instance). Instead, we use a more general ordering relationship: a line x is defined to be to the right of a line y if both endpoints of x are on the same side of y as a point infinitely far to the right, *or* if y is to the left of x, with "left" defined analagously. Thus, in the diagram above, B is to the right of A and B is to the right of C (since C is to the left of B). If x is neither to the left nor to the right of y, then they must intersect. This generalized "line comparison" operation is a simple extension of the *same* procedure of Chapter 24. Except for the use of this function whenever a comparison is needed, the standard binary search tree procedures (even balanced trees, if desired) can be used. For example, the following sequence of diagrams shows the manipulation of the tree for our example between the time that line C is encountered and the time that line D is encountered.

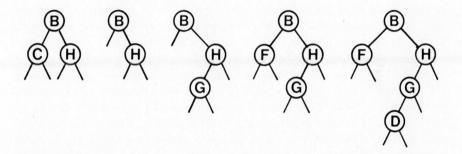

Each "comparison" performed during the tree manipulation procedures is actually a line intersection test: if the binary search tree procedure can't decide to go right or left, then the two lines in question must intersect, and we're finished.

But this is not the whole story, because this generalized comparison operation is not *transitive*. In the example above, F is to the left of B (because B is to the right of F) and B is to the left of D, but F is *not* to the left of D. It is essential to note this, because the binary tree deletion procedure assumes that the comparison operation is transitive: when B is deleted from the last tree in the above sequence, the tree

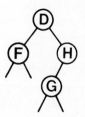

is formed without F and D ever having been explicitly compared. For our intersection-testing algorithm to work correctly, we must explicitly test that comparisons are valid each time we change the tree structure. Specifically, every time we make the left link of node x point to node y, we explicitly test that the line corresponding to x is to the left of the line corresponding to y, according to the above definition, and similarly for the right. Of course, this comparison could result in the detection of an intersection, as it does in our example.

In summary, to test for an intersection among a set of N lines, we use the program above, but with the call to *range* removed, and with the binary tree routines extended to use the generalized comparison as described above. If there is no intersection, we'll start with a null tree and end with a null tree without finding any incomparable lines. If there is an intersection, then the two lines which intersect must be compared against each other at some point during the scanning process and the intersection discovered.

Once we've found an intersection, we can't simply press on and hope to find others, because the two lines that intersect should swap places in the ordering directly after the point of intersection. One way to handle this problem would be to use a priority queue instead of a binary tree for the "y sort": initially put lines on the priority queue according to the y coordinates of their endpoints, then work the scan line up by successively taking the smallest y coordinate from the priority queue and doing a binary tree insert or delete as above. When an intersection is found, new entries are added to the priority queue for each line, using the intersection point as the lower endpoint for each.

Another way to find all intersections, which is appropriate if not too many are expected, is to simply remove one of the intersecting lines when an intersection is found. Then after the scan is completed, we know that all intersecting pairs must involve one of those lines, so we can use a brute force method to enumerate all the intersections.

An interesting feature of the above procedure is that it can be adapted to solve the problem for testing for the existence of an intersecting pair among a set of more general geometric shapes just by changing the generalized comparison procedure. For example, if we implement a procedure which compares two rectangles whose edges are horizontal and vertical according to the trivial rule that rectangle x is to the left of rectangle y if the right edge of x is to the left of the left edge of y, then we can use the above method to test for intersection among a set of such rectangles. For circles, we can use the x coordinates of the centers for the ordering, but explicitly test for intersection (for example, compare the distance between the centers to the sum of the radii). Again, if this comparison procedure is used in the above method, we have an algorithm for testing for intersection among a set of circles. The problem of returning all intersections in such cases is much more complicated, though the brute-force method mentioned in the previous paragraph will always work if few intersections are expected. Another approach that will suffice for many applications is simply to consider complicated objects as sets of lines and to use the line intersection procedure.

360

Exercises

1. How would you determine whether two triangles intersect? Squares? Regular n-gons for $n > 4$?

2. In the horizontal-vertical line intersection algorithm, how many pairs of lines are tested for intersection in a set of lines with no intersections in the worst case? Give a diagram supporting your answer.

3. What happens if the horizontal-vertical line intersection procedure is used on a set of lines with arbitrary slope?

4. Write a program to find the number of intersecting pairs among a set of N random horizontal and vertical lines, each line generated with two random integer coordinates between 0 and 1000 and a random bit to distinguish horizontal from vertical.

5. Give a method for testing whether or not a given polygon is simple (doesn't intersect itself).

6. Give a method for testing whether one polygon is totally contained within another.

7 Describe how you would solve the general line intersection problem given the additional fact that the minimum separation between two lines is greater than the maximum length of the lines.

8. Give the binary tree structure that exists when the line intersection algorithm detects an intersection in the following set of lines:

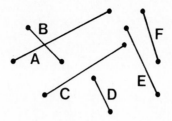

9. Are the comparison procedures for circles and Manhattan rectangles that are described in the text transitive?

10. Write a program to find the number of intersecting pairs among a set of N random lines, each line generated with random integer coordinates between 0 and 1000.

28. Closest Point Problems

Geometric problems involving points on the plane usually involve implicit or explicit treatment of distances between the points. For example, a very natural problem which arises in many applications is the *nearest-neighbor* problem: find the point among a set of given points closest to a given new point. This seems to involve checking the distance from the given point to each point in the set, but we'll see that much better solutions are possible. In this section we'll look at some other distance problems, a prototype algorithm, and a fundamental geometric structure called the *Voronoi diagram* that can be used effectively for a variety of such problems in the plane. Our approach will be to describe a general method for solving closest point problems through careful consideration of a prototype implementation, rather than developing full implementations of programs to solve all of the problems.

Some of the problems that we consider in this chapter are similar to the range-searching problems of Chapter 26, and the grid and 2D tree methods developed there are suitable for solving the nearest-neighbor and other problems. The fundamental shortcoming of those methods is that they rely on randomness in the point set: they have bad worst-case performance. Our aim in this chapter is to examine yet another general approach that has guaranteed good performance for many problems, no matter what the input. Some of the methods are too complicated for us to examine a full implementation, and they involve sufficient overhead that the simpler methods may do better for actual applications where the point set is not large or where it is sufficiently well dispersed. However, we'll see that the study of methods with good worst-case performance will uncover some fundamental properties of point sets that should be understood even if simpler methods turn out to be more suitable.

The general approach that we'll be examining provides yet another example of the use of doubly recursive procedures to intertwine processing along the two coordinate directions. The two previous methods of this type that we've seen (kD trees and line intersection) have been based on binary search trees; in this case the method is based on mergesort.

Closest Pair

The *closest-pair* problem is to find the two points that are closest together among a set of points. This problem is related to the nearest-neighbor problem; though it is not as widely applicable, it will serve us well as a prototype closest-point problem in that it can be solved with an algorithm whose general recursive structure is appropriate for other problems.

It would seem necessary to examine the distances between all pairs of points to find the smallest such distance: for N points this would mean a running time proportional to N^2. However, it turns out that we can use sorting to get by with only examining about $N \log N$ distances between points in the worst case (far fewer on the average) to get a worst-case running time proportional to $N \log N$ (far better on the average). In this section, we'll examine such an algorithm in detail.

The algorithm that we'll use is based on a straightforward "divide-and-conquer" strategy. The idea is to sort the points on one coordinate, say the x coordinate, then use that ordering to divide the points in half. The closest pair in the whole set is either the closest pair in one of the halves or the closest pair with one member in each half. The interesting case, of course, is when the closest pair crosses the dividing line: the closest pair in each half can obviously be found by using recursive calls, but how can all the pairs on either side of the dividing line be checked efficiently?

Since the only information we seek is the closest pair of the point set, we need examine only points within distance *min* of the dividing line, where *min* is the smaller of the distances between the closest pairs found in the two halves. By itself, however, this observation isn't enough help in the worst case, since there could be many pairs of points very close to the dividing line. For example, all the points in each half could be lined up right next to the dividing line.

To handle such situations, it seems necessary to sort the points on y. Then we can limit the number of distance computations involving each point as follows: proceeding through the points in increasing y order, check if each point is inside the vertical strip consisting of all points in the plane within *min* of the dividing line. For each such point, compute the distance between it and any point also in the strip whose y coordinate is less than the y coordinate of the current point, but not more than *min* less. The fact that the distance between all pairs of points in each half is at least *min* means that only a few points are likely to be checked, as demonstrated in our example set of points:

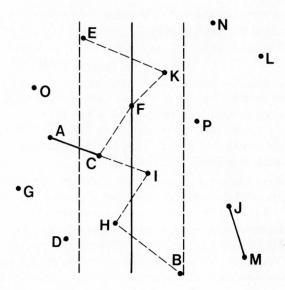

A vertical dividing line just to the right of F has eight points to the left, eight points to the right. The closest pair on the left half is AC (or AO), the closest pair on the right is JM. If we have the points sorted on y, then the closest pair which is split by the line is found by checking the pairs HI, CI, FK, which is the closest pair in the whole point set, and finally EK.

Though this algorithm is simply stated, some care is required to implement it efficiently: for example, it would be too expensive to sort the points on y within our recursive subroutine. We've seen several algorithms with a running time described by the recurrence $T(N) = 2T(N/2)+N$, which implies that $T(N)$ is proportional to $N \log N$; if we were to do the full sort on y, then the recurrence would become $T(N) = 2T(N/2) + N \log N$, and it turns out that this implies that $T(N)$ is proportional to $N \log^2 N$. To avoid this, we need to avoid the sort of y.

The solution to this problem is simple, but subtle. The *mergesort* method from Chapter 12 is based on dividing the elements to be sorted exactly as the points are divided above. We have two problems to solve and the same general method to solve them, so we may as well solve them simultaneously! Specifically, we'll write one recursive routine that both sorts on y *and* finds the closest pair. It will do so by splitting the point set in half, then calling itself recursively to sort the two halves on y and find the closest pair in each half, then merging to complete the sort on y and applying the procedure above to complete the closest pair computation. In this way, we avoid the cost of doing an extra y sort by intermixing the data movement required for the sort with the data movement required for the closest pair computation.

For the y sort, the split in half could be done in any way, but for the closest pair computation, it's required that the points in one half all have smaller x coordinates than the points in the other half. This is easily accomplished by sorting on x before doing the division. In fact, we may as well use the same routine to sort on x! Once this general plan is accepted, the implementation is not difficult to understand.

As mentioned above, the implementation will use the recursive *sort* and *merge* procedures of Chapter 12. The first step is to modify the list structures to hold points instead of keys, and to modify *merge* to check a global variable *pass* to decide how to do its comparison. If *pass=1*, the comparison should be done using the x coordinates of the two points; if *pass=2* we do the y coordinates of the two points. The dummy node z which appears at the end of all lists will contain a "sentinel" point with artificially high x and y coordinates.

The next step is to modify the recursive *sort* of Chapter 12 also to do the closest-point computation when *pass=2*. This is done by replacing the line containing the call to *merge* and the recursive calls to *sort* in that program by the following code:

```
if pass=2 then middle:=b↑.p.x;
c:=merge(sort(a, N div 2), sort(b, N−(N div 2)));
sort:=c;
if pass=2 then
  begin
  a:=c;  p1:=z↑.p;  p2:=z↑.p;  p3:=z↑.p;  p4:=z↑.p;
  repeat
    if abs(a↑.p.x−middle)<min then
      begin
      check(a↑.p, p1);
      check(a↑.p, p2);
      check(a↑.p, p3);
      check(a↑.p, p4);
      p1:=p2;  p2:=p3;  p3:=p4;  p4:=a↑.p
      end;
    a:=a↑.next
  until a=z
  end
```

If *pass=1*, this is straight mergesort: it returns a linked list containing the points sorted on their x coordinates (because of the change to merge). The magic of this implementation comes when *pass=2*. The program not only sorts on y but also completes the closest-point computation, as described in detail below. The procedure *check* simply checks whether the distance between the two points given as arguments is less than the global variable *min*. If so, it resets *min* to that distance and saves the points in the global variables *cp1* and *cp2*. Thus, the global *min* always contains the distance between *cp1* and *cp2*, the closest pair found so far.

First, we sort on x, then we sort on y and find the closest pair by invoking *sort* as follows:

```
new(z);  z↑.next:=z;
z↑.p.x:=maxint;  z↑.p.y:=maxint;
new(h);  h↑.next:=readlist;
min:=maxint;
pass:=1;  h↑.next:=sort(h↑.next, N);
pass:=2;  h↑.next:=sort(h↑.next, N);
```

After these calls, the closest pair of points is found in the global variables *cp1* and *cp2* which are managed by the *check* "find the minimum" procedure.

The crux of the implementation is the operation of *sort* when *pass=2*. *Before* the recursive calls the points are sorted on x: this ordering is used to divide the points in half and to find the x coordinate of the dividing line. *After* the recursive calls the points are sorted on y and the distance between every pair of points in each half is known to be greater than *min*. The ordering on y is used to scan the points near the dividing line; the value of *min* is used to limit the number of points to be tested. Each point within a distance of *min* of the dividing line is *checked* against each of the previous four points found within a distance of *min* of the dividing line. This is guaranteed to find any pair of points closer together than *min* with one member of the pair on either side of the dividing line. This is an amusing geometric fact which the reader may wish to check. (We know that points which fall on the same side of the dividing line are spaced by at least *min*, so the number of points falling in any circle of radius *min* is limited.)

It is interesting to examine the order in which the various vertical dividing lines are tried in this algorithm. This can be described with the aid of the following binary tree:

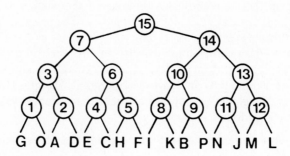

Each node in this tree represents a vertical line dividing the points in the left and right subtree. The nodes are numbered in the order in which the vertical lines are tried in the algorithm. Thus, first the line between G and O is tried and the pair GO is retained as the closest so far. Then the line between A and D is tried, but A and D are too far apart to change *min*. Then the line between O and A is tried and the pairs GD GA and OA all are successively closer pairs. It happens for this example that no closer pairs are found until FK, which is the last pair checked for the last dividing line tried. This diagram reflects the difference between top-down and bottom-up mergesort. A bottom-up version of the closest-pair problem can be developed in the same way as for mergesort, which would be described by a tree like the one above, numbered left to right and bottom to top.

The general approach that we've used for the closest-pair problem can be used to solve other geometric problems. For example, another problem of interest is the *all-nearest-neighbors* problem: for each point we want to find the point nearest to it. This problem can be solved using a program like the one above with extra processing along the dividing line to find, for each point, whether there is a point on the other side closer than its closest point on its own side. Again, the "free" *y* sort is helpful for this computation.

Voronoi Diagrams

The set of all points closer to a given point in a point set than to all other points in the set is an interesting geometric structure called the *Voronoi polygon* for the point. The union of all the Voronoi polygons for a point set is called its *Voronoi diagram*. This is the ultimate in closest-point computations: we'll see that most of the problems involving distances between points that we face have natural and interesting solutions based on the Voronoi diagram. The diagram for our sample point set is comprised of the thick lines in the diagram below:

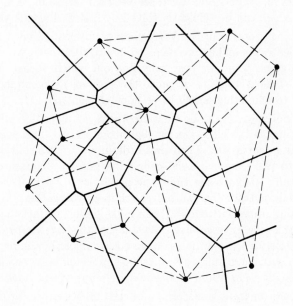

Basically, the Voronoi polygon for a point is made up of the perpendicular bisectors separating the point from those points closest to it. The actual definition is the other way around: the Voronoi polygon is defined to be the set of all points in the plane closer to the given point than to any other point in the point set, and the points "closest to" a point are defined to be those that lead to edges on the Voronoi polygon. The *dual* of the Voronoi diagram makes this correspondence explicit: in the dual, a line is drawn between each point and all the points "closest to" it. Put another way, x and y are connected in the Voronoi dual if their Voronoi polygons have an edge in common. The dual for our example is comprised of the thin dotted lines in the above diagram.

The Voronoi diagram and its dual have many properties that lead to efficient algorithms for closest-point problems. The property that makes these algorithms efficient is that the number of lines in both the diagram and the dual is proportional to a small constant times N. For example, the line connecting the closest pair of points must be in the dual, so the problem of the previous section can be solved by computing the dual and then simply finding the minimum length line among the lines in the dual. Similarly, the line connecting each point to its nearest neighbor must be in the dual, so the all-nearest-neighbors problem reduces directly to finding the dual. The convex hull of the point set is part of the dual, so computing the Voronoi dual is yet another convex hull algorithm. We'll see yet another example in Chapter 31 of a problem which can be efficiently solved by first finding the Voronoi dual.

The defining property of the Voronoi diagram means that it can be used to solve the nearest-neighbor problem: to identify the nearest neighbor in a point set to a given point, we need only find out which Voronoi polygon the point falls in. It is possible to organize the Voronoi polygons in a structure like a 2D tree to allow this search to be done efficiently.

The Voronoi diagram can be computed using an algorithm with the same general structure as the closest-point algorithm above. The points are first sorted on their x coordinate. Then that ordering is used to split the points in half, leading to two recursive calls to find the Voronoi diagram of the point set for each half. At the same time, the points are sorted on y; finally, the two Voronoi diagrams for the two halves are merged together. As before, the merging together (done with $pass=2$) can make use of the fact that the points are sorted on x before the recursive calls and that they are sorted on y and the Voronoi diagrams for the two halves have been built after the recursive calls. However, even with these aids, it is quite a complicated task, and presentation of a full implementation would be beyond the scope of this book.

The Voronoi diagram is certainly the natural structure for closest-point problems, and understanding the characteristics of a problem in terms of the Voronoi diagram or its dual is certainly a worthwhile exercise. However, for many particular problems, a direct implementation based on the general schema given in this chapter may be suitable. This is powerful enough to compute the Voronoi diagram, so it is powerful enough for algorithms based on the Voronoi diagram, and it may admit to simpler, more efficient code, just as we saw for the closest-pair problem.

Exercises

1. Write programs to solve the nearest-neighbor problem, first using the grid method, then using 2D trees.

2. Describe what happens when the closest-pair procedure is used on a set of points that fall on the same horizontal line, equally spaced.

3. Describe what happens when the closest-pair procedure is used on a set of points that fall on the same vertical line, equally spaced.

4. Give an algorithm that, given a set of $2N$ points, half with positive x coordinates, half with negative x coordinates, finds the closest pair with one member of the pair in each half.

5. Give the successive pairs of points assigned to *cp1* and *cp2* when the program in the text is run on the example points, but with A removed.

6. Test the effectiveness of making *min* global by comparing the performance of the implementation given to a purely recursive implementation for some large random point set.

7. Give an algorithm for finding the closest pair from a set of *lines*.

8. Draw the Voronoi diagram and its dual for the points A B C D E F from the sample point set.

9. Give a "brute-force" method (which might require time proportional to N^2) for computing the Voronoi diagram.

10. Write a program that uses the same recursive structure as the closest-pair implementation given in the text to find the convex hull of a set of points.

SOURCES for Geometric Algorithms

Much of the material described in this section has actually been developed quite recently, so there are many fewer available references than for older, more central areas such as sorting or mathematical algorithms. Many of the problems and solutions that we've discussed were presented by M. Shamos in 1975. Shamos' manuscript treats a large number of geometric algorithms, and has stimulated much of the recent research.

For the most part, each of the geometric algorithms that we've discussed is described in its own original reference. The convex hull algorithms treated in Chapter 25 may be found in the papers by Jarvis, Graham, and Eddy. The range searching methods of Chapter 26 come from Bentley and Freidman's survey article, which contains many references to original sources (of particular interest is Bentley's own original article on KD trees, written while he was an undergraduate). The intersection algorithms of Chapter 27 are from Shamos and Hoey's 1975 paper, and the treatment of closest point problems is based on their 1976 paper.

But the best route for someone interested in learning more about geometric algorithms is to implement some, work with them and try to learn about their behavior on different types of point sets. This field is still in its infancy and the best algorithms are yet to be discovered.

J. L. Bentley, "Multidimensional binary search trees used for associative searching," *Communications of the ACM*, **18**, 9 (September, 1975).

J. L. Bentley and J.H. Friedman, "Data structures for range searching," *Computing Surveys*, **11**, 4 (December, 1979).

W. F. Eddy, "A new convex hull algorithm for planar sets," *ACM Transactions on Mathematical Software*, **3** (1977).

R. L. Graham, "An efficient algorithm for determining the convex hull of a finite planar set," *Information Processing Letters*, **1** (1972).

R. A. Jarvis, "On the identification of the convex hull of a finite set of points in the plane," *Information Processing Letters*, **2** (1973).

M. I. Shamos, "Geometric complexity," in *7th Symposium on Theory of Computing*, ACM, 1975.

M. I. Shamos, *Problems in Computational Geometry*, unpublished manuscript, 1975.

M. I. Shamos and D. Hoey, "Closest-point problems," in *16th Annual Symposium on Foundations of Computer Science*, IEEE, 1975.

M. I. Shamos and D. Hoey, "Geometric intersection problems," in *17th Annual Symposium on Foundations of Computer Science*, IEEE, 1976.

GRAPH ALGORITHMS

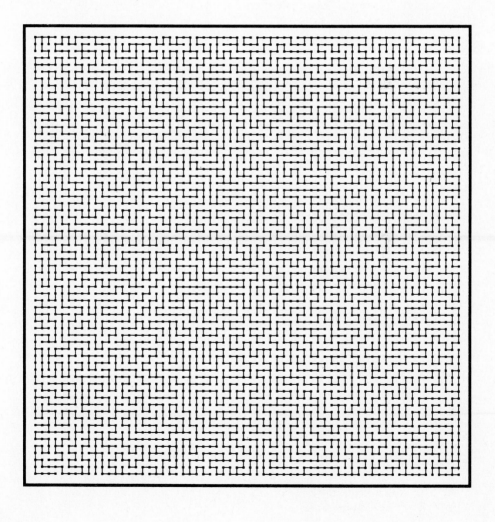

29. Elementary Graph Algorithms

A great many problems are naturally formulated in terms of objects and connections between them. For example, given an airline route map of the eastern U. S., we might be interested in questions like: "What's the fastest way to get from Providence to Princeton?" Or we might be more interested in money than in time, and look for the cheapest way to get from Providence to Princeton. To answer such questions we need only information about interconnections (airline routes) between objects (towns).

Electric circuits are another obvious example where interconnections between objects play a central role. Circuit elements like transistors, resistors, and capacitors are intricately wired together. Such circuits can be represented and processed within a computer in order to answer simple questions like "Is everything connected together?" as well as complicated questions like "If this circuit is built, will it work?" In this case, the answer to the first question depends only on the properties of the interconnections (wires), while the answer to the second question requires detailed information about both the wires and the objects that they connect.

A third example is "job scheduling," where the objects are tasks to be performed, say in a manufacturing process, and interconnections indicate which jobs should be done before others. Here we might be interested in answering questions like "When should each task be performed?"

A *graph* is a mathematical object which accurately models such situations. In this chapter, we'll examine some basic properties of graphs, and in the next several chapters we'll study a variety of algorithms for answering questions of the type posed above.

Actually, we've already encountered graphs in several instances in previous chapters. Linked data structures are actually representations of graphs, and some of the algorithms that we'll see for processing graphs are similar to algorithms that we've already seen for processing trees and other structures.

For example, the finite-state machines of Chapters 19 and 20 are represented with graph structures.

Graph theory is a major branch of combinatorial mathematics and has been intensively studied for hundreds of years. Many important and useful properties of graphs have been proved, but many difficult problems have yet to be resolved. We'll be able here only to scratch the surface of what is known about graphs, covering enough to be able to understand the fundamental algorithms.

As with so many of the problem domains that we've studied, graphs have only recently begun to be examined from an algorithmic point of view. Although some of the fundamental algorithms are quite old, many of the interesting ones have been discovered within the last ten years. Even trivial graph algorithms lead to interesting computer programs, and the nontrivial algorithms that we'll examine are among the most elegant and interesting (though difficult to understand) algorithms known.

Glossary

A good deal of nomenclature is associated with graphs. Most of the terms have straightforward definitions, and it is convenient to put them in one place even though we won't be using some of them until later.

A *graph* is a collection of *vertices* and *edges*. Vertices are simple objects which can have names and other properties; an edge is a connection between two vertices. One can draw a graph by marking points for the vertices and drawing lines connecting them for the edges, but it must be borne in mind that the graph is defined independently of the representation. For example, the following two drawings represent the same graph:

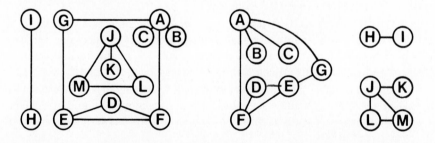

We define this graph by saying that it consists of the set of vertices A B C D E F G H I J K L M and the set of edges between these vertices AG AB AC LM JM JL JK ED FD HI FE AF GE .

For some applications, such as the airline route example above, it might not make sense to rearrange the placement of the vertices as in the diagrams above. But for some other applications, such as the electric circuit application above, it is best to concentrate only on the edges and vertices, independent of any particular geometric placement. And for still other applications, such as the finite-state machines in Chapters 19 and 20, no particular geometric placement of nodes is ever implied. The relationship between graph algorithms and geometric problems is discussed in further detail in Chapter 31. For now, we'll concentrate on "pure" graph algorithms which process simple collections of edges and nodes.

A *path* from vertex x to y in a graph is a list of vertices in which successive vertices are connected by edges in the graph. For example, BAFEG is a path from B to G in the graph above. A graph is *connected* if there is a path from every node to every other node in the graph. Intuitively, if the vertices were physical objects and the edges were strings connecting them, a connected graph would stay in one piece if picked up by any vertex. A graph which is not connected is made up of *connected components*; for example, the graph drawn above has three connected components. A *simple path* is a path in which no vertex is repeated. (For example, BAFEGAC is not a simple path.) A *cycle* is a path which is simple except that the first and last vertex are the same (a path from a point back to itself): the path AFEGA is a cycle.

A graph with no cycles is called a *tree*. There is only one path between any two nodes in a tree. (Note that binary trees and other types of trees that we've built with algorithms are special cases included in this general definition of trees.) A group of disconnected trees is called a *forest*. A *spanning tree* of a graph is a subgraph that contains all the vertices but only enough of the edges to form a tree. For example, below is a spanning tree for the large component of our sample graph.

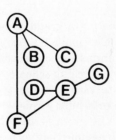

Note that if we add any edge to a tree, it must form a cycle (because there is already a path between the two vertices that it connects). Also, it is easy to prove by induction that a tree on V vertices has exactly $V - 1$ edges.

If a graph with V vertices has less than $V-1$ edges, it can't be connected. If it has more that $V-1$ edges, it must have a cycle. (But if it has exactly $V-1$ edges, it need not be a tree.)

We'll denote the number of vertices in a given graph by V, the number of edges by E. Note that E can range anywhere from 0 to $\frac{1}{2}V(V-1)$. Graphs with all edges present are called *complete* graphs; graphs with relatively few edges (say less than $V\log V$) are called *sparse*; graphs with relatively few of the possible edges missing are called *dense*.

This fundamental dependence on two parameters makes the comparative study of graph algorithms somewhat more complicated than many algorithms that we've studied, because more possibilities arise. For example, one algorithm might take about V^2 steps, while another algorithm for the same problem might take $(E+V)\log E$ steps. The second algorithm would be better for sparse graphs, but the first would be preferred for dense graphs.

Graphs as defined to this point are called *undirected graphs*, the simplest type of graph. We'll also be considering more complicated type of graphs, in which more information is associated with the nodes and edges. In *weighted graphs* integers (*weights*) are assigned to each edge to represent, say, distances or costs. In *directed graphs* , edges are "one-way": an edge may go from x to y but not the other way. Directed weighted graphs are sometimes called *networks*. As we'll discover, the extra information weighted and directed graphs contain makes them somewhat more difficult to manipulate than simple undirected graphs.

Representation

In order to process graphs with a computer program, we first need to decide how to represent them within the computer. We'll look at two commonly used representations; the choice between them depends whether the graph is dense or sparse.

The first step in representing a graph is to map the vertex names to integers between 1 and V. The main reason for doing this is to make it possible to quickly access information corresponding to each vertex, using array indexing. Any standard searching scheme can be used for this purpose: for instance, we can translate vertex names to integers between 1 and V by maintaining a hash table or a binary tree which can be searched to find the integer corresponding to any given vertex name. Since we have already studied these techniques, we'll assume that we have available a function *index* to convert from vertex names to integers between 1 and V and a function *name* to convert from integers to vertex names. In order to make the algorithms easy to follow, our examples will use one-letter vertex names, with the ith letter of the alphabet corresponding to the integer i. Thus, though *name* and *index* are trivial to implement for our examples, their use makes it easy to extend the algorithms to handle graphs with real vertex names using techniques from Chapters 14–17.

The most straightforward representation for graphs is the so-called *adjacency matrix* representation. A V-by-V array of boolean values is maintained, with $a[x, y]$ set to *true* if there is an edge from vertex x to vertex y and *false* otherwise. The adjacency matrix for our example graph is given below.

	A	B	C	D	E	F	G	H	I	J	K	L	M
A	1	1	1	0	0	1	1	0	0	0	0	0	0
B	1	1	0	0	0	0	0	0	0	0	0	0	0
C	1	0	1	0	0	0	0	0	0	0	0	0	0
D	0	0	0	1	1	1	0	0	0	0	0	0	0
E	0	0	0	1	1	1	1	0	0	0	0	0	0
F	1	0	0	1	1	1	0	0	0	0	0	0	0
G	1	0	0	0	1	0	1	0	0	0	0	0	0
H	0	0	0	0	0	0	0	1	1	0	0	0	0
I	0	0	0	0	0	0	0	1	1	0	0	0	0
J	0	0	0	0	0	0	0	0	0	1	1	1	1
K	0	0	0	0	0	0	0	0	0	1	1	0	0
L	0	0	0	0	0	0	0	0	0	1	0	1	1
M	0	0	0	0	0	0	0	0	0	1	0	1	1

Notice that each edge is really represented by two bits: an edge connecting x and y is represented by true values in both $a[x, y]$ and $a[y, x]$. While it is possible to save space by storing only half of this symmetric matrix, it is inconvenient to do so in Pascal and the algorithms are somewhat simpler with the full matrix. Also, it's sometimes convenient to assume that there's an "edge" from each vertex to itself, so $a[x, x]$ is set to 1 for x from 1 to V.

A graph is defined by a set of nodes and a set of edges connecting them. To read in a graph, we need to settle on a format for reading in these sets. The obvious format to use is first to read in the vertex names and then read in pairs of vertex names (which define edges). As mentioned above, one easy way to proceed is to read the vertex names into a hash table or binary search tree and to assign to each vertex name an integer for use in accessing vertex-indexed arrays like the adjacency matrix. The ith vertex read can be assigned the integer i. (Also, as mentioned above, we'll assume for simplicity in our examples that the vertices are the first V letters of the alphabet, so that we can read in graphs by reading V and E, then E pairs of letters from the first

V letters of the alphabet.) Of course, the order in which the edges appear is not relevant. All orderings of the edges represent the same graph and result in the same adjacency matrix, as computed by the following program:

```
program adjmatrix(input, output);
const maxV=50;
var j, x, y, V, E: integer;
    a: array [1..maxV, 1..maxV] of boolean;
begin
readln(V, E);
for x:=1 to V do
  for y:=1 to V do a[x, y]:=false;
for x:=1 to V do a[x, x]:=true;
for j:=1 to E do
  begin
  readln(v1, v2);
  x:=index(v1);  y:=index(v2);
  a[x, y]:=true;  a[y, x]:=true
  end;
end.
```

The types of *v1* and *v2* are omitted from this program, as well as the code for *index*. These can be added in a straightforward manner, depending on the graph input representation desired. (For our examples, *v1* and *v2* could be of type *char* and *index* a simple function which uses the Pascal *ord* function.)

The adjacency matrix representation is satisfactory only if the graphs to be processed are dense: the matrix requires V^2 bits of storage and V^2 steps just to initialize it. If the number of edges (the number of one bits in the matrix) is proportional to V^2, then this may be no problem because about V^2 steps are required to read in the edges in any case, but if the graph is sparse, just initializing this matrix could be the dominant factor in the running time of an algorithm. Also this might be the best representation for some algorithms which require more than V^2 steps for execution. Next we'll look at a representation which is more suitable for graphs which are not dense.

In the *adjacency structure* representation all the vertices connected to each vertex are listed on an *adjacency list* for that vertex. This can be easily accomplished with linked lists, as shown in the program below which builds the adjacency structure for our sample graph.

```
program adjlist(input, output);
const maxV=1000;
type  link=↑node;
      node=record v: integer; next: link end;
var j, x, y, V, E: integer;
    t, z: link;
    adj: array [1..maxV] of link;
begin
readln(V, E);
new(z); z↑.next:=z;
for j:=1 to V do adj[j]:=z;
for j:=1 to E do
  begin
  readln(v1, v2);
  x:=index(v1); y:=index(v2);
  new(t); t↑.v:=x; t↑.next:=adj[y]; adj[y]:=t;
  new(t); t↑.v:=y; t↑.next:=adj[x]; adj[x]:=t;
  end;
end.
```

(As usual, each linked list ends with a link to an artificial node z, which links to itself.) For this representation, the order in which the edges appear in the input is quite relevant: it (along with the list insertion method used) determines the order in which the vertices appear on the adjacency lists. Thus, the same graph can be represented in many different ways in an adjacency list structure. Indeed, it is difficult to predict what the adjacency lists will look like by examining just the sequence of edges, because each edge involves insertions into two adjacency lists.

The order in which edges appear on the adjacency list affects, in turn, the order in which edges are processed by algorithms. That is, the adjacency list structure determines the way that various algorithms that we'll be examining "see" the graph. While an algorithm should produce a correct answer no matter what the order of the edges on the adjacency lists, it might get to that answer by quite different sequences of computations for different orders. And if there is more than one "correct answer," different input orders might lead to different output results.

If the edges appear in the order listed after the first drawing of our sample graph at the beginning of the chapter, the program above builds the following adjacency list structure:

```
A:  F   C   B   G
B:  A
C:  A
D:  F   E
E:  G   F   D
F:  A   E   D
G:  E   A
H:  I
I:  H
J:  K   L   M
K:  J
L:  J   M
M:  J   L
```

Note that again each edge is represented twice: an edge connecting x and y is represented as a node containing x on y's adjacency list and as a node containing y on x's adjacency list. It is important to include both, since otherwise simple questions like "Which nodes are connected directly to node x?" could not be answered efficiently.

Some simple operations are not supported by this representation. For example, one might want to delete a vertex, x, and all the edges connected to it. It's not sufficient to delete nodes from the adjacency list: each node on the adjacency list specifies another vertex whose adjacency list must be searched for a node corresponding to x to be deleted. This problem can be corrected by linking together the two list nodes which correspond to a particular edge and making the adjacency lists doubly linked. Then if an edge is to be removed, both list nodes corresponding to that edge can be deleted quickly. Of course, all these extra links are quite cumbersome to process, and they certainly shouldn't be included unless operations like deletion are needed.

Such considerations also make it plain why we don't use a "direct" representation for graphs: a data structure which exactly models the graph, with vertices represented as allocated records and edge lists containing links to vertices instead of vertex names. How would one add an edge to a graph represented in this way?

Directed and weighted graphs are represented with similar structures. For directed graphs, everything is the same, except that each edge is represented just once: an edge from x to y is represented by a *true* value in a$[x, y]$ in the adjacency matrix or by the appearance of y on x's adjacency list in the adjacency structure. Thus an undirected graph might be thought of as a directed graph with directed edges going both ways between each pair of vertices connected by an edge. For weighted graphs, everything again is the same except that we fill the adjacency matrix with weights instead of boolean values (using some non-existent weight to represent *false*), or we include a field for the edge weight in adjacency list records in the adjacency structure.

It is often necessary to associate other information with the vertices or nodes of a graph to allow it to model more complicated objects or to save bookkeeping information in complicated algorithms. Extra information associated with each vertex can be accommodated by using auxiliary arrays indexed by vertex number (or by making *adj* an array of records in the adjacency structure representation). Extra information associated with each edge can be put in the adjacency list nodes (or in an array *a* of records in the adjacency matrix representation), or in auxiliary arrays indexed by edge number (this requires numbering the edges).

Depth-First Search

At the beginning of this chapter, we saw several natural questions that arise immediately when processing a graph. Is the graph connected? If not, what are its connected components? Does the graph have a cycle? These and many other problems can be easily solved with a technique called *depth-first search*, which is a natural way to "visit" every node and check every edge in the graph systematically. We'll see in the chapters that follow that simple variations on a generalization of this method can be used to solve a variety of graph problems.

For now, we'll concentrate on the mechanics of examining every piece of the graph in an organized way. Below is an implementation of depth-first search which fills in an array *val*[1..V] as it visits every vertex of the graph. The array is initially set to all zeros, so *val*[k]=0 indicates that vertex k has not yet been visited. The goal is to systematically visit all the vertices of the graph, setting the *val* entry for the *now*th vertex visited to *now*, for *now*= $1, 2, \ldots, V$. The program uses a recursive procedure *visit* which visits all the vertices in the same connected component as the vertex given in the argument. To *visit* a vertex, we check all its edges to see if they lead to vertices which haven't yet been visited (as indicated by 0 *val* entries); if so, we *visit* them:

```
procedure dfs;
  var now, k: integer;
      val: array [1..maxV] of integer;
  procedure visit(k: integer);
    var t: link;
    begin
    now:=now+1;  val[k]:=now;
    t:=adj[k];
    while t<>z do
      begin
      if val[t↑.v]=0 then visit(t↑.v);
      t:=t↑.next
      end
    end;
  begin
  now:=0;
  for k:=1 to V do val[k]:=0;
  for k:=1 to V do
    if val[k]=0 then visit(k)
  end;
```

First *visit* is called for the first vertex, which results in nonzero *val* values being set for all the vertices connected to that vertex. Then *dfs* scans through the *val* array to find a zero entry (corresponding to a vertex that hasn't been seen yet) and calls *visit* for that vertex, continuing in this way until all vertices have been visited.

The best way to follow the operation of depth-first search is to redraw the graph as indicated by the recursive calls during the *visit* procedure. This gives the following structure.

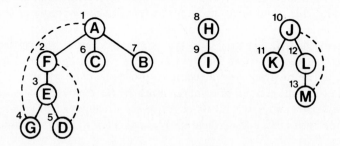

Vertices in this structure are numbered with their *val* values: the vertices are actually visited in the order A F E G D C B H I J K L M. Each connected component leads to a tree, called the *depth-first search tree*. It is important to note that this forest of depth-first search trees is simply another way of drawing the graph; all vertices and edges of the graph are examined by the algorithm.

Solid lines in the diagram indicate that the lower vertex was found by the algorithm to be on the edge list of the upper vertex and had not been visited at that time, so that a recursive call was made. Dotted lines correspond to edges to vertices which had already been visited, so the **if** test in *visit* failed, and the edge was not "followed" with a recursive call. These comments apply to the *first* time each edge is encountered; the **if** test in *visit* also guards against following the edge the *second* time that it is encountered. For example, once we've gone from A to F (on encountering F in A's adjacency list), we don't want to go back from F to A (on encountering A in F's adjacency list). Similarly, dotted links are actually checked twice: even though we checked that A was already visited while at G (on encountering A in G's adjacency list), we'll check that G was already visited later on when we're back at A (on encountering G in A's adjacency list).

A crucial property of these depth-first search trees for undirected graphs is that the dotted links always go from a node to some *ancestor* in the tree (another node in the same tree, that is higher up on the path to the root). At any point during the execution of the algorithm, the vertices divide into three classes: those for which *visit* has finished, those for which *visit* has only partially finished, and those which haven't been seen at all. By definition of *visit*, we won't encounter an edge pointing to any vertex in the first class, and if we encounter an edge to a vertex in the third class, a recursive call will be made (so the edge will be solid in the depth-first search tree). The only vertices remaining are those in the second class, which are precisely the vertices on the path from the current vertex to the root in the same tree, and any edge to any of them will correspond to a dotted link in the depth-first search tree.

The running time of *dfs* is clearly proportional to $V + E$ for any graph. We set each of the V *val* values (hence the V term), and we examine each edge twice (hence the E term).

The same method can be applied to graphs represented with adjacency matrices by using the following *visit* procedure:

```
procedure visit(k: integer);
   var t: integer;
   begin
   now:=now+1; val[k]:=now;
   for t:=1 to V do
     if a[k, t] then
        if val[t]=0 then visit(t);
   end;
```

Traveling through an adjacency list translates to scanning through a row in the adjacency matrix, looking for *true* values (which correspond to edges). As before, any edge to a vertex which hasn't been seen before is "followed" via a recursive call. Now, the edges connected to each vertex are examined in a different order, so we get a different depth-first search forest:

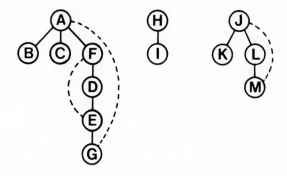

This underscores the point that the depth-first search forest is simply another representation of the graph whose particular structure depends both on the search algorithm and the internal representation used. The running time of *dfs* when this *visit* procedure is used is proportional to V^2, since every bit in the adjacency matrix is checked.

Now, testing if a graph has a cycle is a trivial modification of the above program. A graph has a cycle if and only if a nonzero *val* entry is discovered in *visit*. That is, if we encounter an edge pointing to a vertex that we've already visited, then we have a cycle. Equivalently, all the dotted links in the depth-first search trees belong to cycles.

Similarly, depth-first search finds the connected components of a graph. Each nonrecursive call to *visit* corresponds to a different connected component. An easy way to print out the connected components is to have *visit* print out

the vertex being visited (say, by inserting *write(name(k))* just before exiting),
then print out some indication that a new connected component is to start
just before the call to *visit* in *dfs* (say, by inserting two *writeln* statements).
This technique would produce the following output when *dfs* is used on the
adjacency list representation of our sample graph:

G   D   E   F   C   B   A
I   H
K   M   L   J

Note that the adjacency matrix version of *visit* will compute the same connected components (of course), but that the vertices will be printed out in a
different order.

Extensions to do more complicated processing on the connected components are straightforward. For example, by simply inserting *inval[now]=k*
after *val[k]=now* we get the "inverse" of the *val* array, whose *now*th entry
is the index of the *now*th vertex visited. (This is similar to the inverse heap
that we studied at the end of Chapter 11, though it serves a quite different
purpose.) Vertices in the same connected components are contiguous in this
array, the index of each new connected component given by the value of *now*
each time *visit* is called in *dfs*. These values could be stored, or used to mark
delimiters in *inval* (for example, the first entry in each connected component
could be made negative). The following table would be produced for our
example if the adjacency list version of *dfs* were modified in this way:

k	name(k)	val[k]	inval[k]
1	A	1	−1
2	B	7	6
3	C	6	5
4	D	5	7
5	E	3	4
6	F	2	3
7	G	4	2
8	H	8	−8
9	I	9	9
10	J	10	−10
11	K	11	11
12	L	12	12
13	M	13	13

With such techniques, a graph can be divided up into its connected components for later processing by more sophisticated algorithms.

Mazes

This systematic way of examining every vertex and edge of a graph has a distinguished history: depth-first search was first stated formally hundreds of years ago as a method for traversing mazes. For example, at left in the diagram below is a popular maze, and at right is the graph constructed by putting a vertex at each point where there is more than one path to take, then connecting the vertices according to the paths:

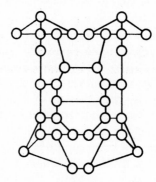

This is significantly more complicated than early English garden mazes, which were constructed as paths through tall hedges. In these mazes, all walls were connected to the outer walls, so that gentlemen and ladies could stroll in and clever ones could find their way out by simply keeping their right hand on the wall (laboratory mice have reportedly learned this trick). When independent inside walls can occur, it is necessary to resort to a more sophisticated strategy to get around in a maze, which leads to depth-first search.

To use depth-first search to get from one place to another in a maze, we use *visit*, starting at the vertex on the graph corresponding to our starting point. Each time *visit* "follows" an edge via a recursive call, we walk along the corresponding path in the maze. The trick in getting around is that we must walk *back* along the path that we used to enter each vertex when *visit* finishes for that vertex. This puts us back at the vertex one step higher up in the depth-first search tree, ready to follow its next edge.

The maze graph given above is an interesting "medium-sized" graph which the reader might be amused to use as input for some of the algorithms in later chapters. To fully capture the correspondence with the maze, a weighted version of the graph should be used, with weights on edges corresponding to distances (in the maze) between vertices.

Perspective

In the chapters that follow we'll consider a variety of graph algorithms largely aimed at determining connectivity properties of both undirected and directed graphs. These algorithms are fundamental ones for processing graphs, but are only an introduction to the subject of graph algorithms. Many interesting and useful algorithms have been developed which are beyond the scope of this book, and many interesting problems have been studied for which good algorithms have not yet been found.

Some very efficient algorithms have been developed which are much too complicated to present here. For example, it is possible to determine efficiently whether or not a graph can be drawn on the plane without any intersecting lines. This problem is called the *planarity* problem, and no efficient algorithm for solving it was known until 1974, when R. E. Tarjan developed an ingenious (but quite intricate) algorithm for solving the problem in linear time, using depth-first search.

Some graph problems which arise naturally and are easy to state seem to be quite difficult, and no good algorithms are known to solve them. For example, no efficient algorithm is known for finding the minimum-cost tour which visits each vertex in a weighted graph. This problem, called the *traveling salesman problem*, belongs to a large class of difficult problems that we'll discuss in more detail in Chapter 40. Most experts believe that no efficient algorithms exist for these problems.

Other graph problems may well have efficient algorithms, though none has been found. An example of this is the *graph isomorphism* problem: determine whether two graphs could be made identical by renaming vertices. Efficient algorithms are known for this problem for many special types of graphs, but the general problem remains open.

In short, there is a wide spectrum of problems and algorithms for dealing with graphs. We certainly can't expect to solve every problem which comes along, because even some problems which appear to be simple are still baffling the experts. But many problems which are relatively easy to solve do arise quite often, and the graph algorithms that we will study serve well in a great variety of applications.

Exercises

1. Which undirected graph representation is most appropriate for determining quickly whether a vertex is isolated (is connected to no other vertices)?

2. Suppose depth-first search is used on a binary search tree and the right edge taken before the left out of each node. In what order are the nodes visited?

3. How many bits of storage are required to represent the adjacency matrix for an undirected graph with V nodes and E edges, and how many are required for the adjacency list representation?

4. Draw a graph which cannot be written down on a piece of paper without two edges crossing.

5. Write a program to delete an edge from a graph represented with adjacency lists.

6. Write a version of *adjlist* that keeps the adjacency lists in sorted order of vertex index. Discuss the merits of this approach.

7. Draw the depth-first search forests that result for the example in the text when *dfs* scans the vertices in reverse order (from V down to 1), for both representations.

8. Exactly how many times is *visit* called in the depth-first search of an undirected graph, in terms of the number of vertices V, the number of edges E, and the number of connected components C?

9. Find the shortest path which connects all the vertices in the maze graph example, assuming each edge to be of length 1.

10. Write a program to generate a "random" graph of V vertices and E edges as follows: for each pair of integers $i < j$ between 1 and V, include an edge from i to j if and only if *randomint*($V*(V-1)$**div** 2) is less than E. Experiment to determine about how many connected components are created for $V = E = 10, 100$, and 1000.

30. Connectivity

The fundamental depth-first search procedure in the previous chapter finds the connected components of a given graph; in this section we'll examine related algorithms and problems concerning other graph connectivity properties.

As a first example of a non-trivial graph algorithm we'll look at a generalization of connectivity called *biconnectivity*. Here we are interested in knowing if there is more than one way to get from one vertex to another in the graph. A graph is *biconnected* if and only if there are at least two different paths connecting each pair of vertices. Thus even if one vertex and all the edges touching it are removed, the graph is still connected. If it is important that a graph *be* connected for some application, it might also be important that it *stay* connected. We'll look at a method for testing whether a graph is biconnected using depth-first search.

Depth-first search is certainly not the only way to traverse the nodes of a graph. Other strategies are appropriate for other problems. In particular, we'll look at *breadth-first search*, a method appropriate for finding the shortest path from a given vertex to any other vertex. This method turns out to differ from depth-first search only in the data structure used to save unfinished paths during the search. This leads to a generalized graph traversal program that encompasses not just depth-first and breadth-first search, but also classical algorithms for finding the minimum spanning tree and shortest paths in the graph, as we'll see in Chapter 31.

One particular version of the connectivity problem which arises frequently involves a dynamic situation where edges are added to the graph one by one, interspersed with queries as to whether or not two particular vertices belong to the same connected component. We'll look at an interesting family of algorithms for this problem. The problem is sometimes called the "union-find" problem, a nomenclature which comes from the application of the algorithms to processing simple operations on sets of elements.

Biconnectivity

It is sometimes reasonable to design more than one route between points on a graph, so as to handle possible failures at the connection points (vertices). For example, we can fly from Providence to Princeton even if New York is snowed in by going through Philadelphia instead. Or the main communications lines in an integrated circuit might be biconnected, so that the rest of the circuit still can function if one component fails. Another application, which is not particularly realistic but which illustrates the concept is to imagine a wartime stituation where we can make it so that an enemy must bomb at least two stations in order to cut our rail lines.

An *articulation point* in a connected graph is a vertex which, if deleted, would break the graph into two or more pieces. A graph with no articulation points is said to be *biconnected*. In a biconnected graph, there are two distinct paths connecting each pair of vertices. If a graph is not biconnected, it divides into *biconnected components*, sets of nodes mutually accessible via two distinct paths. For example, consider the following undirected graph, which is connected but not biconnected:

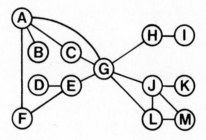

(This graph is obtained from the graph of the previous chapter by adding the edges GC, GH, JG, and LG. In our examples, we'll assume that these fours edges are added in the order given at the end of the input, so that (for example) the adjacency lists are similar to those in the example of the previous chapter with eight new entries added to the lists to reflect the four new edges.) The articulation points of this graph are A (because it connects B to the rest of the graph), H (because it connects I to the rest of the graph), J (because it connects K to the rest of the graph), and G (because the graph would fall into three pieces if G were deleted). There are six biconnected components: ACGDEF, GJLM, and the individual nodes B, H, I, and K.

Determining the articulation points turns out to be a simple extension of depth-first search. To see this, consider the depth-first search tree for this graph (adjacency list representation):

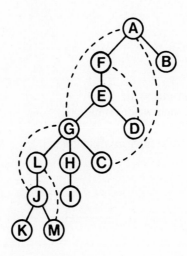

Deleting node E will not disconnect the graph because G and D both have dotted links that point above E, giving alternate paths from them to F (E's father in the tree). On the other hand, deleting G will disconnect the graph because there are no such alternate paths from L or H to E (L's father).

A vertex x is not an articulation point if every son y has some node lower in the tree connected (via a dotted link) to a node higher in the tree than x, thus providing an alternate connection from x to y. This test doesn't quite work for the root of the depth-first search tree, since there are no nodes "higher in the tree." The root is an articulation point if it has two or more sons, since the only path connecting sons of the root goes through the root. These tests are easily incorporated into depth-first search by changing the node-visit procedure into a function which returns the highest point in the tree (lowest *val* value) seen during the search, as follows:

```
function visit(k: integer): integer;
  var t: link;
      m, min: integer;
  begin
  now:=now+1; val[k]:=now; min:=now;
  t:=adj[k];
  while t<>z do
    begin
    if val[t↑.v]=0 then
      begin
      m:=visit(t↑.v);
      if m<min then min:=m;
      if m>=val[k] then write(name(k));
      end
    else if val[t↑.v]<min then min:=val[t↑.v];
    t:=t↑.next
    end
  visit:=min;
  end;
```

This procedure recursively determines the highest point in the tree reachable (via a dotted link) from any descendant of vertex k and uses this information to determine if k is an articulation point. Normally this calculation simply involves testing whether the minimum value reachable from a son is higher up in the tree, but we need an extra test to determine whether k is the root of a depth-first search tree (or, equivalently, whether this is the first call to *visit* for the connected component containing k), since we're using the same recursive program for both cases. This test is properly performed outside the recursive *visit*, so it does not appear in the code above.

The program above simply prints out the articulation points. Of course, as before, it is easily extended to do additional processing on the articulation points and biconnected components. Also, since it is a depth-first search procedure, the running time is proportional to $V + E$.

Besides the "reliability" sort of application mentioned above, biconnectedness can be helpful in decomposing large graphs into manageable pieces. It is obvious that a very large graph may be processed one connected component at a time for many applications; it is somewhat less obvious but sometimes as useful that a graph can sometimes be processed one biconnected component at a time.

Graph Traversal Algorithms

Depth-first search is a member of a family of graph traversal algorithms that are quite natural when viewed nonrecursively. Any one of these methods can be used to solve the simple connectivity problems posed in the last chapter. In this section, we'll see how one program can be used to implement graph traversal methods with quite different characteristics, merely by changing the value of one variable. This method will be used to solve several graph problems in the chapters which follow.

Consider the analogy of traveling through a maze. Any time we face a choice of several vertices to visit, we can only go along a path to one of them, so we must "save" the others to visit later. One way to implement a program based on this idea is to remove the recursion from the recursive depth-first algorithm given in the previous chapter. This would result in a program that saves, on a stack, the point in the adjacency list of the vertex being visited at which the search should resume after all vertices connected to previous vertices on the adjacency list have been visited. Instead of examining this implementation in more detail, we'll look at a more general framework which encompasses several algorithms.

We begin by thinking of the vertices as being divided into three classes: *tree* (or *visited*) vertices, those connected together by paths that we've traversed; *fringe* vertices, those adjacent to tree vertices but not yet visited; and *unseen* vertices, those that haven't been encountered at all yet. To search a connected component of a graph systematically (implement the *visit* procedure of the previous chapter), we begin with one vertex on the fringe, all others unseen, and perform the following step until all vertices have been visited: "move one vertex (call it x) from the fringe to the tree, and put any unseen vertices adjacent to x on the fringe." Graph traversal methods differ in how it is decided which vertex should be moved from the fringe to the tree.

For depth-first search, we always want to choose the vertex from the fringe that was most recently encountered. This can be implemented by always moving the first vertex on the fringe to the tree, then putting the unseen vertices adjacent to that vertex (x) at the front of the fringe and moving vertices adjacent to x which happen to be already on the fringe to the front. (Note carefully that a completely different traversal method results if we leave untouched the vertices adjacent to x which are already on the fringe.)

For example, consider the undirected graph given at the beginning of this chapter. The following table shows the contents of the fringe each time a vertex is moved to the tree; the corresponding search tree is shown at the right:

```
          A
A     G   B   C   F
G     E   C   H   J   L   B   F
E     D   F   C   H   J   L   B
D     F   C   H   J   L   B
F     C   H   J   L   B
C     H   J   L   B
H     I   J   L   B
I     J   L   B
J     M   L   K   B
M     L   K   B
L     K   B
K     B
B
```

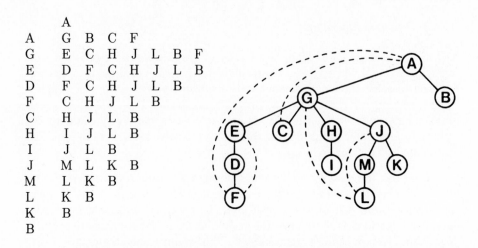

In this algorithm, the fringe essentially operates as a pushdown stack: we remove a vertex (call it x) from the beginning of the fringe, then go through x's edge list, adding unseen vertices to the beginning of the fringe, and moving fringe vertices to the beginning. Note that this is not strictly a stack, since we use the additional operation of moving a vertex to the beginning. The algorithm can be efficiently implemented by maintaining the fringe as a linked list and keeping pointers into this list in an array indexed by vertex number: we'll omit this implementation in favor of a program that can implement other traversal strategies as well.

This gives a different depth-first search tree than the one drawn in the biconnectivity section above because x's edges are put on the stack one at a time, so their order is reversed when the stack is popped. The same tree as for recursive depth-first search would result if we were to append all of x's edge list on the front of the stack, then go through and delete from the stack any other occurrences of vertices from x's edge list. The reader might find it interesting to compare this process with the result of directly removing the recursion from the *visit* procedure of the previous chapter.

By itself, the fringe table does not give enough information to reconstruct the search tree. In order to actually build the tree, it is necessary to store, with each node in the fringe table, the name of its father in the tree. This is available when the node is entered into the table (it's the node that caused the entry), it is changed whenever the node moves to the front of the fringe, and it is needed when the node is removed from the table (to determine where in the tree to attach it).

A second classic traversal method derives from maintaining the fringe as a *queue*: always pick the *least* recently encountered vertex. This can be maintained by putting the unseen vertices adjacent to x at the *end* of the fringe

in the general strategy above. This method is called *breadth-first search*: first
we visit a node, then all the nodes adjacent to it, then all the nodes adjacent
to those nodes, etc. This leads to the following fringe table and search tree
for our example graph:

	A				
A	F	C	B	G	
F	C	B	G	E	D
C	B	G	E	D	
B	G	E	D		
G	E	D	L	J	H
E	D	L	J	H	
D	L	J	H		
L	J	H	M		
J	H	M	K		
H	M	K	I		
M	K	I			
K	I				
I					

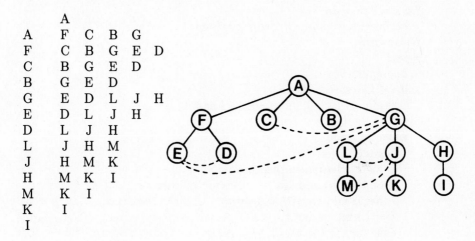

We remove a vertex (call it x) from the beginning of the fringe, then go
through x's edge list, putting unseen vertices at the end of the fringe. Again,
an efficient implementation is available using a linked list representation for
the fringe, but we'll omit this in favor of a more general method.

A fundamental feature of this general graph traversal strategy is that the
fringe is actually operating as a priority queue: the vertices can be assigned
priorities with the property that the "highest priority" vertex is the one moved
from the fringe to the tree. That is, we can directly use the priority queue
routines from Chapter 11 to implement a general graph searching program.
For the applications in the next chapter, it is convenient to assign the highest
priority to the lowest value, so we assume that the inequalities are switched
in the programs from Chapter 11. The following program is a "priority first
search" routine for a graph represented with adjacency lists (so it is most
appropriate for sparse graphs).

```
procedure sparsepfs;
  var now, k: integer;
      t: link;
  begin
  now:=0;
  for k:=1 to V do
    begin val[k]:=unseen; dad[k]:=0 end;
  pqconstruct;
  repeat
    k:=pqremove;
    if val[k]=unseen then
      begin val[k]:=0; now:=now+1 end
    t:=adj[k];
    while t<>z do
      begin
      if val[t↑.v]=unseen then now:=now+1;
      if onpq(t↑.v) and (val[t↑.v]>priority) then
        begin pqchange(t↑.v, priority); dad[t↑.v]:=k end;
      t:=t↑.next
      end
    until pqempty;
  end;
```

(The functions *onpq* and *pqempty* are priority queue utility routines which are easily implemented additions to the set of programs given in Chapter 11: *pqempty* returns *true* if the priority queue is empty; *onpq* returns *true* if the given vertex is currently on the priority queue. Below and in Chapter 31, we'll see how the substitution of various expressions for *priority* in this program yields several classical graph traversal algorithms. Specifically, the program operates as follows: first, we give all the vertices the sentinel value *unseen* (which could be *maxint*) and initialize the *dad* array, which is used to store the search tree. Next we construct an indirect priority queue containing all the vertices (this construction is trivial because the values are initially all the same). In the terminology above, tree vertices are those which are not on the priority queue, unseen vertices are those on the priority queue with value *unseen*, and fringe vertices are the others on the priority queue. With these conventions understood, the operation of the program is straightforward: it repeatedly removes the highest priority vertex from the queue and puts it on the tree, then updates the priorities of all fringe or unseen vertices connected to that vertex.

If all vertices on the priority queue are unseen, then no vertex previously encountered is connected to any vertex on the queue: that is, we're entering a new connected component. This is automatically handled by the priority queue mechanism, so there is no need for a separate *visit* procedure inside a main procedure. But note that maintaining the proper value of *now* is more complicated than for the recursive depth-first search program of the previous chapter. The convention of this program is to leave the *val* entry *unseen* and zero for the root of the depth-first search tree for each connected component: it might be more convenient to set it to zero or some other value (for example, *now*) for various applications.

Now, recall that *now* increases from 1 to V during the execution of the algorithm so it can be used to assign unique priorities to the vertices. If we change the two occurrences of *priority* in *sparsepfs* to $V-now$, we get depth-first search, because newly encountered nodes have the highest priority. If we use *now* for *priority* we get breadth-first search, because old nodes have the highest priority. These priority assignments make the priority queues operate like stacks and queues as described above. (Of course, if we were only interested in using depth-first or breadth-first search, we would use a direct implementation for stacks or queues, not priority queues as in *sparsepfs*.) In the next chapter, we'll see that other priority assignments lead to other classical graph algorithms.

The running time for graph traversal when implemented in this way depends on the method of implementing the priority queue. In general, we have to do a priority queue operation for each edge and for each vertex, so the worst case running time should be proportional to $(E+V)\log V$ if the priority queue is implemented as a heap as indicated. However, we've already noted that for both depth-first and breadth-first search we can take advantage of the fact that each new priority is the highest or the lowest so far encountered to get a running time proportional to $E+V$. Also, other priority queue implementations might sometimes be appropriate: for example if the graph is dense then we might as well simply keep the priority queue as an unordered array. This gives a worst case running time proportional to $E+V^2$ (or just V^2), since each edge simply requires setting or resetting a priority, but each vertex now requires searching through the whole queue to find the highest priority vertex. An implementation which works in this way is given in the next chapter.

The difference between depth-first and breadth-first search is quite evident when a large graph is considered. The diagram at left below shows the edges and nodes visited when depth-first search is halfway through the maze graph of the previous chapter starting at the upper left corner; the diagram at right is the corresponding picture for breadth-first search:

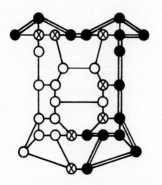 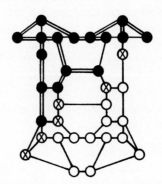

Tree nodes are blackened in these diagrams, fringe nodes are crossed, and unseen nodes are blank. Depth-first search "explores" the graph by looking for new vertices far away from the start point, taking closer vertices only when dead ends are encountered; breadth-first search completely covers the area close to the starting point, moving farther away only when everything close has been looked at. Depth-first search is appropriate for one person looking for something in a maze because the "next place to look" is always close by; breadth-first search is more like a group of people looking for something by fanning out in all directions.

Beyond these operational differences, it is interesting to reflect on the fundamental differences in the implementations of these methods. Depth-first search is very simply expressed recursively (because its underlying data structure is a stack), and breadth-first search admits to a very simple nonrecursive implementation (because its underlying data structure is a queue). But we've seen that the true underlying data structure for graph algorithms is a priority queue, and this admits a wealth of interesting properties to consider. Again, we'll see more examples in the next chapter.

Union-Find Algorithms

In some applications we wish to know simply whether a vertex x is connected to a vertex y in a graph; the actual path connecting them may not be relevant. This problem has been carefully studied in recent years; some efficient algorithms have been developed which are of independent interest because they can also be used for processing *sets* (collections of objects).

Graphs correspond to sets of objects in a natural way, with vertices corresponding to objects and edges have the meaning "is in the same set as." Thus, the sample graph in the previous chapter corresponds to the sets {A B C D E F G}, {H I} and {J K L M}. Each connected component corresponds to a different set. For sets, we're interested in the fundamental question "is x in the same set as y?" This clearly corresponds to the fundamental graph question "is vertex x connected to vertex y?"

Given a set of edges, we can build an adjacency list representation of the corresponding graph and use depth-first search to assign to each vertex the index of its connected component, so the questions of the form "is x connected to y?" can be answered with just two array accesses and a comparison. The extra twist in the methods that we consider here is that they are *dynamic*: they can accept new edges arbitrarily intermixed with questions and answer the questions correctly using the information received. From the correspondence with the set problem, the addition of a new edge is called a *union* operation, and the queries are called *find* operations.

Our objective is to write a function which can check if two vertices x and y are in the same set (or, in the graph representation, the same connected component) and, if not, can put them in the same set (put an edge between them in the graph). Instead of building a direct adjacency list or other representation of the graph, we'll gain efficiency by using an internal structure specifically oriented towards supporting the *union* and *find* operations. The internal structure that we will use is a forest of trees, one for each connected component. We need to be able to find out if two vertices belong to the same tree and to be able to combine two trees to make one. It turns out that both of these operations can be implemented efficiently.

To illustrate the way this algorithm works, we'll look at the forest constructed when the edges from the sample graph that we've been using in this chapter are processed in some arbitrary order. Initially, all nodes are in separate trees. Then the edge AG causes a two node tree to be formed, with A at the root. (This choice is arbitrary —we could equally well have put G at the root.) The edges AB and AC add B and C to this tree in the same way, leaving

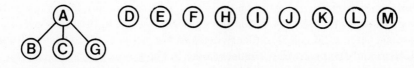

The edges LM, JM, JL, and JK build a tree containing J, K, L, and M that has a slightly different structure (note that JL doesn't contribute anything, since LM and JM put L and J in the same component), and the edges ED, FD, and HI build two more trees, leaving the forest:

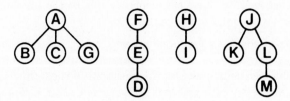

This forest indicates that the edges processed to this point describe a graph with four connected components, or, equivalently, that the set union operations processed to this point have led to four sets {A B C G}, {J K L M}, {D E F} and {H I}. Now the edge FE doesn't contribute anything to the structure, since F and E are in the same component, but the edge AF combines the first two trees; then GC doesn't contribute anything, but GH and JG result in everything being combined into one tree:

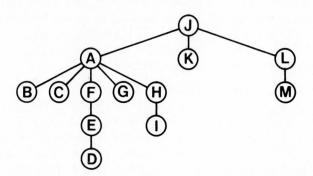

It must be emphasized that, unlike depth-first search trees, the only relationship between these union-find trees and the underlying graph with the given edges is that they divide the vertices into sets in the same way. For example, there is no correspondence between the paths that connect nodes in the trees and the paths that connect nodes in the graph.

We know that we can represent these trees in exactly the same way that we represented graph search trees: we keep an array of integers $dad[1..V]$ which contains, for each vertex, the index of its father (with a 0 entry for nodes which are at the root of a tree). To find the father of a vertex j, we simply set $j:=dad[j]$, and to find the root of the tree to which j belongs, we repeat this operation until reaching 0. The *union* and *find* operations are then very easily implemented:

```
function find(x, y: integer; union: boolean): boolean;
  var i, j: integer;
  begin
  i:=x;  while dad[i]>0 do i:=dad[i];
  j:=y;  while dad[j]>0 do j:=dad[j];
  if union and (i<>j) then dad[j]:=i;
  find:=(i<>j)
  end;
```

This function returns *true* if the two given vertices are in the same component. In addition, if they are not in the same component and the *union* flag is set, they are put into the same component. The method used is simple: Use the *dad* array to get to the root of the tree containing each vertex, then check to see if they are the same. To merge the tree rooted at *j* with the tree rooted at *i*, we simply set $dad[j]=i$ as shown in the following table:

```
     A B C D E F G H I J K L M
AG:              A
AB:  A           A
AC:  A A         A
LM:  A A         A         L
JM:  A A         A       J L
JL:  A A         A       J L *
JK:  A A         A     J J L
ED:  A A E       A     J J L
FD:  A A E F     A     J J L
HI:  A A E F     A   H J J L
FE:  A A E F     A   H J J L *
AF:  A A E F A A     H J J L
GE:  A A E F A A     H J J L *
GC:  A A E F A A     H J J L *
GH:  A A E F A A A A H J J L
JG:  J A A E F A A A A H J J L
LG:  J A A E F A A A A H J J L *
```

An asterisk at the right indicates that the vertices are already in the same component at the time the edge is processed. As usual, we are assuming that we have available functions *index* and *name* to translate between vertex names and integers between 1 and V: each table entry is the name of the corresponding *dad* array entry. Also, for example, the function call *find*($index(x), index(y), false$) would be used to test whether a vertex named x is in the same component as a vertex named y (without introducing an edge between them).

The algorithm described above has bad worst-case performance because the trees formed could be degenerate. For example, taking the edges AB BC CD DE EF FG GH HI IJ ... YZ in that order will produce a long chain with Z pointing to Y, Y pointing to X, etc. This kind of structure takes time proportional to V^2 to build, and has time proportional to V for an average equivalence test.

Several methods have been suggested to deal with this problem. One natural method, which may have already occurred to the reader, is to try to do the "right" thing when merging two trees rather than arbitrarily setting $dad[j]=i$. When a tree rooted at i is to be merged with a tree rooted at j, one of the nodes must remain a root and the other (and all its descendants) must go one level down in the tree. To minimize the distance to the root for the most nodes, it makes sense to take as the root the node with more descendants. This idea, called *weight balancing*, is easily implemented by maintaining the size of each tree (number of descendants of the root) in the *dad* array entry for each root node, encoded as a nonpositive number so that the root node can be detected when traveling up the tree in *find*.

Ideally, we would like every node to point directly to the root of its tree. No matter what strategy we use, achieving this ideal would require examining at least all the nodes in the smaller of the two trees to be merged, and this could be quite a lot compared to the relatively few nodes on the path to the root that *find* usually examines. But we can approach the ideal by making all the nodes that we do examine point to the root! This seems like a drastic step at first blush, but it is relatively easy to do, and it must be remembered that there is nothing sacrosanct about the structure of these trees: if they can be modified to make the algorithm more efficient, we should do so. This method, called *path compression*, is easily implemented by making another pass through each tree after the root has been found, and setting the *dad* entry of each vertex encountered along the way to point to the root.

The combination of weight balancing and path compression ensures that the algorithms will run very quickly. The following implementation shows that the extra code involved is a small price to pay to guard against degenerate cases.

```
function fastfind(x, y: integer; union: boolean): boolean;
var i, j, t: integer;
begin
i:=x; while dad[i]>0 do i:=dad[i];
j:=y; while dad[j]>0 do j:=dad[j];
while dad[x]>0 do
   begin t:=x; x:=dad[x]; dad[t]:=i end;
while dad[y]>0 do
   begin t:=y; y:=dad[y]; dad[t]:=j end;
if union and (i<>j) then
   if dad[j]<dad[i]
      then begin dad[j]:=dad[j]+dad[i]−1; dad[i]:=j end
      else  begin dad[i]:=dad[i]+dad[j]−1; dad[j]:=i end;
fastfind:=(i<>j)
end;
```

The *dad* array is assumed to be initialized to 0. (We'll assume in later chapters that this is done in a separate procedure *findinit*.) For our sample set of edges, this program builds the following trees from the first ten edges:

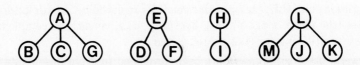

The trees containing DEF and JKLM are flatter than before because of path compression. Now FE doesn't contribute anything, as before, then AF combines the first two trees (with A at the root, since its tree is larger), the GC doesn't contribute anything, then GH puts the HI tree below A, then JG puts the JKLM tree below A, resulting in the following tree:

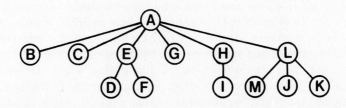

The following table gives the contents of the *dad* array as this forest is constructed:

		A	B	C	D	E	F	G	H	I	J	K	L	M	
AG:	1	0	0	0	0	0	A	0	0	0	0	0	0		
AB:	2	A	0	0	0	0	A	0	0	0	0	0	0		
AC:	3	A	A	0	0	0	A	0	0	0	0	0	0		
LM:	3	A	A	0	0	0	A	0	0	0	0	1	L		
JM:	3	A	A	0	0	0	A	0	0	L	0	2	L		
JL:	3	A	A	0	0	0	A	0	0	L	0	2	L	*	
JK:	3	A	A	0	0	0	A	0	0	L	L	3	L		
ED:	3	A	A	E	1	0	A	0	0	L	L	3	L		
FD:	3	A	A	E	2	E	A	0	0	L	L	3	L		
HI:	3	A	A	E	2	E	A	1	H	L	L	3	L		
FE:	3	A	A	E	2	E	A	1	H	L	L	3	L	*	
AF:	6	A	A	E	A	E	A	1	H	L	L	3	L		
GE:	6	A	A	E	A	E	A	1	H	L	L	3	L	*	
GC:	6	A	A	E	A	E	A	1	H	L	L	3	L	*	
GH:	8	A	A	E	A	E	A	A	H	L	L	3	L		
JG:	12	A	A	E	A	E	A	A	H	L	L	A	L		
LG:	12	A	A	E	A	E	A	A	H	L	L	A	L	*	

For clarity in this table, each positive entry i is replaced by the ith letter of the alphabet (the name of the father), and each negative entry is complemented to give a positive integer (the weight of the tree).

Several other techniques have been developed to avoid degenerate structures. For example, path compression has the disadvantage that it requires another pass up through the tree. Another technique, called *halving*, is to make each node point to its granddad on the way up the tree. Still another technique, *splitting*, is like halving, but is applied only to every other node on the search path. Either of these can be used in combination with weight balancing or with *height balancing*, which is similar but uses tree height instead of tree size to decide which way to merge trees.

How is one to choose from among all these methods? And exactly how "flat" are the trees produced? Analysis for this problem is quite difficult because the performance depends not only on the V and E parameters, but also on the number of *find* operations and, what's worse, on the order in which the *union* and *find* operations appear. Unlike sorting, where the actual files that appear in practice are quite often close to "random," it's hard to see how to model graphs and request patterns that might appear in practice. For this reason, algorithms which do well in the worst case are normally preferred for union-find (and other graph algorithms), though this may be an overly conservative approach.

Even if only the worst case is being considered, the analysis of union-find algorithms is extremely complex and intricate. This can be seen even from the nature of the results, which do give us clear indications of how the algorithms will perform in a practical situation. If either weight balancing or height balancing is used in combination with either path compression, halving, or splitting, then the total number of operations required to build up a structure with E edges is proportional to $E\alpha(E)$, where $\alpha(E)$ is a function that is so slowly growing that $\alpha(E) < 4$ unless E is so large that taking $\lg E$, then taking $\lg$ of the result, then taking $\lg$ of that result, and continuing 16 times still gives a number bigger than 1. This is a stunningly large number; for all practical purposes, it is safe to assume that the average amount of time to execute each *union* and *find* operation is constant. This result is due to R. E. Tarjan, who further showed that *no* algorithm for this problem (from a certain general class) can do better that $E\alpha(E)$, so that this function is intrinsic to the problem.

An important practical application of union-find algorithms is that they can be used to determine whether a graph with V vertices and E edges is connected in space proportional to V (and almost linear time). This is an advantage over depth-first search in some situations: here we don't need to ever store the edges. Thus connectivity for a graph with thousands of vertices and millions of edges can be determined with one quick pass through the edges.

Exercises

1. Give the articulation points and the biconnected components of the graph formed by deleting GJ and adding IK to our sample graph.

2. Write a program to print out the biconnected components of a graph.

3. Give adjacency lists for one graph where breadth-first search would find a cycle before depth first search would, and another graph where depth-first search would find the cycle first.

4. Draw the search tree that results if, in the depth-first search we ignore nodes already on the fringe (as in breadth-first search).

5. Draw the search tree that results if, in the breadth-first search we change the priority of nodes already on the fringe (as in depth-first search).

6. Draw the union-find forest constructed for the example in the text, but assuming that *find* is changed to set $a[i]=j$ rather than $a[j]=i$.

7. Solve the previous problem, assuming further that path compression is used.

8. Draw the union-find forests constructed for the edges AB BC CD DE EF ... YZ, assuming first that weight balancing without path compression is used, then that path compression without weight balancing is used, then that both are used.

9. Implement the union-find variants described in the text, and empirically determine their comparative performance for 1000 *union* operations with both arguments random integers between 1 and 100.

10. Write a program to generate a random connected graph on V vertices by generating random pairs of integers between 1 and V. Estimate how many edges are needed to produce a connected graph as a function of V.

31. Weighted Graphs

It is often necessary to model practical problems using graphs in which *weights* or *costs* are associated with each edge. In an airline map where edges represent flight routes, these weights might represent distances or fares. In an electric circuit where edges represent wires, the length or cost of the wire are natural weights to use. In a job-scheduling chart, weights could represent time or cost of performing tasks or of waiting for tasks to be performed.

Questions entailing minimizing costs naturally arise for such situations. In this chapter, we'll examine algorithms for two such problems in detail: "find the lowest-cost way to connect all of the points," and "find the lowest-cost path between two given points." The first, which is obviously useful for graphs representing something like an electric circuit, is called the *minimum spanning tree* problem; the second, which is obviously useful for graphs representing something like an airline route map, is called the *shortest path* problem. These problems are representative of a variety of problems that arise on weighted graphs.

Our algorithms involve searching through the graph, and sometimes our intuition is supported by thinking of the weights as distances: we speak of "the closest vertex to x," etc. In fact, this bias is built into the nomenclature for the shortest path problem. Despite this, it is important to remember that the weights need not be proportional to any distance at all; they might represent time or cost or something else entirely different. When the weights actually *do* represent distances, other algorithms may be appropriate. This issue is discussed in further detail at the end of the chapter.

A typical weighted undirected graph is diagramed below, with edges comprising a minimum spanning tree drawn with double lines. Note that the shortest paths in the graph do not necessarily use edges of the minimum spanning tree: for example, the shortest path from vertex A to vertex G is AFEG.

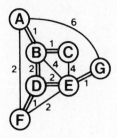

It is obvious how to represent weighted graphs: in the adjacency matrix representation, the matrix can contain edge weights rather than boolean values, and in the adjacency structure representation, each list element (which represents an edge) can contain a weight.

We'll start by assuming that all of the weights are positive. Some of the algorithms can be adapted to handle negative weights, but they become significantly more complicated. In other cases, negative weights change the nature of the problem in an essential way, and require far more sophisticated algorithms than those considered here. For an example of the type of difficulty that can arise, suppose that we have a situation where the sum of the weights of the edges around a cycle is negative: an infinitely short path could be generated by simply spinning around the cycle.

Minimum Spanning Tree

A *minimum spanning tree* of a weighted graph is a collection of edges that connects all the vertices such that the sum of the weights of the edges is at least as small as the sum of the weights of any other collection of edges that connects all the vertices. The minimum spanning tree need not be unique: for example, the following diagram shows three other minimum spanning trees for our sample graph.

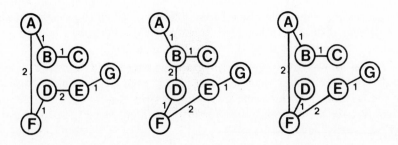

It's easy to prove that the "collection of edges" referred to in the definition above must form a spanning tree: if there's any cycle, some edge in the cycle can be deleted to give a collection of edges which still connects the vertices but has a smaller weight.

We've seen in previous chapters that many graph traversal procedures compute a spanning tree for the graph. How can we arrange things for a weighted graph so that the tree computed is the one with the lowest total weight? The answer is simple: always visit next the vertex which can be connected to the tree using the edge of lowest weight. The following sequence of diagrams illustrates the sequence in which the edges are visited when this strategy is used for our example graph.

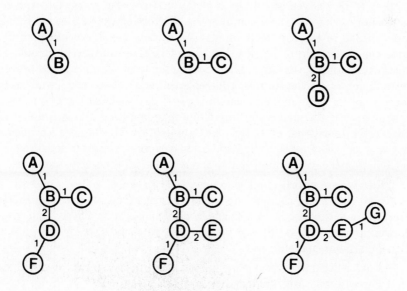

The implementation of this strategy is a trivial application of the priority graph search procedure in the previous chapter: we simply add a weight field to the *edge* record (and modify the input code to read in weights as well), then use $t{\uparrow}.weight$ for *priority* in that program. Thus we always visit next the vertex in the fringe which is closest to the tree. The traversal is diagramed as above for comparison with a completely different method that we'll examine below; we can also redraw the graph in our standard search tree format:

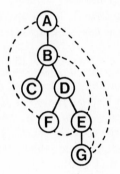

This method is based on the following fundamental property of minimum spanning trees: "Given any division of the vertices of a graph into two sets, the minimum spanning tree contains the shortest of the edges connecting a vertex in one of the sets to a vertex in the other set." For example if we divide the vertices into the sets ABCD and EFG in our sample graph, this says that DF must be in any minimum spanning tree. This is easy to prove by contradiction. Call the shortest edge connecting the two sets s, and assume that s is not in the minimum spanning tree. Then consider the graph formed by adding s to the purported minimum spanning tree. This graph has a cycle; furthermore, that cycle must have some other edge besides s connecting the two sets. Deleting this edge and adding s gives a shorter spanning tree, contradicting the assumption that s is not in the minimum spanning tree.

When we use priority-first searching, the two sets of nodes in question are the visited nodes and the unvisited ones. At each step, we pick the shortest edge from a visited node to a fringe node (there are no edges from visited nodes to unseen nodes). By the property above every edge that is picked is on the minimum spanning tree.

As described in the previous chapter, the priority graph traversal algorithm has a worst-case running time proportional to $(E + V)\log V$, though a different implementation of the priority queue can give a V^2 algorithm, which is appropriate for dense graphs. Later in this chapter, we'll examine this implementation of the priority graph traversal for dense graphs in full detail. For minimum spanning trees, this reduces to a method discovered by R. Prim in 1956 (and independently by E. Dijkstra soon thereafter). Though the methods are the same in essence (just the graph representation and implementation of priority queues differ), we'll refer to the *sparsepfs* program of the previous chapter with *priority* replaced by $t\uparrow.weight$ as the "priority-first search solution" to the minimum spanning tree problem and the adjacency matrix version given later in this chapter (for dense graphs) as "Prim's algorithm." Note that Prim's algorithm takes time proportional to V^2 even for sparse graphs (a factor of about $V^2/E\log V$ slower than the priority-first search solution, and that the priority-first search solution is a factor of $\log V$ slower than Prim's algorithm for dense graphs.

A completely different approach to finding the minimum spanning tree is to simply add edges one at a time, at each step using the shortest edge that does not form a cycle. This algorithm gradually builds up the tree one edge at a time from disconnected components, as illustrated in the following sequence of diagrams for our sample graph:

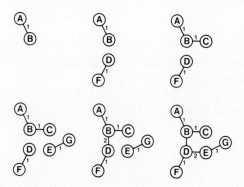

The correctness of this algorithm also follows from the general property of minimum spanning trees that is proved above.

The code for this method can be pieced together from programs that we've already seen. A priority queue is obviously the data structure to use to consider the edges in order of their weight, and the job of testing for cycles can be obviously done with union-find structures. The appropriate data structure to use for the graph is simply an array *edge* with one entry for each edge. The indirect priority queue procedures *pqbuild* and *pqremove* from Chapter 11 can be used to maintain the priority queue, using the *weight* fields in the *edge* array for priorities. Also, the program uses the *findinit* and *fastfind* procedures from Chapter 30. The program simply prints out the edges which comprise the spanning tree; with slightly more work a *dad* array or other representation could be computed:

```
program kruskal(input, output);
const maxV=50; maxE=2500;
type edge=record x, y, weight: integer end;
var i, j, m, x, y, V, E: integer;
    edges: array [0..maxE] of edge;
begin
readln(V, E);
for j:=1 to E do
  begin
  readln(c, d, edges[j].weight);
  edges[j].x:=index(c);
  edges[j].y:=index(d);
  end;
findinit; pqconstruct  i:=0;
repeat
  m:=pqremove; x:=edges[m].x; y:=edges[m].y;
  if not fastfind(x, y, true) then
    begin
    writeln(name(x), name(y), edges[m].weight);
    i:=i+1
    end
until i=V−1;
end.
```

The running time of this program is dominated by the time spent processing edges in the priority queue. Suppose that the graph consists of two clusters of vertices all connected together by very short edges, and only one edge which is very long connecting the two clusters. Then the longest edge in the graph is in the minimum spanning tree, but it will be the last edge out of the priority queue. This shows that the running time could be proportional to $E \log E$ in the worst case, although we might expect it to be much smaller for typical graphs (though it always takes time proportional to E to build the priority queue initially).

An alternate implementation of the same strategy is to sort the edges by weight initially, then simply process them in order. Also, the cycle testing can be done in time proportional to $E \log E$ with a much simpler strategy than union-find, to give a minimum spanning tree algorithm that always takes $E \log E$ steps. This method was proposed by J. Kruskal in 1956, even earlier than Prim's algorithm. We'll refer to the modernized version above, which uses priority queues and union-find structures, as "Kruskal's algorithm."

The performance characteristics of these three methods indicate that the priority-first search method will be faster for some graphs, Prim's for some others, Kruskal's for still others. As mentioned above, the worst case for the priority-first search method is $(E + V)\log V$ while the worst case for Prim's is V^2 and the worst case for Kruskal's is $E\log E$. But it is unwise to choose between the algorithms on the basis of these formulas because "worst-case" graphs are unlikely to occur in practice. In fact, the priority-first search method and Kruskal's method are both likely to run in time proportional to E for graphs that arise in practice: the first because most edges do not really require a priority queue adjustment that takes $\log V$ steps and the second because the longest edge in the minimum spanning tree is probably sufficiently short that not many edges are taken off the priority queue. Of course, Prim's method also runs in time proportional to about E for dense graphs (but it shouldn't be used for sparse graphs).

Shortest Path

The *shortest path* problem is to find the path in a weighted graph connecting two given vertices x and y with the property that the sum of the weights of all the edges is minimized over all such paths.

If the weights are all 1, then the problem is still interesting: it is to find the path containing the minimum number of edges which connects x and y. Moreover, we've already considered an algorithm which solves the problem: breadth-first search. It is easy to prove by induction that breadth-first search starting at x will first visit all vertices which can be reached from x with 1 edge, then all vertices which can be reached from x with 2 edges, etc., visiting all vertices which can be reached with k edges before encountering any that require $k + 1$ edges. Thus, when y is first encountered, the shortest path from x has been found (because no shorter paths reached y).

In general, the path from x to y could touch all the vertices, so we usually consider the problem of finding the shortest paths connecting a given vertex x with *each* of the other vertices in the graph. Again, it turns out that the problem is simple to solve with the priority graph traversal algorithm of the previous chapter.

If we draw the shortest path from x to each other vertex in the graph, then we clearly get no cycles, and we have a spanning tree. Each vertex leads to a different spanning tree; for example, the following three diagrams show the shortest path spanning trees for vertices A, B, and E in the example graph that we've been using.

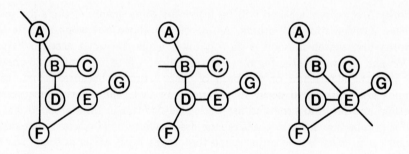

The priority-first search solution to this problem is very similar to the solution for the minimum spanning tree: we build the tree for vertex x by adding, at each step, the vertex on the fringe which is closest to x (before we added the one closest to the *tree*). To find which fringe vertex is closest to x, we use the *val* array: for each tree vertex k, *val*$[k]$ will be the distance from that vertex to x, using the shortest path (which must be comprised of tree nodes). For each vertex on the fringe, we keep track of the shortest way to get to x by going to some tree vertex k, then going a distance *val*$[k]$ to the root. The closest fringe vertex to x will be the one for which this distance is smallest. With this observation, the algorithm is trivially implemented by taking *val*$[k]+t\uparrow$.*weight* for *priority* in the priority graph traversal program. This is because *val*$[k]$ is the shortest path to k during the execution of the algorithm, and $t\uparrow$.*weight* is the distance from k to t, so their sum is the best path found so far to t. The following sequence of diagrams shows the construction of the shortest path search tree for vertex A in our example.

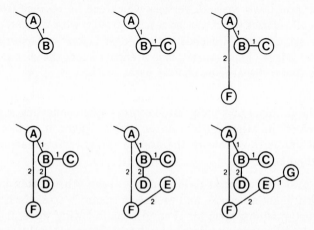

First we visit the closest vertex to A, which is B. Then both C and F are distance 2 from A, so we visit them next (in whatever order the priority queue returns them). Then D can be attached at F or at B to get a path of distance 3 to A. (The algorithm attaches it to B because it was put on the tree before F, so D was already on the fringe when F was put on the tree and F didn't provide a shorter path to A.) Finally, E and G are visited. As usual, the tree is represented by the *dad* array of father links. The following table shows the array computed by the priority graph traversal procedure for our example:

	A	B	C	D	E	F	G
dad:		A	B	B	F	A	E
val:	0	1	2	3	4	2	5

Thus the shortest path from A to G has a total weight of 5 (found in the *val* entry for G) and goes from A to F to E to G (found by tracing backwards in the *dad* array, starting at G). Note that the correct operation of this program depends on the *val* entry for the root being zero, the convention that we adopted for *sparsepfs*.

As before, the priority graph traversal algorithm has a worst-case running time proportional to $(E + V) \log V$, though a different implementation of the priority queue can give a V^2 algorithm, which is appropriate for dense graphs. Below, we'll examine this implementation of priority graph traversal for dense graphs in full detail. For the shortest path problem, this reduces to a method discovered by E. Dijkstra in 1956. Though the methods are the same in essence, we'll refer to the *sparsepfs* program of the previous chapter with *priority* replaced by $val[k] + t\uparrow.weight$ as the "priority-first search solution" to the shortest paths problem and the adjacency matrix version given below as "Dijkstra's algorithm."

Dense Graphs

As we've discussed, when a graph is represented with a adjacency matrix, it is best to use an unordered array representation for the priority queue in order to achieve a V^2 running time for any priority graph traversal algorithm. That is, this provides a linear algorithm for the priority first search (and thus the minimum spanning tree and shortest path problems) for dense graphs.

Specifically, we maintain the priority queue in the *val* array just as in *sparsepfs* but we implement the priority queue operations directly rather than using heaps. First, we adopt the convention that the priority values in the *val* array will be negated, so that the sign of a *val* entry tells whether the corresponding vertex is on the tree or the priority queue. To change the

priority of a vertex, we simply assign the new priority to the *val* entry for
that vertex. To remove the highest priority vertex, we simply scan through
the *val* array to find the vertex with the largest negative (closest to 0) *val* value
(then complement its *val* entry). After making these mechanical changes to
the *sparsepfs* program of the previous chapter, we are left with the following
compact program.

```
procedure densepfs;
  var k, min, t: integer;
  begin
  for k:=1 to V do
    begin val[k]:=unseen; dad[k]:=0 end;
  val[0]:=-(unseen+1);
  min:=1;
  repeat
    k:=min; val[k]:=-val[k]; min:=0;
    if val[k]=unseen then val[k]:=0;
    for t:=1 to V do
      if val[t]<0 then
        begin
        if (a[k,t]<>0) and (val[t]<-priority) then
          begin val[k]:=-priority; dad[t]:=k end;
        if val[t]>val[min] then min:=t;
        end
    until min=0;
  end;
```

Note that, the loop to update the priorities and the loop to find the minimum
are combined: each time we remove a vertex from the fringe, we pass through
all the vertices, updating their priority if necessary, and keeping track of the
minimum value found. (Also, note that *unseen* must be slightly less than
maxint since a value one higher is used as a sentinel to find the minimum,
and the negative of this value must be representable.)

If we use $a[k, t]$ for *priority* in this program, we get Prim's algorithm
for finding the minimum spanning tree; if we use $val[k]+a[k, t]$ for *priority*
we get Dijkstra's algorithm for the shortest path problem. As in Chapter
30, if we include the code to maintain *now* as the number of vertices so far
searched and use $V-now$ for *priority*, we get depth-first search; if we use *now*
we get breadth-first search. This program differs from the *sparsepfs* program
of Chapter 30 only in the graph representation used (adjacency matrix instead
of adjacency list) and the priority queue implementation (unordered array

instead of indirect heap). These changes yield a worst-case running time proportional to V^2, as opposed to $(E + V) \log V$ for *sparsepfs*. That is, the running time is linear for dense graphs (when E is proportional to V^2), but *sparsepfs* is likely to be much faster for sparse graphs.

Geometric Problems

Suppose that we are given N points in the plane and we want to find the shortest set of lines connecting all the points. This is a geometric problem, called the *Euclidean minimum spanning tree* problem. It can be solved using the graph algorithm given above, but it seems clear that the geometry provides enough extra structure to allow much more efficient algorithms to be developed.

The way to solve the Euclidean problem using the algorithm given above is to build a complete graph with N vertices and $N(N - 1)/2$ edges, one edge connecting each pair of vertices weighted with the distance between the corresponding points. Then the minimum spanning tree can be found with the algorithm above for dense graphs in time proportional to N^2.

It has been proven that it is possible to do better. The point is that the geometric structure makes most of the edges in the complete graph irrelevant to the problem, and we can eliminate most of the edges before even starting to construct the minimum spanning tree. In fact, it has been proven that the minimum spanning tree is a subset of the graph derived by taking only the edges from the dual of the Voronoi diagram (see Chapter 28). We know that this graph has a number of edges proportional to N, and both Kruskal's algorithm and the priority-first search method work efficiently on such sparse graphs. In principle, then, we could compute the Voronoi dual (which takes time proportional to $N \log N$), then run either Kruskal's algorithm or the priority-first search method to get a Euclidean minimum spanning tree algorithm which runs in time proportional to $N \log N$. But writing a program to compute the Voronoi dual is quite a challenge even for an experienced programmer.

Another approach which can be used for random point sets is to take advantage of the distribution of the points to limit the number of edges included in the graph, as in the grid method used in Chapter 26 for range searching. If we divide up the plane into squares such that each square is likely to contain about 5 points, and then include in the graph only the edges connecting each point to the points in the neighboring squares, then we are very likely (though not guaranteed) to get all the edges in the minimum spanning tree, which would mean that Kruskal's algorithm or the priority-first search method would efficiently finish the job.

It is interesting to reflect on the relationship between graph and geometric algorithms brought out by the problem posed in the previous paragraphs. It is certainly true that many problems can be formulated either as geometric problems or as graph problems. If the actual physical placement of objects is a dominating characteristic, then the geometric algorithms of the previous section may be appropriate, but if interconnections between objects are of fundamental importance, then the graph algorithms of this section may be better. The Euclidean minimum spanning tree seems to fall at the interface between these two approaches (the input involves geometry and the output involves interconnections) and the development of simple, straightforward methods for this and related problems remains an important though elusive goal.

Another example of the interaction between geometric and graph algorithms is the problem of finding the shortest path from x to y in a graph whose vertices are points in the plane and whose edges are lines connecting the points. For example, the maze graph at the end of Chapter 29 might be viewed as such a graph. The solution to this problem is simple: use priority first searching, setting the priority of each fringe vertex encountered to the distance in the tree from x to the fringe vertex (as in the algorithm given) *plus* the Euclidean distance from the fringe vertex to y. Then we stop when y is added to the tree. This method will very quickly find the shortest path from x to y by always going towards y, while the standard graph algorithm has to "search" for y. Going from one corner to another of a large maze graph like that one at the end of Chapter 29 might require examining a number of nodes proportional to $\sqrt{V}$, while the standard algorithm has to examine virtually all the nodes.

Exercises

1. Give another minimum spanning tree for the example graph at the beginning of the chapter.

2. Give an algorithm to find the *minimum spanning forest* of a connected graph (each vertex must be touched by some edge, but the resulting graph doesn't have to be connected).

3. Is there a graph with V vertices and E edges for which the priority-first solution to the minimum spanning tree problem algorithm could require time proportional to $(E + V) \log V$? Give an example or explain your answer.

4. Suppose we maintained the priority queue as a sorted list in the general graph traversal implementations. What would be the worst-case running time, to within a constant factor? When would this method be appropriate, if at all?

5. Give counterexamples which show why the following "greedy" strategy doesn't work for either the shortest path or the minimum spanning tree problems: "at each step visit the unvisited vertex closest to the one just visited."

6. Give the shortest path trees for the other nodes in the example graph.

7. Find the shortest path from the upper right-hand corner to the lower left-hand corner in the maze graph of Chapter 29, assuming all edges have weight 1.

8. Write a program to generate random connected graphs with V vertices, then find the minimum spanning tree and shortest path tree for some vertex. Use random weights between 1 and V. How do the weights of the trees compare for different values of V?

9. Write a program to generate random complete weighted graphs with V vertices by simply filling in an adjacency matrix with random numbers between 1 and V. Run empirical tests to determine which method finds the minimum spanning tree faster for $V = 10, 25, 100$: Prim's or Kruskal's.

10. Give a counterexample to show why the following method for finding the Euclidean minimum spanning tree doesn't work: "Sort the points on their x coordinates, then find the minimum spanning trees of the first half and the second half, then find the shortest edge that connects them."

32. Directed Graphs

Directed graphs are graphs in which edges connecting nodes are one-way; this added structure makes it more difficult to determine various properties. Processing such graphs is akin to traveling around in a city with many one-way streets or to traveling around in a country where airlines rarely run round-trip routes: getting from one point to another in such situations can be a challenge indeed.

Often the edge direction reflects some type of precedence relationship in the application being modeled. For example, a directed graph might be used to model a manufacturing line, with nodes corresponding to jobs to be done and with an edge from node x to node y if the job corresponding to node x must be done before the job corresponding to node y. How do we decide when to perform each of the jobs so that none of these precedence relationships are violated?

In this chapter, we'll look at depth-first search for directed graphs, as well as algorithms for computing the *transitive closure* (which summarizes connectivity information) and for *topological sorting* and for computing *strongly connected components* (which have to do with precedence relationships).

As mentioned in Chapter 29, representations for directed graphs are simple extensions of representations for undirected graphs. In the adjacency list representation, each edge appears only once: the edge from x to y is represented as a list node containing y in the linked list corresponding to x. In the adjacency matrix representation, we need to maintain a full V-by-V matrix, with a 1 bit in row x and column y (but not necessarily in row y and column x) if there is an edge from x to y.

A directed graph similar to the undirected graph that we've been considering is drawn below. This graph consists of the edges AG AB CA LM JM JL JK ED DF HI FE AF GE GC HG GJ LG IH ML.

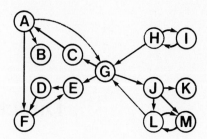

The order in which the edges appear is now significant: the notation AG describes an edge which points from A to G, but *not* from G to A. But it is possible to have two edges between two nodes, one in either direction (we have both HI and IH and both LM and ML in the above graph).

Note that, in these representations, no difference could be perceived between an undirected graph and a directed graph with two opposite directed edges for each edge in the undirected graph. Thus, some of algorithms in this chapter can be considered generalizations of algorithms in previous chapters.

Depth-First Search

The depth-first search algorithm of Chapter 29 works properly for directed graphs exactly as given. In fact, its operation is a little more straightforward than for undirected graphs because we don't have to be concerned with double edges between nodes unless they're explicitly included in the graph. However, the search trees have a somewhat more complicated structure. For example, the following depth-first search structure describes the operation of the recursive algorithm of Chapter 29 on our sample graph.

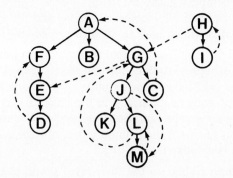

As before, this is a redrawn version of the graph, with solid edges corresponding to those edges that were actually used to visit vertices via recursive calls and dotted edges corresponding to those edges pointing to vertices that had already been visited at the time the edge was considered. The nodes are visited in the order A F E D B G J K L M C H I.

Note that the directions on the edges make this depth-first search forest quite different from the depth-first search forests that we saw for undirected graphs. For example, even though the original graph was connected, the depth-first search structure defined by the solid edges is not connected: it is a forest, not a tree.

For undirected graphs, we had only one kind of dotted edge, one that connected a vertex with some ancestor in the tree. For directed graphs, there are three kinds of dotted edges: *up* edges, which point from a vertex to some ancestor in the tree, *down* edges, which point from a vertex to some descendant in the tree, and *cross* edges, which point from a vertex to some vertex which is neither a descendant nor an ancestor in the tree.

As with undirected graphs, we're interested in connectivity properties of directed graphs. We would like to be able to answer questions like "Is there a *directed path* from vertex x to vertex y (a path which only follows edges in the indicated direction)?" and "Which vertices can we get to from vertex x with a directed path?" and "Is there a directed path from vertex x to vertex y *and* a directed path from y to x?" Just as with undirected graphs, we'll be able to answer such questions by appropriately modifying the basic depth-first search algorithm, though the various different types of dotted edges make the modifications somewhat more complicated.

Transitive Closure

In undirected graphs, simple connectivity gives the vertices that can be reached from a given vertex by traversing edges from the graph: they are all those in the same connected component. Similarly, for directed graphs, we're often interested in the set of vertices which can be reached from a given vertex by traversing edges from the graph in the indicated direction.

It is easy to prove that the recursive *visit* procedure from the depth-first search method in Chapter 29 visits all the nodes that can be reached from the start node. Thus, if we modify that procedure to print out the nodes that it is visiting (say, by inserting *write(name(k))* just upon entering), we are printing out all the nodes that can be reached from the start node. But note carefully that it is *not* necessarily true that each tree in the depth-first search forest contains all the nodes that can be reached from the root of that tree (in our example, all the nodes in the graph can be reached from H, not just I). To get all the nodes that can be visited from each node, we simply call *visit V* times, once for each node:

```
for k:=1 to V do
   begin
   now:=0;
   for j:=1 to V do val[j]:=0;
   visit(k);
   writeln
   end;
```

This program produces the following output for our sample graph:

```
A F E D B G J K L M C
B
C A F E D B G J K L M
D F E
E D F
F E D
G J K L M C A F E D B
H G J K L M C A F E D B I
I H G J K L M C A F E D B
J K L G C A F E D B M
K
L G J K M C A F E D B
M L G J K C A F E D B
```

For undirected graphs, this computation would produce a table with the property that each line corresponding to the nodes in a connected component lists all the nodes in that component. The table above has a similar property: certain of the lines list identical sets of nodes. Below we shall examine the generalization of connectedness that explains this property.

As usual, we could add code to do extra processing rather than just writing out the table. One operation we might want to perform is to add an edge directly from x to y if there is some way to get from x to y. The graph which results from adding all edges of this form to a directed graph is called the *transitive closure* of the graph. Normally, a large number of edges will be added and the transitive closure is likely to be dense, so an adjacency matrix representation is called for. This is an analogue to connected components in an undirected graph; once we've performed this computation once, then we can quickly answer questions like "is there a way to get from x to y?"

Using depth-first search to compute the transitive closure requires V^3 steps in the worst case, since we may have to examine every bit of the adjacency matrix for the depth-first search from each vertex. There is a remarkably simple nonrecursive program for computing the transitive closure of a graph represented with an adjacency matrix:

```
for y:=1 to V do
  for x:=1 to V do
    if a[x, y] then
      for j:=1 to V do
        if a[y, j] then a[x, j]:=true;
```

S. Warshall invented this method in 1962, using the simple observation that "if there's a way to get from node x to node y and a way to get from node y to node j then there's a way to get from node x to node j." The trick is to make this observation a little stronger, so that the computation can be done in only one pass through the matrix, to wit: "if there's a way to get from node x to node y *using only nodes with indices less than* x and a way to get from node y to node j then there's a way to get from node x to node j *using only nodes with indices less than* $x+1$." The above program is a direct implementation of this.

Warshall's method converts the adjacency matrix for our sample graph, given at left in the table below, into the adjacency matrix for its transitive closure, given at the right:

```
   A B C D E F G H I J K L M          A B C D E F G H I J K L M
A  1 1 0 0 0 1 1 0 0 0 0 0 0      A   1 1 1 1 1 1 1 0 0 1 1 1 1
B  0 1 0 0 0 0 0 0 0 0 0 0 0      B   0 1 0 0 0 0 0 0 0 0 0 0 0
C  1 0 1 0 0 0 0 0 0 0 0 0 0      C   1 1 1 1 1 1 1 0 0 1 1 1 1
D  0 0 0 1 0 1 0 0 0 0 0 0 0      D   0 0 0 1 1 1 0 0 0 0 0 0 0
E  0 0 0 1 1 0 0 0 0 0 0 0 0      E   0 0 0 1 1 1 0 0 0 0 0 0 0
F  0 0 0 0 1 1 0 0 0 0 0 0 0      F   0 0 0 1 1 1 0 0 0 0 0 0 0
G  0 0 1 0 1 0 1 0 0 1 0 0 0      G   1 1 1 1 1 1 1 0 0 1 1 1 1
H  0 0 0 0 0 0 1 1 1 0 0 0 0      H   1 1 1 1 1 1 1 1 1 1 1 1 1
I  0 0 0 0 0 0 0 1 1 0 0 0 0      I   1 1 1 1 1 1 1 1 1 1 1 1 1
J  0 0 0 0 0 0 0 0 0 1 1 1 1      J   1 1 1 1 1 1 1 0 0 1 1 1 1
K  0 0 0 0 0 0 0 0 0 0 1 0 0      K   0 0 0 0 0 0 0 0 0 0 1 0 0
L  0 0 0 0 0 0 1 0 0 0 0 1 1      L   1 1 1 1 1 1 1 0 0 1 1 1 1
M  0 0 0 0 0 0 0 0 0 0 0 1 1      M   1 1 1 1 1 1 1 0 0 1 1 1 1
```

For very large graphs, this computation can be organized so that the operations on bits can be done a computer word at a time, which will lead to significant savings in many environments. (As we've seen, it is not intended that such optimizations be tried with Pascal.)

Topological Sorting

For many applications involving directed graphs, cyclic graphs do arise. If, however, the graph above modeled a manufacturing line, then it would imply, say, that job A must be done before job G, which must be done before job C, which must be done before job A. But such a situation is inconsistent: for this and many other applications, directed graphs with no *directed cycles* (cycles with all edges pointing the same way) are called for. Such graphs are called *directed acyclic graphs*, or just *dags* for short. Dags may have many cycles if the directions on the edges are not taken into account; their defining property is simply that one should never get in a cycle by following edges in the indicated direction. A dag similar to the directed graph above, with a few edges removed or directions switched in order to remove cycles, is given below.

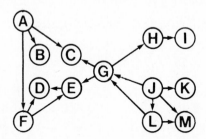

The edge list for this graph is the same as for the connected graph of Chapter 30, but here, again, the order in which the vertices are given when the edge is specified makes a difference.

Dags really are quite different objects from general directed graphs: in a sense, they are part tree, part graph. We can certainly take advantage of their special structure when processing them. Viewed from any vertex, a dag looks like a tree; put another way, the depth-first search forest for a dag has no up edges. For example, the following depth-first search forest describes the operation of *dfs* on the example dag above.

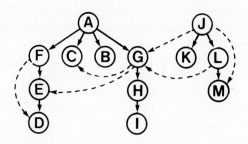

A fundamental operation on dags is to process the vertices of the graph in such an order that no vertex is processed before any vertex that points to it. For example, the nodes in the above graph could be processed in the following order:

J   K   L   M   A   G   H   I   F   E   D   B   C

If edges were to be drawn with the vertices in these positions, all the edges would go from left to right. As mentioned above, this has obvious application, for example, to graphs which represent manufacturing processes, for it gives a specific way to proceed within the constraints represented by the graph. This operation is called *topological sorting*, because it involves ordering the vertices of the graph.

In general, the vertex order produced by a topological sort is not unique. For example, the order

A   J   G   F   K   L   E   M   B   H   C   I   D

is a legal topological ordering for our example (and there are many others). In the manufacturing application mentioned, this situation occurs when one job has no direct or indirect dependence on another and thus they can be performed in either order.

It is occasionally useful to interpret the edges in a graph the other way around: to say that an edge directed from x to y means that vertex x "depends" on vertex y. For example, the vertices might represent terms to be defined in a programming language manual (or a book on algorithms!) with an edge from x to y if the definition of x uses y. In this case, it would be useful to find an ordering with the property that every term is defined before it is used in another definition. This corresponds to positioning the vertices in a line so that edges would all go from right to left. A *reverse topological order* for our sample graph is:

D   E   F   C   B   I   H   G   A   K   M   L   J

The distinction here is not crucial: performing a reverse topological sort on a graph is equivalent to performing a topological sort on the graph obtained by reversing all the edges.

But we've already seen an algorithm for reverse topological sorting, the standard recursive depth-first search procedure of Chapter 29! Simply changing *visit* to print out the vertex visited just before *exiting*, for example by inserting *write(name[k])* right at the end, causes *dfs* to print out the vertices in reverse topological order, when the input graph is a dag. A simple induction argument proves that this works: we print out the name of each vertex *after* we've printed out the names of all the vertices that it points to. When *visit* is changed in this way and run on our example, it prints out the vertices in the reverse topological order given above. Printing out the vertex name on exit from this recursive procedure is exactly equivalent to putting the vertex name on a stack on entry, then popping it and printing it on exit. It would be ridiculous to use an explicit stack in this case, since the mechanism for recursion provides it automatically; we mention this because we do need a stack for the more difficult problem to be considered next.

Strongly Connected Components

If a graph contains a *directed cycle*, (if we can get from a node back to itself by following edges in the indicated direction), then it it is not a dag and it can't be topologically sorted: whichever vertex on the cycle is printed out first will have another vertex which points to it which hasn't yet been printed out. The nodes on the cycle are mutually accessible in the sense that there is a way to get from every node on the cycle to another node on the cycle and back. On the other hand, even though a graph may be connected, it is not likely to be true that any node can be reached from any other via a connected path. In fact, the nodes divide themselves into sets called *strongly connected components* with the property that all nodes within a component are mutually accessible, but there is no way to get from a node in one component to a node in another component and back. The strongly connected components of the directed graph at the beginning of this chapter are two single nodes B and K, one pair of nodes H I, one triple of nodes D E F, and one large component with six nodes A C G J L M. For example, vertex A is in a different component from vertex F because though there is a path from A to F, there is no way to get from F to A.

The strongly connected components of a directed graph can be found using a variant of depth-first search, as the reader may have learned to expect. The method that we'll examine was discovered by R. E. Tarjan in 1972. Since it is based on depth-first search, it runs in time proportional to $V + E$, but it is actually quite an ingenious method. It requires only a few simple modifications to our basic *visit* procedure, but before Tarjan presented the method, no linear time algorithm was known for this problem, even though many people had worked on it.

The modified version of depth first search that we use to find the strongly connected components of a graph is quite similar to the program that we studied in Chapter 30 for finding biconnected components. The recursive *visit* function given below uses the same *min* computation to find the highest vertex reachable (via an up link) from any descendant of vertex *k*, but uses the value of *min* in a slightly different way to write out the strongly connected components:

```
function visit(k: integer): integer;
    var t: link;
        m, min: integer;
    begin
    now:=now+1; val[k]:=now; min:=now;
    stack[p]:=k; p:=p+1;
    t:=adj[k];
    while t<>z do
      begin
      if val[t↑.v]=0 then
        begin
        m:=visit(t↑.v);
        if m<min then min:=m;
        end
      else if val[t↑.v]<min then min:=val[t↑.v];
      t:=t↑.next
      end;
    if min=val[k] then
      begin
      repeat p:=p-1; write(name(stack[p])) until stack[p]=k;
      writeln
      end;
    visit:=min;
    val[k]:=V+1;
    end;
```

This program pushes the vertex names onto a stack on entry to *visit*, then pops them and prints them on exit from visiting the last member of each strongly connected component. The point of the computation is the test whether *min=val[k]* at the end: if so, all vertices encountered since entry (except those already printed out) belong to the same strongly connected component as *k*.

The strongly connected components are printed out on separate lines: the point of the computation is the test whether *min=val*[*k*] in the next-to-last line of the procedure to determine whether or not to do a *writeln* (begin a new component). As usual, this program could easily be modified to do more sophisticated processing than simply writing out the components.

The method is based on two observations that we've actually already made in other contexts. First, once we reach the end of a call to *visit* for a vertex, then we won't encounter any more vertices in the same strongly connected component (because all the vertices which can be reached from that vertex have been processed, as we noted above for topological sorting). Second, the "up" links in the tree provide a second path from one vertex to another and bind together the strong components. As with the algorithm in Chapter 30 for finding articulation points, we keep track of the highest ancestor reachable via one "up" link from all descendants of each node. Now, if a vertex x has no descendants or "up" links in the depth-first search tree, or if it has a descendant in the depth-first search tree with an "up" link that points to x, and no descendants with "up" links that point higher up in the tree, then it and all its descendants (except those vertices which are descendants of a vertex satisfying the same property) comprise a strongly connected component. In the depth-first search tree at the beginning of the chapter, nodes B and K satisfy the first condition (so they represent strongly connected components themselves) and nodes F(representing F E D), H (representing H I), and A (representing A G J L M C) satisfy the second condition. The members of the component represented by A are found by deleting B K F and their descendants (they appear in previously discovered components). Every descendant y of x that does not satisfy this same property has some descendant that has an "up" link that points higher than y in the tree. There is a path from x to y down through the tree; and a path from y to x can be found by going down from y to the vertex with the "up" link that reaches past y, then continuing the same process until x is reached. A crucial extra twist is that once we're done with a vertex, we give it a high *val*, so that "cross" links to that vertex will be ignored.

This program provides a deceptively simple solution to a relatively difficult problem. It is certainly testimony to the subtleties involved in searching directed graphs, subtleties which can be handled (in this case) by a carefully crafted recursive program.

Exercises

1. Give the adjacency matrix for the transitive closure of the example dag given in this chapter.

2. What would be the result of running the transitive closure algorithms on an undirected graph which is represented with an adjacency matrix?

3. Write a program to determine the number of edges in the transitive closure of a given directed graph, using the adjacency list representation.

4. Discuss how Warshall's algorithm compares with the transitive closure algorithm derived from using the depth-first search technique described in the text, but using the adjacency matrix form of *visit* and removing the recursion.

5. Give the topological ordering produced for the example dag given in the text when the suggested method is used with an adjacency matrix representation, but *dfs* scans the vertices in reverse order (from V down to 1) when looking for unvisited vertices.

6. Does the shortest path algorithm from Chapter 31 work for directed graphs? Explain why or give an example for which it fails.

7. Write a program to determine whether or not a given directed graph is a dag.

8. How many strongly connected components are there in a dag? In a graph with a directed cycle of size V?

9. Use your programs from Chapters 29 and 30 to produce large random directed graphs with V vertices. How many strongly connected components do such graphs tend to have?

10. Write a program that is functionally analogous to *find* from Chapter 30, but maintains *strongly* connected components of the *directed* graph described by the input edges. (This is not an easy problem: you certainly won't be able to get as efficient a program as *find*.)

33. Network Flow

Weighted directed graphs are useful models for several types of applications involving commodities flowing through an interconnected network. Consider, for example, a network of oil pipes of varying sizes, interconnected in complex ways, with switches controlling the direction of flow at junctions. Suppose further that the network has a single source (say, an oil field) and a single destination (say, a large refinery) to which all of the pipes ultimately connect. What switch settings will maximize the amount of oil flowing from source to destination? Complex interactions involving material flow at junctions make this problem, called the *network flow problem*, a nontrivial problem to solve.

This same general setup can be used to describe traffic flowing along highways, materials flowing through factories, etc. Many different versions of the problem have been studied, corresponding to many different practical situations where it has been applied. There is clearly strong motivation to find an efficient algorithm for these problems.

This type of problem lies at the interface between computer science and the field of *operations research*. Operations researchers are generally concerned with mathematical modeling of complex systems for the purpose of (preferably optimal) decision-making. Network flow is a typical example of an operations research problem; we'll briefly touch upon some others in Chapters 37–40.

As we'll see, the modeling can lead to complex mathematical equations that require computers for solution. For example, the classical solution to the network flow problem given below is closely related to the graph algorithms that we have been examining. But this problem is one which is still actively being studied: unlike many of the problems that we've looked at, the "best" solution has not yet been found and good new algorithms are still being discovered.

The Network Flow Problem

Consider the following rather idealized drawing of a small network of oil pipes:

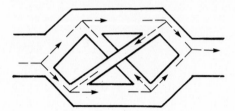

The pipes are of fixed capacity proportional to their size and oil can flow in them only in the direction indicated (perhaps they run downhill or have unidirectional pumps). Furthermore, switches at each junction control how much of the oil goes in each direction. No matter how the switches are set, the system reaches a state of equilibrium when the amount of oil flowing into the system at the left is equal to the amount flowing out of the system at the right (this is the quantity that we want to maximize) and when the amount of oil flowing in at each junction is equal to the amount of oil flowing out. We measure both flow and pipe capacity in terms of integral "units" (say, gallons per second).

It is not immediately obvious that the switch settings can really affect the total flow: the following example will illustrate that they can. First, suppose that all switches are open so that the two diagonal pipes and the top and bottom pipes are full. This gives the following configuration:

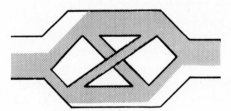

The total flow into and out of the network in this case is less than half the capacity of the input pipe, only slightly more than half the capacity of the output pipe. Now suppose that the upward diagonal pipe is shut off. This shuts flow equal to its capacity out of the bottom, and the top is unaffected because there's room to replace its flow from the input pipe; thus we have:

The total flow into and out of the network is increased to substantially.

This situation can obviously be modeled using a directed graph, and it turns out that the programs that we have studied can apply. Define a *network* as a weighted directed graph with two distinguished vertices: one with no edges pointing in (the *source*); one with no edges pointing out (the *sink*). The weights on the edges, which we assume to be non-negative, are called the *edge capacities*. Now, a *flow* is defined as another set of weights on the edges such that the flow on each edge is equal to or less than the capacity, and the flow into each vertex is equal to the flow out of that vertex. The *value* of the flow is the flow into the source (or out of the sink). The *network flow problem* is to find a flow with maximum value for a given network.

Networks can obviously be represented using either the adjacency matrix or adjacency list representations that we have used for graphs in previous chapters. Instead of a single weight, two weights are associated with each edge, the *size* and the *flow*. These could be represented as two fields in an adjacency list node, as two matrices in the adjacency matrix representation, or as two fields within a single record in either representation. Even though networks are directed graphs, the algorithms that we'll be examining need to traverse edges in the "wrong" direction, so we use an undirected graph representation: if there is an edge from x to y with size s and flow f, we also keep an edge from y to x with size $-s$ and flow $-f$. In an adjacency list representation, it is necessary to maintain links connecting the two list nodes which represent each edge, so that when we change the flow in one we can update it in the other.

Ford-Fulkerson Method

The classical approach to the network flow problem was developed by L. R. Ford and D. R. Fulkerson in 1962. They gave a method to improve any legal flow (except, of course, the maximum). Starting with a zero flow, we apply the method repeatedly; as long as the method can be applied, it produces an increased flow; if it can't be applied, the maximum flow has been found.

Consider any directed path through the network (from source to sink). Clearly, the flow can be increased by at least the smallest amount of unused capacity on any edge on the path, by increasing the flow in all edges on the

path by that amount. In our example, this rule could be applied along the
path ADEBCF:

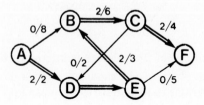

then along the path ABCDEF:

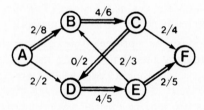

and then along the path ABCF:

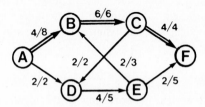

Now all directed paths through the network have at least one edge which
is filled to capacity. But there is another way to increase the flow: we can
consider arbitrary paths through the network which can contain edges which
point the "wrong way" (from sink to source along the path). The flow can

be increased along such a path by *increasing* the flow on edges from source to sink and *decreasing* the flow on edges from sink to source by the same amount. To simplify terminology, we'll call edges which flow from source to sink along a particular path *forward* edges and edges which flow from sink to source *backward* edges. For example, the flow in the network above can be increased by 2 along the path ABEF.

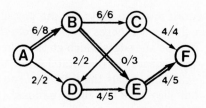

This corresponds to shutting off the oil on the pipe from E to B; this allows 2 units to be redirected from E to F without losing any flow at the other end, because the 2 units which used to come from E to B can be replaced by 2 units from A.

Notice that the amount by which the flow can be increased is limited by the minimum of the unused capacities in the forward edges and the minimum of the flows in the backward edges. Put another way, in the new flow, at least one of the forward edges along the path becomes full or at least one of the backward edges along the path becomes empty. Furthermore, the flow can't be increased on any path containing a full forward edge or an empty backward edge.

The paragraph above gives a method for increasing the flow on any network, *provided* that a path with no full forward edges or empty backward edges can be found. The crux of the Ford-Fulkerson method is the observation that if no such path can be found then the flow is maximal. The proof of this fact goes as follows: if every path from the source to the sink has a full forward edge or an empty backward edge, then go through the graph and identify the first such edge on every path. This set of edges *cuts* the graph in two parts, as shown in the diagram below for our example.

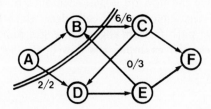

For any cut of the network into two parts, we can measure the flow "across" the cut: the total of the flow on the edges which go from the source to the sink minus the total of the flow on the edges which go the other way. Our example cut has a value of 8, which is equal to the total flow for the network. It turns out that whenever the cut flow equals the total flow, we know not only that the flow is maximal, but also that the cut is minimal (that is, every other cut has at least as high a flow "across"). This is called the *maxflow-mincut theorem*: the flow couldn't be any larger (otherwise the cut would have to be larger also); and no smaller cuts exist (otherwise the flow would have to be smaller also).

Network Searching

The Ford-Fulkerson method described above may be summarized as follows: "start with zero flow everywhere and increase the flow along any path from source to sink with no full forward edges or empty backward edges, continuing until there are no such paths in the network." But this is not an *algorithm* in the sense to which we have become accustomed, since the method for finding paths is not specified. The example above is based on the intuition that the longer the path, the more the network is filled up, and thus that long paths should be preferred. But the following classical example shows that some care should be exercised:

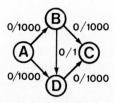

Now, if the first path chosen is ABDC, then the flow is increased by only one. Then the second path chosen might be ADBC, again increasing the flow by one, and leaving an situation identical to the initial situation, except that the flows on the outside edges are increased only by 1. Any algorithm which chose those two paths (for example, one that looks for long paths) would continue to do so, thus requiring 1000 pairs of iterations before the maximum flow is found. If the numbers on the sides were a billion, then a billion iterations would be used. Obviously, this is an undesirable situation, since the paths ABC and ADC would give the maximum flow in just two steps. For the algorithm to be useful, we must avoid having the running time so dependent on the magnitude of the capacities.

Fortunately, the problem is easily eliminated. It was proven by Edmonds and Karp that if *breadth-first* search were used to find the path, then the number of paths used before the maximum flow is found in a network of V vertices and E edges must be less than VE. (This is a worst-case bound: a typical network is likely to require many fewer steps.) In other words, simply use the shortest available path from source to sink in the Ford-Fulkerson method.

With the priority graph traversal method of Chapters 30 and 31, we can implement another method suggested by Edmonds and Karp: find the path through the network which increases the flow by the largest amount. This can be achieved simply by using a variable for *priority* (whose value is set appropriately) in either the adjacency list *sparsepfs* of Chapter 30 or the adjacency matrix *densepfs* of Chapter 31. For example, the following statements compute the priority assuming a matrix representation:

---

if *size*[k, t]>0 then *priority*:=*size*[k, t]$-$*flow*[k, t]
else *priority*:=$-$*flow*[k, t];
if *priority*>*val*[k] then *priority*:=*val*[k];

Then, since we want to take the node with the *highest* priority value, we must either reorient the priority queue mechanisms in those programs to return the maximum instead of the minimum or use them as is with *priority* set to *maxint*$-1-$*priority* (and the process reversed when the value is removed). Also, we modify the priority first search procedure to take the source and sink as arguments, then to start at the source and stop when a path to the sink has been found. If such a path is not found, the partial priority search tree defines a mincut for the network. Finally, the *val* for the source should be set to *maxint* before the search is started, to indicate that any amount of flow can be achieved at the source (though this is immediately restricted by the total capacity of all the pipes leading directly out of the source).

With *densepfs* implemented as described in the previous paragraph, finding the maximum flow is actually quite simple, as shown by the following program:

```
repeat
    densepfs(1, V);
    y:=V; x:=dad[V];
    while x<>0 do
      begin
      flow[x, y]:=flow[x, y]+val[V];
      flow[y, x]:=-flow[x, y];
      y:=x; x:=dad[y]
      end
    until val[V]=unseen
```

This program assumes that the adjacency matrix representation is being used for the network. As long as *densepfs* can find a path which increases the flow (by the maximum amount), we trace back through the path (using the *dad* array constructed by *densepfs*) and increase the flow as indicated. If V remains unseen after some call to *densepfs*, then a mincut has been found and the algorithm terminates.

For our example network, the algorithm first increases the flow along the path ABCF, then along ADEF, then along ABCDEF. No backwards edges are used in this example, since the algorithm does not make the unwise choice ADEBCF that we used to illustrate the need for backwards edges. In the next chapter we'll see a graph for which this algorithm uses backwards edges to find the maximum flow.

Though this algorithm is easily implemented and is likely to work well for networks that arise in practice, the analysis of this method is quite complicated. First, as usual, *densepfs* requires V^2 steps in the worst case, or we could use *sparsepfs* to run in time proportional to $(E + V) \log V$ per iteration, though the algorithm is likely to run somewhat faster than this, since it stops when it reaches the sink. But how many iterations are required? Edmonds and Karp show the worst case to be $1 + \log_{M/M-1} f^*$ where f^* is the cost of the flow and M is the maximum number of edges in a cut of the network. This is certainly complicated to compute, but it is not likely to be large for real networks. This formula is included to give an indication not of how long the algorithm might take on an actual network, but rather of the complexity of the analysis. Actually, this problem has been quite widely studied, and complicated algorithms with much better worst-case time bounds have been developed.

The network flow problem can be extended in several ways, and many variations have been studied in some detail because they are important in actual applications. For example, the *multicommodity flow problem* involves introducing multiple sources, sinks, and types of material in the network. This makes the problem much more difficult and requires more advanced algorithms than those considered here: for example, no analogue to the max-flow min-cut theorem is known to hold for the general case. Other extensions to the network flow problem include placing capacity constraints on vertices (easily handled by introducing artificial edges to handle these capacities), allowing undirected edges (also easily handled by replacing undirected edges by pairs of directed edges), and introducing lower bounds on edge flows (not so easily handled). If we make the realistic assumption that pipes have associated costs as well as capacities, then we have the *min-cost* flow problem, a quite difficult problem from operations research.

☐

Exercises

1. Give an algorithm to solve the network flow problem for the case that the network forms a tree if the sink is removed.

2. What paths are traced by the algorithm given in the text when finding the maximum flow in the following network?

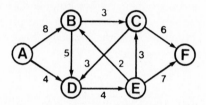

3. Draw the priority search trees computed on each call to *densepfs* for the example discussed in the text.

4. True or false: No algorithm can find the maximum flow without examining every edge in the network.

5. What happens to the Ford-Fulkerson method in the case that the network has a directed cycle?

6. Give an example where a shortest-path traversal would produce a different set of paths than the method given in the text.

7. Give a counterexample which shows why depth-first search is not appropriate for the network flow problem.

8. Find an assignment of sizes that would make the algorithm given in the text use a backward edge on the example graph, or prove that none exists.

9. Implement the breadth-first search solution to the network flow problem, using *sparsepfs*.

10. Write a program to find maximum flows in random networks with V nodes and about $10V$ edges. How many calls to *sparsepfs* are made, for $V = 25, 50, 100$?

34. Matching

A problem which often arises is to "pair up" objects according to preference relationships which are likely to conflict. For example, a quite complicated system has been set up in the U. S. to place graduating medical students into hospital residence positions. Each student lists several hospitals in order of preference, and each hospital lists several students in order of preference. The problem is to assign students to positions in a fair way, respecting all the stated preferences. A sophisticated algorithm is required because the best students are likely to be preferred by several hospitals, and the best hospital positions are likely to be preferred by several students. It's not even clear that each hospital position can be filled by a student that the hospital has listed and each student can be assigned to a position that the student has listed, let alone respect the order in the preference lists. Actually this frequently occurs: after the algorithm has done the best that it can, there is a last minute scramble among unmatched hospitals and students to complete the process.

This example is a special case of a difficult fundamental problem on graphs that has been widely studied. Given a graph, a *matching* is a subset of the edges in which no vertex appears more than once. That is, each vertex touched by one of the edges in the matching is paired with the other vertex of that edge, but some vertices may be left unmatched. Even if we insist that there should be no edges connecting unmatched vertices, different ways of choosing the edges could lead to different numbers of leftover (unmatched) vertices.

Of particular interest is a *maximum matching*, which contains as many edges as possible or, equivalently, which minimizes the number of unmatched vertices. The best that we could hope to do would be to have a set of edges in which each vertex appears exactly once (such a matching in a graph with $2V$ vertices would have V edges), but it is not always possible to achieve this.

For example, consider our sample undirected graph:

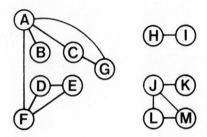

The edges AF DE CG HI JK LM make a maximum matching for this graph, which is the best that can be done, but there's no three-edge matching for the subgraph consisting of just the first six vertices and the edges connecting them.

For the medical student matching problem described above, the students and hospitals would correspond to nodes in the graph; their preferences to edges. If they assign values to their preferences (perhaps using the time-honored "1–10" scale), then we have the *weighted matching* problem: given a weighted graph, find a set of edges in which no vertex appears more than once such that the sum of the weights on the edges in the set chosen is maximized. Below we'll see another alternative, where we respect the order in the preferences, but do not require (arbitrary) values to be assigned to them.

The matching problem has attracted attention because of its intuitive nature and its wide applicability. Its solution in the general case involves intricate and beautiful combinatorial mathematics beyond the scope of this book. Our intent here is to provide the reader with an appreciation for the problem by considering some interesting special cases while at the same time developing some useful algorithms.

Bipartite Graphs

The example mentioned above, matching medical students to residencies, is certainly representative of many other matching applications. For example, we might be matching men and women for a dating service, job applicants to available positions, courses to available hours, or congressmen to committee assignments. The graphs resulting from modeling such cases are called *bipartite graphs*, which are defined to be graphs in which all edges go between two sets of nodes (that is, the nodes divide into two sets and no edges connect two nodes in the same set). Obviously, we wouldn't want to "match" one job applicant to another or one committee assignment to another.

The reader might be amused to search for a maximum matching in the typical bipartite graph drawn below:

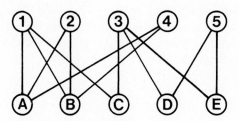

In an adjacency matrix representation for bipartite graphs, one can achieve obvious savings by including only rows for one set and only columns for the other set. In an adjacency list representation, no particular saving suggests itself, except naming the vertices intelligently so that it is easy to tell which set a vertex belongs to.

In our examples, we use letters for nodes in one set, numbers for nodes in the other. The maximum matching problem for bipartite graphs can be simply expressed in this representation: "Find the largest subset of a set of letter-number pairs with the property that no two pairs have the same letter or number." Finding the maximum matching for our example bipartite graph corresponds to solving this puzzle on the pairs E5 A2 A1 C1 B4 C3 D3 B2 A4 D5 E3 B1.

It is an interesting exercise to attempt to find a direct solution to the matching problem for bipartite graphs. The problem seems easy at first glance, but subtleties quickly become apparent. Certainly there are far too many pairings to try all possibilities: a solution to the problem must be clever enough to try only a few of the possible ways to match the vertices.

The solution that we'll examine is an indirect one: to solve a particular instance of the matching problem, we'll construct an instance of the network flow problem, use the algorithm from the previous chapter, then use the solution to the network flow problem to solve the matching problem. That is, we *reduce* the matching problem to the network flow problem. Reduction is a method of algorithm design somewhat akin to the use of a library subroutine by a systems programmer. It is of fundamental importance in the theory of advanced combinatorial algorithms (see Chapter 40). For the moment, reduction will provide us with an efficient solution to the bipartite matching problem.

The construction is straightforward: given an instance of bipartite matching, construct an instance of network flow by creating a source vertex with edges pointing to all the members of one set in the bipartite graph, then make all the edges in the bipartite graph point from that set to the other, then add a sink vertex pointed to by all the members of the other set. All of the edges in the resulting graph are given a capacity of 1. For example, the bipartite graph given above corresponds to the network below: the darkened edges show the first four paths found when the network flow algorithm of the previous chapter is run on this graph.

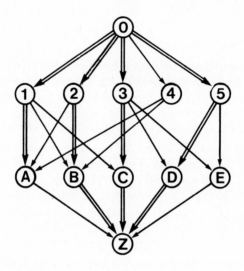

Note that the bipartite property of the graph, the direction of the flow, and the fact that all capacities are 1 force each path through the network to correspond to an edge in a matching: in the example, the paths found so far correspond to the partial matching A1 B2 C3 D5. Each time the network flow algorithm calls *pfs* it either finds a path which increases the flow by one or terminates.

Now all forward paths through the network are full, and the algorithm must use backward edges. The path found in this example is the path 04A1C3EZ. This path clearly increases the flow in the network, as described in the previous chapter. In the present context, we can think of the path as a set of instructions to create a new partial matching (with one more edge) from the current one. This construction follows in a natural way from tracing through the path in order: "4A" means to add A4 to the matching, which requires that "A1" be deleted; "1C" means to add C1 to the matching, which requires that "C3" be deleted; "3E" means to add E3 to the matching. Thus, after this path is processed, we have the matching A4 B2 C1 D5 E3; equivalently, the flow in the network is given by full pipes in the edges connecting those nodes, and all pipes leaving 0 and entering Z full.

The proof that the matching is exactly those edges which are filled to capacity by the maxflow algorithm is straightforward. First, the network flow always gives a legal matching: since each vertex has an edge of capacity 1 either coming in (from the sink) or going out (to the source), at most one unit of flow can go through each vertex, which implies that each vertex will be included at most once in the matching. Second, no matching can have more edges, since any such matching would lead directly to a better flow than that produced by the maxflow algorithm.

Thus, to compute the maximum matching for a bipartite graph we simply format the graph so as to be suitable for input to the network flow algorithm of the previous chapter. Of course, the graphs presented to the network flow algorithm in this case are much simpler than the general graphs the algorithm is designed to handle, and it turns out that the algorithm is somewhat more efficient for this case. The construction ensures that each call to *pfs* adds one edge to the matching, so we know that there are at most $V/2$ calls to *pfs* during the execution of the algorithm. Thus, for example, the total time to find the maximum matching for a dense bipartite graph with V vertices (using the adjacency matrix representation) is proportional to V^3.

Stable Marriage Problem

The example given at the beginning of this chapter, involving medical students and hospitals, is obviously taken quite seriously by the participants. But the method that we'll examine for doing the matching is perhaps better understood in terms of a somewhat whimsical model of the situation. We assume that we have N men and N women who have expressed mutual preferences (each man must say exactly how he feels about each of the N women and vice versa). The problem is to find a set of N marriages that respects everyone's preferences.

How should the preferences be expressed? One method would be to use the "1–10" scale, each side assigning an absolute score to certain members of the other side. This makes the marriage problem the same as the weighted matching problem, a relatively difficult problem to solve. Furthermore, use of absolute scales in itself can lead to inaccuracies, since peoples' scales will be inconsistent (one woman's 10 might be another woman's 7). A more natural way to express the preferences is to have each person list in order of preference all the people of the opposite sex. The following two tables might show preferences among a set of five women and five men. As usual (and to protect the innocent!) we assume that hashing or some other method has been used to translate actual names to single digits for women and single letters for men:

A:	2	5	1	3	4		1:	E	A	D	B	C
B:	1	2	3	4	5		2:	D	E	B	A	C
C:	2	3	5	4	1		3:	A	D	B	C	E
D:	1	3	2	4	5		4:	C	B	D	A	E
E:	5	3	2	1	4		5:	D	B	C	E	A

Clearly, these preferences often conflict: for example, both A and C list 2 as their first choice, and nobody seems to want 4 very much (but someone must get her). The problem is to engage all the women to all the men in such a way as to respect all their preferences as much as possible, then perform N marriages in a grand ceremony. In developing a solution, we must assume that anyone assigned to someone less than their first choice will be disappointed and will always prefer anyone higher up on the list. A set of marriages is called *unstable* if two people who are not married both prefer each other to their spouses. For example, the assignment A1 B3 C2 D4 E5 is unstable because A prefers 2 to 1 and 2 prefers A to C. Thus, acting according to their preferences, A would leave 1 for 2 and 2 would leave C for A (leaving 1 and C with little choice but to get together).

Finding a stable configuration seems on the face of it to be a difficult problem, since there are so many possible assignments. Even determining whether a configuration is stable is not simple, as the reader may discover by looking (before reading the next paragraph) for the unstable couple in the example above after the new matches A2 and C1 have been made. In general, there are many different stable assignments for a given set of preference lists, and we only need to find one. (Finding *all* stable assignments is a much more difficult problem.)

One possible algorithm for finding a stable configuration might be to remove unstable couples one at a time. However, not only is this slow because of the time required to determine stability, but also the process does not even necessarily terminate! For example, after A2 and C1 have been matched in the example above, B and 2 make an unstable couple, which leads to the configuration A3 B2 C1 D4 E5. In this arrangement, B and 1 make an unstable couple, which leads to the configuration A3 B1 C2 D4 E5. Finally, A and 1 make an unstable configuration which leads back to the original configuration. An algorithm which attempts to solve the stable marriage problem by removing stable pairs one by one is bound to get caught in this type of loop.

Instead, we'll look at an algorithm which tries to build stable pairings systematically using a method based on what might happen in the somewhat idealized "real-life" version of the problem. The idea is to have each man, in turn, become a "suitor" and seek a bride. Obviously, the first step in his quest is to propose to the first woman on his list. If she is already engaged to a man whom she prefers, then our suitor must try the next woman on his list, continuing until he finds a woman who is not engaged or who prefers him to her current fiancee. If this women is not engaged, then she becomes engaged to the suitor and the next man becomes the suitor. If she is engaged, then she breaks the engagement and becomes engaged to the suitor (whom she prefers). This leaves her old fiancee with nothing to do but become the suitor once again, starting where he left off on his list. Eventually he finds a new fiancee, but another engagement may need to be broken. We continue in this way, breaking engagements as necessary, until some suitor finds a woman who has not yet been engaged.

This method may model what happens in some 19th-century novels, but some careful examination is required to show that it produces a stable set of assignments. The diagram below shows the sequence of events for the initial stages of the process for our example. First, A proposes to 2 (his first choice) and is accepted; then B proposes to 1 (his first choice) and is accepted; then C proposes to 2, is turned down, and proposes to 3 and is accepted, as depicted in the third diagram:

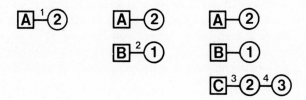

Each diagram shows the sequence of events when a new man sets out as the suitor to seek a fiancee. Each line gives the "used" preference list for the corresponding man, with each link labeled with an integer telling when that link was used by that man to propose to that woman. This extra information is useful in tracking the sequence of proposals when D and E become the suitor, as shown in the following figure:

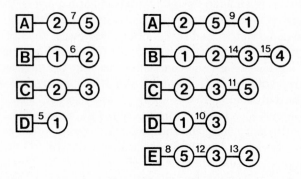

When D proposes to 1, we have our first broken engagement, since 1 prefers D to B. Then B becomes the suitor and proposes to 2, which gives our second broken engagement, since 2 prefers B to A. Then A becomes the suitor and proposes to 5, which leaves a stable situation. The reader might wish to trace through the sequence of proposals made when E becomes the suitor. Things don't settle down until after eight proposals are made. Note that E takes on the suitor role twice in the process.

To begin the implementation, we need data structures to represent the preference lists. Different structures are appropriate for the men and the women, since they use the preference lists in different ways. The men simply go through their preference lists in order, so a straightforward implementation as a two-dimensional array is called for: we'll maintain an two-dimensional array for the preference list so that, for example, $prefer[m, w]$ will be the wth woman in the preference list of the mth man. In addition, we need to keep track of how far each man has progressed on his list. This can be handled with a one-dimensional array $next$, initialized to zero, with $next[m]+1$ the index of the next woman on man m's preference list: her identifier is found in $prefer[m, next[m]+1]$.

For each woman, we need to keep track of her fiancee ($fiancee[w]$ will be the man engaged to woman w) and we need to be able to answer the question "Is man s preferable to $fiancee[w]$?" This could be done by searching the preference list sequentially until either s or $fiancee[w]$ is found, but this method would be rather inefficient if they're both near the end. What is called for is the "inverse" of the preference list: $rank[w, s]$ is the index of man s on woman w's preference list. For the example above this array is

```
1:  2  4  5  3  1
2:  4  3  5  1  2
3:  1  3  4  2  5
4:  4  2  1  3  5
5:  5  2  3  1  4
```

The suitability of suitor s can be very quickly tested by the statement **if** $rank[w, s] < rank[w, fiancee[w]]$ .... These arrays are easily constructed directly from the preference lists. To get things started, we use a "sentinel" man 0 as the initial suitor, and put him at the end of all the women's preference lists.

With the data structures initialized in this way, the implementation as described above is straightforward:

```
for m:=1 to N do
   begin
   s:=m;
   repeat
      next[s]:=next[s]+1;  w:=prefer[s, next[s]];
      if rank[w, s]<rank[w, fiancee[w]] then
         begin t:=fiancee[w]; fiancee[w]:=s; s:=t end;
   until s=0;
   end;
```

Each iteration starts with an unengaged man and ends with an unengaged woman. The **repeat** loop must terminate because every man's list contains every woman and each iteration of the loop involves incrementing some man's list, and thus an unengaged woman must be encountered before any man's list is exhausted. The set of engagements produced by the algorithm is stable because every woman whom any man prefers to his fiancee is engaged to someone that she prefers to him.

There are several obvious built-in biases in this algorithm. First, the men go through the women on their lists in order, while the women must wait for the "right man" to come along. This bias may be corrected (in a somewhat easier manner than in real life) by interchanging the order in which the preference lists are input. This produces the stable configuration 1E 2D 3A 4C 5B, where every women gets her first choice except 5, who gets her second. In general, there may be many stable configurations: it can be shown that this one is "optimal" for the women, in the sense that no other stable configuration will give any woman a better choice from her list. (Of course, the first stable configuration for our example is optimal for the men.)

Another feature of the algorithm which seems to be biased is the order in which the men become the suitor: is it better to be the first man to propose (and therefore be engaged at least for a little while to your first choice) or the last (and therefore have a reduced chance to suffer the indignities of a broken engagement)? The answer is that this is not a bias at all: it doesn't matter in what order the men become the suitor. As long as each man makes proposals and each woman accepts according to their lists, the same stable configuration results.

Advanced Algorithms

The two special cases that we've examined give some indication of how complicated the matching problem can be. Among the more general problems that have been studied in some detail are: the maximum matching problem for general (not necessarily bipartite) graphs; weighted matching for bipartite graphs, where edges have weights and a matching with maximum total weight is sought; and weighted matching for general graphs. Treating the many techniques that have been tried for matching on general graphs would fill an entire volume: it is one of the most extensively studied problems in graph theory.

Exercises

1. Find all the matchings with five edges for the given sample bipartite graph.

2. Use the algorithm given in the text to find maximum matchings for random bipartite graphs with 50 vertices and 100 edges. About how many edges are in the matchings?

3. Construct a bipartite graph with six nodes and eight edges which has a three-edge matching, or prove that none exists.

4. Suppose that vertices in a bipartite graph represent jobs and people and that each person is to be assigned to *two* jobs. Will reduction to network flow give an algorithm for this problem? Prove your answer.

5. Modify the network flow program of Chapter 33 to take advantage of the special structure of the 0-1 networks which arise for bipartite matching.

6. Write an efficient program for determining whether an assignment for the marriage problem is stable.

7. Is it possible for two men to get their last choice in the stable marriage algorithm? Prove your answer.

8. Construct a set of preference lists for $N = 4$ for the stable marriage problem where everyone gets their second choice, or prove that no such set exists.

9. Give a stable configuration for the stable marriage problem for the case where the preference lists for men and women are all the same: in ascending order.

10. Run the stable marriage program for $N = 50$, using random permutations for preference lists. About how many proposals are made during the execution of the algorithm?

SOURCES for Graph Algorithms

There are several textbooks on graph algorithms, but the reader should be forewarned that there is a great deal to be learned about graphs, that they still are not fully understood, and that they are traditionally studied from a mathematical (as opposed to an algorithmic) standpoint. Thus, many references have more rigorous and deeper coverage of much more difficult topics than our treatment here.

Many of the topics that we've treated here are covered in the book by Even, for example, our network flow example in Chapter 33. Another source for further material is the book by Papadimitriou and Steiglitz. Though most of that book is about much more advanced topics (for example, there is a full treatment of matching in general graphs), it has up-to-date coverage of many of the algorithms that we've discussed, including pointers to further reference material.

The application of depth-first search to solve graph connectivity and other problems is the work of R. E. Tarjan, whose original paper merits further study. The many variants on algorithms for the union-find problem of Chapter 30 are ably categorized and compared by van Leeuwen and Tarjan. The algorithms for shortest paths and minimum spanning trees in dense graphs in Chapter 31 are quite old, but the original papers by Dijkstra, Prim, and Kruskal still make interesting reading. Our treatment of the stable marriage problem in Chapter 34 is based on the entertaining account given by Knuth.

E. W. Dijkstra, "A note on two problems in connexion with graphs," *Numerishe Mathematik*, **1** (1959).

S. Even, *Graph Algorithms*, Computer Science Press, Rockville, MD, 1980.

D. E. Knuth, *Marriages stables*, Les Presses de L'Universite de Montreal, Montreal, 1976.

J. R. Kruskal Jr., "On the shortest spanning subtree of a graph and the traveling salesman problem," *Proceedings AMS*, **7**, 1 (1956).

C. H. Papadimitriou and K. Steiglitz, *Combinatorial Optimization: Algorithms and Complexity*, Prentice-Hall, Englewood Cliffs, NJ, 1982.

R. C. Prim, "Shortest connection networks and some generalizations," *Bell System Technical Journal*, **36** (1957).

R. E. Tarjan, "Depth-first search and linear graph algorithms," *SIAM Journal on Computing*, **1**, 2 (1972).

J. van Leeuwen and R. E. Tarjan, "Worst-case analysis of set-union algorithms," *Journal of the ACM*, to appear.

ADVANCED TOPICS

35. Algorithm Machines

The algorithms that we have studied are, for the most part, remarkably robust in their applicability. Most of the methods that we have seen are a decade or more old and have survived many quite radical changes in computer hardware and software. New hardware designs and new software capabilities certainly can have a significant impact on specific algorithms, but good algorithms on old machines are, for the most part, good algorithms on new machines.

One reason for this is that the fundamental design of "conventional" computers has changed little over the years. The design of the vast majority of computing systems is guided by the same underlying principle, which was developed by the mathematician J. von Neumann in the early days of modern computing. When we speak of the *von Neumann model of computation*, we refer to a view of computing in which instructions and data are stored in the same memory and a single processor fetches instructions from the memory and executes them (perhaps operating on the data), one by one. Elaborate mechanisms have been developed to make computers cheaper, faster, smaller (physically), and larger (logically), but the architecture of most computer systems can be viewed as variations on the von Neumann theme.

Recently, however, radical changes in the cost of computing components have made it plausible to consider radically different types of machines, ones in which a large number of instructions can be executed at each time instant or in which the instructions are "wired in" to make special-purpose machines capable of solving only one problem or in which a large number of smaller machines can cooperate to solve the same problem. In short, rather than having a machine execute just one instruction at each time instant, we can think about having a large number of actions being performed simultaneously. In this chapter, we shall consider the potential effect of such ideas on some of the problems and algorithms we have been considering.

General Approaches

Certain fundamental algorithms are used so frequently or for such large problems that there is always pressure to run them on bigger and faster computers. One result of this has been a series of "supercomputers" which embody the latest technology; they make some concessions to the fundamental von Neumann concept but still are designed to be general-purpose and useful for all programs. The common approach to using such a machine for the type of problem we have been studying is to start with the algorithms that are best on conventional machines and adapt them to the particular features of the new machine. This approach encourages the persistence of old algorithms and old architecture in new machines.

Microprocessors with significant computing capabilities have become quite inexpensive. An obvious approach is to try to use a large number of processors together to solve a large problem. Some algorithms can adapt well to being "distributed" in this way; others simply are not appropriate for this kind of implementation.

The development of inexpensive, relatively powerful processors has involved the appearance of general-purpose tools for use in designing and building new processors. This has led to increased activity in the development of special-purpose machines for particular problems. If no machine is particularly well-suited to execute some important algorithm, then we can design and build one that is! For many problems, an appropriate machine can be designed and built that fits on one (very-large-scale) integrated circuit chip.

A common thread in all of these approaches is *parallelism*: we try to have as many different things as possible happening at any instant. This can lead to chaos if it is not done in an orderly manner. Below, we'll consider two examples which illustrate some techniques for achieving a high degree of parallelism for some specific classes of problems. The idea is to assume that we have not just one but M processors on which our program can run. Thus, if things work out well, we can hope to have our program run M times faster than before.

There are several immediate problems involved in getting M processors to work together to solve the same problem. The most important is that they must communicate in some way: there must be wires interconnecting them and specific mechanisms for sending data back and forth along those wires. Furthermore, there are physical limitations on the type of interconnection allowed. For example, suppose that our "processors" are integrated circuit chips (these can now contain more circuitry than small computers of the past) which have, say, 32 pins to be used for interconnection. Even if we had 1000 such processors, we could connect each to at most 32 others. The choice of how to interconnect the processors is fundamental in parallel computing.

Moreover, it's important to remember that this decision must be made ahead of time: a program can change the way in which it does things depending on the particular instance of the problem being solved, but a machine generally can't change the way its parts are wired together.

This general view of parallel computation in terms of independent processors with some fixed interconnection pattern applies in each of the three domains described above: a supercomputer has very specific processors and interconnection patterns that are integral to its architecture (and affect many aspects of its performance); interconnected microprocessors involve a relatively small number of powerful processors with simple interconnections; and very-large-scale integrated circuits themselves involve a very large number of simple processors (circuit elements) with complex interconnections.

Many other views of parallel computation have been studied extensively since von Neumann, with renewed interest since inexpensive processors have become available. It would certainly be beyond the scope of this book to treat all the issues involved. Instead, we'll consider two specific machines that have been proposed for some familiar problems. The machines that we consider illustrate the effects of machine architecture on algorithm design and vice versa. There is a certain symbiosis at work here: one certainly wouldn't design a new computer without some idea of what it will be used for, and one would like to use the best available computers to execute the most important fundamental algorithms.

Perfect Shuffles

To illustrate some of the issues involved in implementing algorithms as machines instead of programs, we'll look at an interesting method for merging which is suitable for hardware implementation. As we'll see, the same general method can be developed into a design for an "algorithm machine" which incorporates a fundamental interconnection pattern to achieve parallel operation of M processors for solving several problems in addition to merging.

As mentioned above, a fundamental difference between writing a program to solve a problem and designing a machine is that a program can *adapt* its behavior to the particular instance of the problem being solved, while the machine must be "wired" ahead of time always to perform the same sequence of operations. To see the difference, consider the first sorting program that we studied, *sort3* from Chapter 8. No matter what three numbers appear in the data, the program always performs the same sequence of three fundamental "compare-exchange" operations. None of the other sorting algorithms that we studied have this property. They all perform a sequence of comparisons that depends on the outcome of previous comparisons, which presents severe problems for hardware implementation.

Specifically, if we have a piece of hardware with two input wires and two output wires that can compare the two numbers on the input and exchange them if necessary for the output, then we can wire three of these together as follows to produce a sorting machine with three inputs (at the top in the figure) and three outputs (at the bottom):

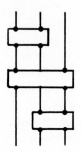

Thus, for example, if C B A were to appear at the top, the first box would exchange the C and the B to give B C A, then the second box would exchange the B and the A to give A C B, then the third box would exchange the C and the B to produce the sorted result.

Of course, there are many details to be worked out before an actual sorting machine based on this scheme can be built. For example, the method of encoding the inputs is left unspecified: one way would be to think of each wire in the diagram above as a "bus" of enough wires to carry the data with one bit per wire; another way is to have the compare-exchangers read their inputs one bit at a time along a single wire (most significant bit first). Also left unspecified is the timing: mechanisms must be included to ensure that no compare-exchanger performs its operation before its input is ready. We clearly won't be able to delve much deeper into such circuit design questions; instead we'll concentrate on the higher level issues concerning interconnecting simple processors such as compare-exchangers for solving larger problems.

To begin, we'll consider an algorithm for merging together two sorted files, using a sequence of "compare-exchange" operations that is independent of the particular numbers to be merged and is thus suitable for hardware implementation. Suppose that we have two sorted files of eight keys to be merged together into one sorted file. First write one file below the other, then compare those that are vertically adjacent and exchange them if necessary to put the larger one below the smaller one.

```
A  E  G  G  I  M  N  R        A  B  E  E  I  M  N  R
A  B  E  E  L  M  P  X        A  E  G  G  L  M  P  X
```

Next, split each line in half and interleave the halves, then perform the same compare-exchange operations on the numbers in the second and third lines. (Note that comparisons involving other pairs of lines are not necessary because of previous sorting.)

```
A  B  E  E            A  B  E  E
I  M  N  R            A  E  G  G
A  E  G  G            I  M  N  R
L  M  P  X            L  M  P  X
```

This leaves both the rows and the columns of the table sorted. This fact is a fundamental property of this method: the reader may wish to check that it is true, but a rigorous proof is a trickier exercise than one might think. It turns out that this property is preserved by the same operation: split each line in half, interleave the halves, and do compare-exchanges between items now vertically adjacent that came from different lines.

```
A  B            A  B
E  E            A  E
A  E            E  E
G  G            G  G
I  M            I  M
N  R            L  M
L  M            N  R
P  X            P  X
```

We have doubled the number of rows, halved the number of columns, and still kept the rows and the columns sorted. One more step completes the merge:

```
          A        A
          B        A
          A        B
          E        E
          E        E
          E        E
          G        G
          G        G
          I        I
          M        L
          L        M
          M        M
          N        N
          R        P
          P        R
          X        X
```

At last we have 16 rows and 1 column, which is sorted. This method obviously extends to merge files of equal lengths which are powers of two. Other sizes can be handled by adding dummy keys in a straightforward manner, though the number of dummy keys can get large (if N is just larger than a power of 2).

The basic "split each line in half and interleave the halves" operation in the above description is easy to visualize on paper, but how can it be translated into wiring for a machine? There is a surprising and elegant answer to this question which follows directly from writing the tables down in a different way. Rather than writing them down in a two-dimensional fashion, we'll write them down as a simple (one-dimensional) list of numbers, organized in *column-major* order: first put the elements in the first column, then put the elements in the second column, etc. Since compare-exchanges are only done between vertically adjacent items, this means that each stage involves a group of compare-exchange boxes, wired together according to the "split-and-interleave" operation which is necessary to bring items together into the compare-exchange boxes.

This leads to the following diagram, which corresponds precisely to the description using tables above, except that the tables are all written in column-major order (including an initial 1 by 16 table with one file, then the other). The reader should be sure to check the correspondence between this diagram and the tables given above. The compare-exchange boxes are drawn explicitly, and explicit lines are drawn showing how elements move in the "split-and-interleave" operation:

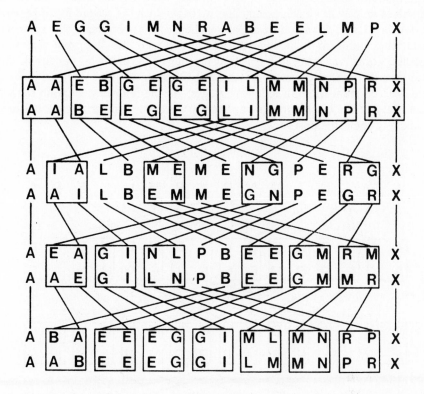

Surprisingly, in this representation each "split-and-interleave" operation reduces to precisely the same interconnection pattern. This pattern is called
the *perfect shuffle* because the wires are exactly interleaved, in the same way
that cards from the two halves would be interleaved in an ideal mix of a deck
of cards.

This method was named the *odd-even merge* by K. E. Batcher, who
invented it in 1968. The essential feature of the method is that all of the
compare-exchange operations in each stage can be done in parallel. It clearly
demonstrates that two files of N elements can be merged together in $\log N$
parallel steps (the number of rows in the table is halved at every step), using
less than $N \log N$ compare-exchange boxes. From the description above, this
might seem like a straightforward result: actually, the problem of finding such
a machine had stumped researchers for quite some time.

Batcher also developed a closely related (but more difficult to understand)
merging algorithm, the *bitonic merge*, which leads to an even simpler machine.

This method can be described in terms of the "split-and-interleave" operation on tables exactly as above, except that we begin with the second file in *reverse* sorted order and always do compare-exchanges between vertically adjacent items that came from the *same* lines. We won't go into the proof that this method works: our interest in it is that it removes the annoying feature in the odd-even merge that the compare-exchange boxes in the first stage are shifted one position from those in following stages. As the following diagram shows, each stage of the bitonic merge has exactly the same number of comparators, in exactly the same positions:

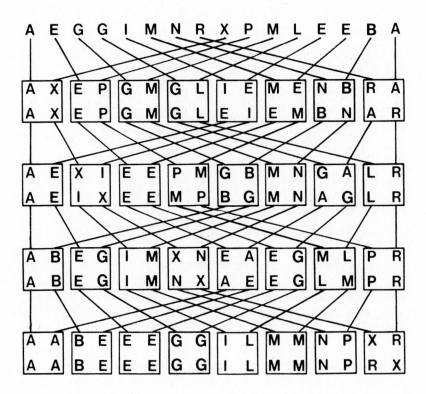

Now there is regularity not only in the interconnections but in the positions of the compare-exchange boxes. There are more compare-exchange boxes than for the odd-even merge, but this is not a problem, since the same number of parallel steps is involved. The importance of this method is that it leads directly to a way to do the merge using only N compare-exchange boxes. The idea is to simply collapse the rows in the table above to just one pair of rows, and thus produce a cycling machine wired together as follows:

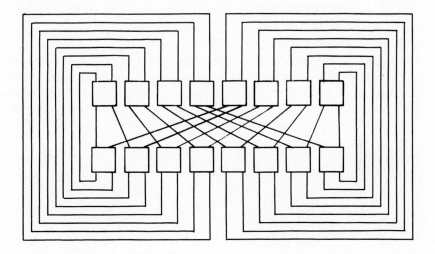

Such a machine can do $\log N$ compare-exchange-shuffle "cycles," one for each of the stages in the figure above.

Note carefully that this is not quite "ideal" parallel performance: since we can merge together two files of N elements using one processor in a number of steps proportional to N, we would hope to be able to do it in a constant number of steps using N processors. In this case, it has been proven that it is not possible to achieve this ideal and that the above machine achieves the best possible parallel performance for merging using compare-exchange boxes.

The perfect shuffle interconnection pattern is appropriate for a variety of other problems. For example, if a 2^n-by-2^n square matrix is kept in row-major order, then n perfect shuffles will transpose the matrix (convert it to column-major order). More important examples include the fast Fourier transform (which we'll examine in the next chapter); sorting (which can be developed by applying either of the methods above recursively); polynomial evaluation; and a host of others. Each of these problems can be solved using a cycling perfect shuffle machine with the same interconnections as the one diagramed above but with different (somewhat more complicated) processors. Some researchers have even suggested the use of the perfect shuffle interconnection for "general-purpose" parallel computers.

Systolic Arrays

One problem with the perfect shuffle is that the wires used for inter-connection are long. Furthermore, there are many wire crossings: a shuffle with N wires involves a number of crossings proportional to N^2. These two properties turn out to create difficulties when a perfect shuffle machine is actually constructed: long wires lead to time delays and crossings make the interconnection expensive and inconvenient.

A natural way to avoid both of these problems is to insist that processors be connected only to processors which are physically adjacent. As above, we operate the processors synchronously: at each step, each processor reads inputs from its neighbors, does a computation, and writes outputs to its neighbors. It turns out that this is not necessarily restrictive, and in fact H. T. Kung showed in 1978 that arrays of such processors, which he termed *systolic arrays* (because the way data flows within them is reminiscent of a heartbeat), allow very efficient use of the processors for some fundamental problems.

As a typical application, we'll consider the use of systolic arrays for matrix-vector multiplication. For a particular example, consider the matrix operation

$$\begin{pmatrix} 1 & 3 & -4 \\ 1 & 1 & -2 \\ -1 & -2 & 5 \end{pmatrix} \begin{pmatrix} 1 \\ 5 \\ 2 \end{pmatrix} = \begin{pmatrix} 8 \\ 2 \\ -1 \end{pmatrix}$$

This computation will be carried out on a row of simple processors each of which has three input lines and two output lines, as depicted below:

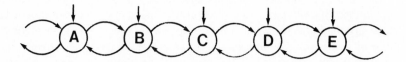

Five processors are used because we'll be presenting the inputs and reading the outputs in a carefully timed manner, as described below.

During each step, each processor reads one input from the *left*, one from the *top*, and one from the *right*; performs a simple computation; and writes one output to the *left* and one output to the *right*. Specifically, the *right* output gets whatever was on the *left* input, and the *left* output gets the result computed by multiplying together the *left* and *top* inputs and adding the *right* input. A crucial characteristic of the processors is that they always perform a dynamic transformation of inputs to outputs; they never have to "remember" computed values. (This is also true of the processors in the perfect shuffle machine.) This is a ground rule imposed by low-level constraints on the hardware design, since the addition of such a "memory" capability can be (relatively) quite expensive.

The paragraph above gives the "program" for the systolic machine; to complete the description of the computation, we need to also describe exactly how the input values are presented. This timing is an essential feature of the systolic machine, in marked contrast to the perfect shuffle machine, where all the input values are presented at one time and all the output values are available at some later time.

The general plan is to bring in the matrix through the *top* inputs of the processors, reflected about the main diagonal and rotated forty-five degrees, and the vector through the *left* input of processor A, to be passed on to the other processors. Intermediate results are passed from right to left in the array, with the output eventually appearing on the *left* output of processor A. The specific timing for our example is shown in the following table, which gives the values of the *left*, *top*, and *right* inputs for each processor at each step:

	A	B	C	D	E	A	B	C	D	E	A	B	C	D
1	1													
2		1												
3	5	1						1						
4		5	1				3	1				1		
5	2	5	1			-4	1	-1			16	1		
6		2	5				-2	-2			8	6	-1	
7			2	5				5				2	-11	
8			2									2	-1	
9				2									-1	
10													-1	

The input vector is presented to the *left* input of processor A at steps 1, 3, and 5 and passed right to the other processors in subsequent steps. The input matrix is presented to the *top* inputs of the processors starting at steps 3, skewed so the right-to-left diagonals of the matrix are presented in successive steps. The output vector appears as the *left* output of processor A at steps 6, 8, and 10. (In the table, this appears as the *right* input of an imaginary processor to the left of A, which is collecting the answer.)

The actual computation can be traced by following the *right* inputs (*left* outputs) which move from right to left through the array. All computations produce a zero result until step 3, when processor C has 1 for its *left* input and 1 for its *top* input, so it computes the result 1, which is passed along as processor B's *right* input for step 4. At step 4, processor B has non-zero values for all three of its inputs, and it computes the value 16, to be passed on to processor A for step 5. Meanwhile, processor D computes a value 1 for processor C's use at step 5. Then at step 5, processor A computes the value 8, which is presented as the first output value at step 6; C computes the value 6 for B's use at step 6, and E computes its first nonzero value (−1) for use by D at step 6. The computation of the second output value is completed by B at step 6 and passed through A for output at step 8, and the computation of the third output value is completed by C at step 7 and passed through B and A for output at step 10.

Once the process has been checked at a detailed level as in the previous paragraph, the method is better understood at a somewhat higher level. The numbers in the middle part of the table above are simply a copy of the input matrix, rotated and reflected as required for presentation to the *top* inputs of the processors. If we check the numbers in the corresponding positions at the left part of the table, we find three copies of the input vector, located in exactly the right positions and at the right times for multiplication against the rows of the matrix. And the corresponding positions on the right give the intermediate results for each multiplication of the input vector with each matrix row. For example, the multiplication of the input vector with the middle matrix row requires the partial computations $1*1 = 1$, $1 + 1*5 = 6$, and $6 + (-2)*2 = 2$, which appear in the entries 1 6 2 in the reflected middle row on the right-hand side of the table. The systolic machine manages to time things so that each matrix element "meets" the proper input vector entry and the proper partial computation at the processor where it is input, so that it can be incorporated into the partial result.

The method extends in an obvious manner to multiply an N-by-N matrix by an N-by-1 vector using $2N - 1$ processors in $4N - 2$ steps. This does come close to the ideal situation of having every processor perform useful work at every step: a quadratic algorithm is reduced to a linear algorithm using a linear number of processors.

One can appreciate from this example that systolic arrays are at once simple and powerful. The output vector at the edge appears almost as if by magic! However, each individual processor is just performing the simple computation described above: the magic is in the interconnection and the timed presentation of the inputs. As before, we've only described a general method of parallel computation. Many details in the logical design need to be worked out before such a systolic machine can be constructed.

As with perfect shuffle machines, systolic arrays may be used in many different types of problems, including string matching and matrix multiplication among others. Some researchers have even suggested the use of this interconnection pattern for "general-purpose" parallel machines.

Certainly, the study of the perfect shuffle and systolic machines illustrates that hardware design can have a significant effect on algorithm design, suggesting changes that can provide interesting new algorithms and fresh challenges for the algorithm designer. While this is an interesting and fruitful area for further research, we must conclude with a few sobering notes. First, a great deal of engineering effort is required to translate general schemes for parallel computation such as those sketched above to actual algorithm machines with good performance. For many applications, the resource expenditure required is simply not justified, and a simple "algorithm machine" consisting of a conventional (inexpensive) microprocessor running a conventional algorithm will do quite well. For example, if one has many instances of the same problem to solve and several microprocessors with which to solve them, then ideal parallel performance can be achieved by having each microprocessor (using a conventional algorithm) working on a different instance of the problem, with no interconnection at all required. If one has N files to sort and N processors available on which to sort them, why not simply use one processor for each sort, rather than having all N processors labor together on all N sorts? Techniques such as those discussed in this chapter are currently justified only for applications with very special time or space requirements. In studying various parallel computation schemes and their effects on the performance of various algorithms, we can look forward to the development of general-purpose parallel computers that will provide improved performance for a wide variety of algorithms.

Exercises

1. Outline two possible ways to use parallelism in Quicksort.

2. Write a conventional Pascal program which merges files using Batcher's bitonic method.

3. Write a conventional Pascal program which merges files using Batcher's bitonic method, but doesn't actually do any shuffles.

4. Does the program of the previous exercise have any advantage over conventional merging?

5. How many perfect shuffles will bring all the elements in an array of size 2^n back to their original positions?

6. Draw a table like the one given in the text to illustrate the operation of the systolic matrix-vector multiplier for the following problem:

$$\begin{pmatrix} 2 & 1 & 4 \\ 3 & 0 & 1 \\ 1 & -1 & 3 \end{pmatrix} \begin{pmatrix} 3 \\ 1 \\ -1 \end{pmatrix} = \begin{pmatrix} 3 \\ 8 \\ -1 \end{pmatrix}$$

7. Write a conventional Pascal program which simulates the operation of the systolic array for multiplying a N-by-N matrix by an N-by-1 vector.

8. Show how to use a systolic array to transpose a matrix.

9. How many processors and how many steps would be required for a systolic machine which can multiply an M-by-N matrix by an N-by-1 vector?

10. Give a simple parallel scheme for matrix-vector multiplication using processors which have the capability to "remember" computed values.

36. The Fast Fourier Transform

One of the most widely used arithmetic algorithms is the *fast Fourier transform*, which (among many other applications) can provide a substantial reduction in the time required to multiply two polynomials. The Fourier transform is of fundamental importance in mathematical analysis and is the subject of volumes of study. The emergence of an efficient algorithm for this computation was a milestone in the history of computing.

It would be beyond the scope of this book to outline the mathematical basis for the Fourier transform or to survey its many applications. Our purpose is to learn the characteristics of a fundamental algorithm within the context of some of the other algorithms that we've been studying. In particular, we'll examine how to use the algorithm for polynomial multiplication, a problem that we studied in Chapter 4. Only a very few elementary facts from complex analysis are needed to show how the Fourier transform can be used to multiply polynomials, and it is possible to appreciate the fast Fourier transform algorithm without fully understanding the underlying mathematics. The divide-and-conquer technique is applied in a way similar to other important algorithms that we've seen.

Evaluate, Multiply, Interpolate

The general strategy of the improved method for polynomial multiplication that we'll be examining takes advantage of the fact that a polynomial of degree $N - 1$ is completely determined by its value at N different points. When we multiply two polynomials of degree $N - 1$ together, we get a polynomial of degree $2N - 2$: if we can find that polynomial's value at $2N - 1$ points, then it is completely determined. But we can find the value of the result at any point simply by evaluating the two polynomials to be multiplied at that point and then multiplying those numbers. This leads to the following general scheme for multiplying two polynomials of degree $N - 1$:

Evaluate the input polynomials at $2N-1$ distinct points.

Multiply the two values obtained at each point.

Interpolate to find the unique result polynomial that has the given value at the given points.

For example, to compute $r(x) = p(x)q(x)$ with $p(x) = 1+x+x^2$ and $q(x) = 2-x+x^2$, we can evaluate $p(x)$ and $q(x)$ at any five points, say $-2,-1,0,1,2$, to get the values

$$[p(-2), p(-1), p(0), p(1), p(2)] = [3, 1, 1, 3, 7],$$
$$[q(-2), q(-1), q(0), q(1), q(2)] = [8, 4, 2, 2, 4].$$

Multiplying these together term-by-term gives enough values for the product polynomial,

$$[r(-2), r(-1), r(0), r(1), r(2)] = [24, 4, 2, 6, 28],$$

that its coefficients can be found by interpolation. By the Lagrange formula,

$$
\begin{aligned}
r(x) = {}& 24\frac{x+1}{-2+1}\frac{x-0}{-2-0}\frac{x-1}{-2-1}\frac{x-2}{-2-2} \\
& + 4\frac{x+2}{-1+2}\frac{x-0}{-1-0}\frac{x-1}{-1-1}\frac{x-2}{-1-2} \\
& + 2\frac{x+2}{0+2}\frac{x+1}{0+1}\frac{x-1}{0-1}\frac{x-2}{0-2} \\
& + 6\frac{x+2}{1+2}\frac{x+1}{1+1}\frac{x-0}{1-0}\frac{x-2}{1-2} \\
& + 28\frac{x+2}{2+2}\frac{x+1}{2+1}\frac{x-0}{2-0}\frac{x-1}{2-1},
\end{aligned}
$$

which simplifies to the result

$$r(x) = 2 + x + 2x^2 + x^4.$$

As described so far, this method is not a good algorithm for polynomial multiplication since the best algorithms we have so far for both evaluation (repeated application of Horner's method) and interpolation (Lagrange formula) require N^2 operations. However, there is some hope of finding a better algorithm because the method works for any choice of $2N-1$ distinct points whatsoever, and it is reasonable to expect that evaluation and interpolation will be easier for some sets of points than for others.

Complex Roots of Unity

It turns out that the most convenient points to use for polynomial interpolation and evaluation are complex numbers, in fact, a particular set of complex numbers called the *complex roots of unity*.

A brief review of some facts about complex analysis is necessary. The number $i = \sqrt{-1}$ is an *imaginary* number: though $\sqrt{-1}$ is meaningless as a real number, it is convenient to give it a name, i, and perform algebraic manipulations with it, replacing i^2 with -1 whenever it appears. A *complex number* consists of two parts, real and imaginary, usually written as $a + bi$, where a and b are reals. To multiply complex numbers, apply the usual rules, but replace i^2 with -1 whenever it appears. For example,

$$(a + bi)(c + di) = (ac - bd) + (ad + bc)i.$$

Sometimes the real or imaginary part can cancel out when a complex multiplication is performed. For example,

$$(1 - i)(1 - i) = -2i,$$

$$(1 + i)^4 = -4,$$

$$(1 + i)^8 = 16.$$

Scaling this last equation by dividing through by $16 = \sqrt{2}^8$), we find that

$$\left(\frac{1}{\sqrt{2}} + \frac{i}{\sqrt{2}} \right)^8 = 1.$$

In general, there are many complex numbers that evaluate to 1 when raised to a power. These are the so-called complex roots of unity. In fact, it turns out that for each N, there are exactly N complex numbers z with $z^N = 1$. One of these, named w_N, is called the *principal Nth root of unity*; the others are obtained by raising w_N to the kth power, for $k = 0, 1, 2, \ldots, N - 1$. For example, we can list the eighth roots of unity as follows:

$$w_8^0, w_8^1, w_8^2, w_8^3, w_8^4, w_8^5, w_8^6, w_8^7.$$

The first root, w_N^0, is 1 and the second, w_N^1, is the principal root. Also, for N even, the root $w_N^{N/2}$ is -1 (because $(w_N^{N/2})^2 = 1$).

The precise values of the roots are unimportant for the moment. We'll be using only simple properties which can easily be derived from the basic fact that the Nth power of any Nth root of unity must be 1.

Evaluation at the Roots of Unity

The crux of our implementation will be a procedure for evaluating a polynomial of degree $N - 1$ at the Nth roots of unity. That is, this procedure
transforms the N coefficients which define the polynomial into the N values
resulting from evaluating that polynomial at all of the Nth roots of unity.

This may not seem to be exactly what we want, since for the first step of
the polynomial multiplication procedure we need to evaluate polynomials of
degree $N-1$ at $2N-1$ points. Actually, this is no problem, since we can view
a polynomial of degree $N - 1$ as a polynomial of degree $2N - 2$ with $N - 1$
coefficients (those for the terms of largest degree) which are 0.

The algorithm that we'll use to evaluate a polynomial of degree $N - 1$
at N points simultaneously will be based on a simple divide-and-conquer
strategy. Rather than dividing the polynomials in the middle (as in the
multiplication algorithm in Chapter 4) we'll divide them into two parts by
putting alternate terms in each part. This division can easily be expressed in
terms of polynomials with half the number of coefficients. For example, for
$N = 8$, the rearrangement of terms is as follows:

$$p(x) = p_0 + p_1 x + p_2 x^2 + p_3 x^3 + p_4 x^4 + p_5 x^5 + p_6 x^6 + p_7 x^7$$
$$= (p_0 + p_2 x^2 + p_4 x^4 + p_6 x^6) + x(p_1 + p_3 x^2 + p_5 x^4 + p_7 x^6)$$
$$\equiv p_e(x^2) + x p_o(x^2).$$

The Nth roots of unity are convenient for this decomposition because if you
square a root of unity, you get another root of unity. In fact, even more is
true: for N even, if you square an Nth root of unity, you get an $\frac{1}{2}N$th root of
unity (a number which evaluates to 1 when raised to the $\frac{1}{2}N$th power). This
is exactly what is needed to make the divide-and-conquer method work. To
evaluate a polynomial with N coefficients on N points, we split it into two
polynomials with $\frac{1}{2}N$ coefficients. These polynomials need only be evaluated
on $\frac{1}{2}N$ points (the $\frac{1}{2}N$th roots of unity) to compute the values needed for the
full evaluation.

To see this more clearly, consider the evaluation of a degree-7 polynomial
$p(x)$ on the eighth roots of unity

$$W_8 : w_8^0, w_8^1, w_8^2, w_8^3, w_8^4, w_8^5, w_8^6, w_8^7.$$

Since $w_8^4 = -1$, this is the same as the sequence

$$W_8 : w_8^0, w_8^1, w_8^2, w_8^3, -w_8^0, -w_8^1, -w_8^2, -w_8^3.$$

Squaring each term of this sequence gives two copies of the sequence $\{W_4\}$ of
the fourth roots of unity:

$$W_8^2 : w_4^0, w_4^1, w_4^2, w_4^3, w_4^0, w_4^1, w_4^2, w_4^3.$$

Now, our equation

$$p(x) = p_e(x^2) + xp_o(x^2)$$

tells us immediately how to evaluate $p(x)$ at the eighth roots of unity from these sequences. First, we evaluate $p_e(x)$ and $p_o(x)$ at the fourth roots of unity. Then we substitute each of the eighth roots of unity for x in the equation above, which requires adding the appropriate p_e value to the product of the appropriate p_o value and the eighth root of unity:

$$
\begin{aligned}
p(w_8^0) &= p_e(w_4^0) + w_8^0 p_o(w_4^0), \\
p(w_8^1) &= p_e(w_4^1) + w_8^1 p_o(w_4^1), \\
p(w_8^2) &= p_e(w_4^2) + w_8^2 p_o(w_4^2), \\
p(w_8^3) &= p_e(w_4^3) + w_8^3 p_o(w_4^3), \\
p(w_8^4) &= p_e(w_4^0) - w_8^0 p_o(w_4^0), \\
p(w_8^5) &= p_e(w_4^1) - w_8^1 p_o(w_4^1), \\
p(w_8^6) &= p_e(w_4^2) - w_8^2 p_o(w_4^2), \\
p(w_8^7) &= p_e(w_4^3) - w_8^3 p_o(w_4^3).
\end{aligned}
$$

In general, to evaluate $p(x)$ on the Nth roots of unity, we recursively evaluate $p_e(x)$ and $p_o(x)$ on the $\frac{1}{2}N$th roots of unity and perform the N multiplications as above. This only works when N is even, so we'll assume from now on that N is a power of two, so that it remains even throughout the recursion. The recursion stops when $N = 2$ and we have $p_0 + p_1 x$ to be evaluated at 1 and -1, with the results $p_0 + p_1$ and $p_0 - p_1$.

The number of multiplications used satisfies the "fundamental divide-and-conquer" recurrence

$$M(N) = 2M(N/2) + N,$$

which has the solution $M(N) = N \lg N$. This is a substantial improvement over the straightforward N^2 method for interpolation but, of course, it works only at the roots of unity.

This gives a method for transforming a polynomial from its representation as N coefficients in the conventional manner to its representation in terms of its values at the roots of unity. This conversion of the polynomial from the first representation to the second is the Fourier transform, and the efficient recursive calculation procedure that we have described is called the "fast" Fourier transform (FFT). (These same techniques apply to more general functions than polynomials. More precisely we're doing the "discrete" Fourier transform.)

Interpolation at the Roots of Unity

Now that we have a fast way to evaluate polynomials at a specific set of points, all that we need is a fast way to interpolate polynomials at those same points, and we will have a fast polynomial multiplication method. Surprisingly, it works out that, for the complex roots of unity, running the evaluation program on a particular set of points will do the interpolation! This is a specific instance of a fundamental "inversion" property of the Fourier transform, from which many important mathematical results can be derived.

For our example with $N = 8$, the interpolation problem is to find the polynomial

$$r(x) = r_0 + r_1 x + r_2 x^2 + r_3 x^3 + r_4 x^4 + r_5 x^5 + r_6 x^6 + r_7 x^7$$

which has the values

$$r(w_8^0) = s_0, \quad r(w_8^1) = s_1, \quad r(w_8^2) = s_2, \quad r(w_8^3) = s_3,$$

$$r(w_8^4) = s_4, \quad r(w_8^5) = s_5, \quad r(w_8^6) = s_6, \quad r(w_8^7) = s_7.$$

As we've said before, the interpolation problem is the "inverse" of the evaluation problem. When the points under consideration are the complex roots of unity, this is literally true. If we let

$$s(x) = s_0 + s_1 x + s_2 x^2 + s_3 x^3 + s_4 x^4 + s_5 x^5 + s_6 x^6 + s_7 x^7$$

then we can get the coefficients

$$r_0, r_1, r_2, r_3, r_4, r_5, r_6, r_7$$

just by evaluating the polynomial $s(x)$ at the inverses of the complex roots of unity

$$W_8^{-1} : w_8^0, w_8^{-1}, w_8^{-2}, w_8^{-3}, w_8^{-4}, w_8^{-5}, w_8^{-6}, w_8^{-7}$$

which is the same sequence as the complex roots of unity, but in a different order:

$$W_8^{-1} : w_8^0, w_8^7, w_8^6, w_8^5, w_8^4, w_8^3, w_8^2, w_8^1.$$

In other words, we can use exactly the same routine for interpolation as for evaluation: only a simple rearrangement of the points to be evaluated is required.

The proof of this fact requires some elementary manipulations with finite sums: those unfamiliar with such manipulations may wish to skip to the end of this paragraph. Evaluating $s(x)$ at the inverse of the tth Nth root of unity gives

$$s(w_N^{-t}) = \sum_{0 \le j < N} s_j(w_N^{-t})^j$$

$$= \sum_{0 \le j < N} r(w_N^j)(w_N^{-t})^j$$

$$= \sum_{0 \le j < N} \sum_{0 \le i < N} r_i(w_N^j)^i(w_N^{-t})^j$$

$$= \sum_{0 \le j < N} \sum_{0 \le i < N} r_i w_N^{j(i-t)}$$

$$= \sum_{0 \le i < N} r_i \sum_{0 \le j < N} w_N^{j(i-t)} = N r_t.$$

Nearly everything disappears in the last term because the inner sum is trivially N if $i = t$: if $i \ne t$ then it evaluates to

$$\sum_{0 \le j < N} w_N^{j(i-t)} = \frac{w_N^{(i-t)N} - 1}{w_N^{(i-t)} - 1} = 0.$$

Note that an extra scaling factor of N arises. This is the "inversion theorem" for the discrete Fourier transform, which says that the same method will convert a polynomial both ways: between its representation as coefficients and its representation as values at the complex roots of unity.

While the mathematics may seem complicated, the results indicated are quite easy to apply: to interpolate a polynomial on the Nth roots of unity, use the same procedure as for evaluation, using the interpolation values as polynomial coefficients, then rearrange and scale the answers.

Implementation

Now we have all the pieces for a divide-and-conquer algorithm to multiply two polynomials using only about $N \lg N$ operations. The general scheme is to:

Evaluate the input polynomials at the $(2N - 1)$st roots of unity.

Multiply the two values obtained at each point.

Interpolate to find the result by evaluating the polynomial defined by the numbers just computed at the $(2N - 1)$st roots of unity.

The description above can be directly translated into a program which uses a procedure that can evaluate a polynomial of degree $N - 1$ at the Nth roots of unity. Unfortunately, all the arithmetic in this algorithm is to be complex arithmetic, and Pascal has no built-in type *complex*. While it is possible

to have a user-defined type for the complex numbers, it is then necessary
to also define procedures or functions for all the arithmetic operations on
the numbers, and this obscures the algorithm unnecessarily. The following
implementation assumes a type *complex* for which the obvious arithmetic
functions are defined:

```
eval(p, outN, 0);
eval(q, outN, 0);
for i:=0 to outN do r[i]:=p[i]*q[i];
eval(r, outN, 0);
for i:=1 to N do
    begin t:=r[i]; r[i]:=r[outN+1−i]; r[outN+1−i]:=t end;
for i:=0 to outN do r[i]:=r[i]/(outN+1);
```

This program assumes that the global variable *outN* has been set to $2N-1$,
and that p, q, and r are arrays indexed from 0 to $2N - 1$ which hold complex
numbers. The two polynomials to be multiplied, p and q are of degree $N - 1$,
and the other coefficients in those arrays are initially set to 0. The procedure
eval replaces the coefficients of the polynomial given as the first argument by
the values obtained when the polynomial is evaluated at the roots of unity.
The second argument specifies the degree of the polynomial (one less than the
number of coefficients and roots of unity) and the third argument is described
below. The above code computes the product of p and q and leaves the result
in r.

Now we are left with the implementation of *eval*. As we've seen before,
recursive programs involving arrays can be quite cumbersome to implement. It
turns out that for this algorithm it is possible to get around the usual storage
management problem by reusing the storage in a clever way. What we would
like to do is have a recursive procedure that takes as input a contiguous array
of $N + 1$ coefficients and returns the $N + 1$ values in the same array. But
the recursive step involves processing two noncontiguous arrays: the odd and
even coefficients. On reflection, the reader will see that the "perfect shuffle"
of the previous chapter is exactly what is needed here. We can get the odd
coefficients in a contiguous subarray (the first half) and the even coefficients
in a contiguous subarray (the second half) by doing a "perfect unshuffle" of
the input, as diagramed below for $N = 15$:

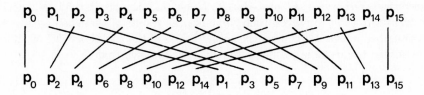

This leads to the following implementation of the FFT:

```
procedure eval(var p: poly; N, k: integer);
   var i, j: integer;
   begin
   if N=1 then
      begin
      t:=p[k];  p1:=p[k+1];
      p[k]:=t+p1;  p[k+1]:=t−p1
      end
   else
      begin
        for i:=0 to N div 2 do
           begin
           j:=k+2*i;
           t[i]:=p[j];  t[i+1+N div 2]:=p[j+1]
           end;
        for i:=0 to N do p[k+i]:=t[i];
        eval(p, N div 2, k);
        eval(p, N div 2, k+1+N div 2);
        j:=(outN+1) div (N+1);
        for i:=0 to N div 2 do
           begin
           t:=w[i*j]*p[k+(N div 2)+1+i];
           t[i]:=p[k+i]+t;  t[i+(N div 2)+1]:=p[k+i]−t
           end;
        for i:=0 to N do p[k+i]:=t[i]
        end
   end;
```

This program transforms the polynomial of degree N inplace in the subarray $p[k..k+N]$ using the recursive method outlined above. (For simplicity, the code assumes that $N+1$ is a power of two, though this dependence is not hard to remove.) If $N = 1$, then the easy computation to evaluate at 1 and -1 is

performed. Otherwise the procedure first shuffles, then recursively calls itself
to transform the two halves, then combines the results of these computations
as described above. Of course, the actual values of the complex roots of unity
are needed to do the implementation. It is well known that

$$w_N^j = \cos\left(\frac{2\pi j}{N+1}\right) + i\sin\left(\frac{2\pi j}{N+1}\right);$$

these values are easily computed using conventional trigonometric functions.
In the above program, the array w is assumed to hold the ($outN+1$)st roots of
unity. To get the roots of unity needed, the program selects from this array
at an interval determined by the variable i. For example, if $outN$ were 15,
the fourth roots of unity would be found in $w[0]$, $w[4]$, $w[8]$, and $w[12]$. This
eliminates the need to recompute roots of unity each time they are used.

As mentioned at the outset, the scope of applicability of the FFT is far
greater than can be indicated here; and the algorithm has been intensively
used and studied in a variety of domains. Nevertheless, the fundamental
principles of operation in more advanced applications are the same as for the
polynomial multiplication problem that we've discussed here. The FFT is
a classic example of the application of the "divide-and-conquer" algorithm
design paradigm to achieve truly significant computational savings.

Exercises

1. Explain how you would improve the simple evaluate-multiply-interpolate algorithm for multiplying together two polynomials $p(x)$ and $q(x)$ with known roots $p_0, p_1, \ldots, p_{N-1}$ and $q_0, q_1, \ldots, q_{N-1}$.

2. Find a set of N real numbers at which a polynomial of degree N can be evaluated using substantially fewer than N^2 operations.

3. Find a set of N real numbers at which a polynomial of degree N can be interpolated using substantially fewer than N^2 operations.

4. What is the value of w_N^M for $M > N$?

5. Would it be worthwhile to multiply sparse polynomials using the FFT?

6. The FFT implementation has three calls to *eval*, just as the polynomial multiplication procedure in Chapter 4 has three calls to *mult*. Why is the FFT implementation more efficient?

7. Give a way to multiply two complex numbers together using fewer than four integer multiplication operations.

8. How much storage would be used by the FFT if we didn't circumvent the "storage management problem" with the perfect shuffle?

9. Why can't some technique like the perfect shuffle be used to avoid the problems with dynamically declared arrays in the polynomial multiplication procedure of Chapter 4?

10. Write an efficient program to multiply a polynomial of degree N by a polynomial of degree M (not necessarily powers of two).

37. Dynamic Programming

The principle of *divide-and-conquer* has guided the design of many of the algorithms we've studied: to solve a large problem, break it up into smaller problems which can be solved independently. In *dynamic programming* this principle is carried to an extreme: when we don't know exactly which smaller problems to solve, we simply solve them all, then store the answers away to be used later in solving larger problems.

There are two principal difficulties with the application of this technique. First, it may not always be possible to combine the solutions of two problems to form the solution of a larger one. Second, there may be an unacceptably large number of small problems to solve. No one has precisely characterized which problems can be effectively solved with dynamic programming; there are certainly many "hard" problems for which it does not seem to be applicable (see Chapters 39 and 40), as well as many "easy" problems for which it is less efficient than standard algorithms. However, there is a certain class of problems for which dynamic programming is quite effective. We'll see several examples in this section. These problems involve looking for the "best" way to do something, and they have the general property that any decision involved in finding the best way to do a small subproblem remains a good decision even when that subproblem is included as a piece of some larger problem.

Knapsack Problem

Suppose that a thief robbing a safe finds N items of varying size and value that he could steal, but has only a small knapsack of capacity M which he can use to carry the goods. The *knapsack problem* is to find the combination of items which the thief should choose for his knapsack in order to maximize the total take. For example, suppose that he has a knapsack of capacity 17 and the safe contains many items of each of the following sizes and values:

name	A	B	C	D	E
size	3	4	7	8	9
value	4	5	10	11	13

(As before, we use single letter names for the items in the example and integer indices in the programs, with the knowledge that more complicated names could be translated to integers using standard searching techniques.) Then the thief could take five A's (but not six) for a total take of 20, or he could fill up his knapsack with a D and an E for a total take of 24, or he could try many other combinations.

Obviously, there are many commercial applications for which a solution to the knapsack problem could be important. For example, a shipping company might wish to know the best way to load a truck or cargo plane with items for shipment. In such applications, other variants to the problem might arise: for example, there might be a limited number of each kind of item available. Many such variants can be handled with the same approach that we're about to examine for solving the basic problem stated above.

In a dynamic programming solution to the knapsack problem, we calculate the best combination for *all* knapsack sizes up to M. It turns out that we can perform this calculation very efficiently by doing things in an appropriate order, as in the following program:

```
for j:=1 to N do
  begin
  for i:=1 to M do
    if i−size[j]>=0 then
      if cost[i]<(cost[i−size[j]]+val[j]) then
        begin
        cost[i]:=cost[i−size[j]]+val[j];
        best[i]:=j
        end;
  end;
```

In this program, $cost[i]$ is the highest value that can be achieved with a knapsack of capacity i and $best[i]$ is the last item that was added to achieve that maximum (this is used to recover the contents of the knapsack, as described below). First, we calculate the best that we can do for all knapsack sizes when only items of type A are taken, then we calculate the best that we can do when only A's and B's are taken, etc. The solution reduces to a simple calculation for $cost[i]$. Suppose an item j is chosen for the knapsack: then the best value that could be achieved for the total would be $val[j]$ (for the item) plus $cost[i-size[j]]$ (to fill up the rest of the knapsack). If this value exceeds the best value that can be achieved *without* an item j, then we update $cost[i]$ and $best[i]$; otherwise we leave them alone. A simple induction proof shows that this strategy solves the problem.

The following table traces the computation for our example. The first pair of lines shows the best that can be done (the contents of the *cost* and *best* arrays) with only A's, the second pair of lines shows the best that can be done with only A's and B's, etc.:

1	2	3	4	5	6	7	8	9	10	11	12	13	14	15	16	17
0	0	4	4	4	8	8	8	12	12	12	16	16	16	20	20	20
		A	A	A	A	A	A	A	A	A	A	A	A	A	A	A
0	0	4	5	5	8	9	10	12	13	14	16	17	18	20	21	22
		A	B	B	A	B	B	A	B	B	A	B	B	A	B	B
0	0	4	5	5	8	10	10	12	14	15	16	18	20	20	22	24
		A	B	B	A	C	B	A	C	C	A	C	C	A	C	C
0	0	4	5	5	8	10	11	12	14	15	16	18	20	21	22	24
		A	B	B	A	C	D	A	C	C	A	C	C	D	C	C
0	0	4	5	5	8	10	11	13	14	15	17	18	20	21	23	24
		A	B	B	A	C	D	E	C	C	E	C	C	D	E	C

Thus the highest value that can be achieved with a knapsack of size 17 is 24. In order to compute this result, we also solved many smaller subproblems. For example, the highest value that can be achieved with a knapsack of size 16 using only A's B's and C's is 22.

The actual contents of the optimal knapsack can be computed with the aid of the *best* array. By definition, $best[M]$ is included, and the remaining contents are the same as for the optimal knapsack of size $M-size[best[M]]$. Therefore, $best[M-size[best[M]]]$ is included, and so forth. For our example, $best[17]=C$, then we find another type C item at size 10, then a type A item at size 3.

It is obvious from inspection of the code that the running time of this algorithm is proportional to NM. Thus, it will be fine if M is not large, but could become unacceptable for large capacities. In particular, a crucial point that should not be overlooked is that the method does not work at all if M and the sizes or values are, for example, real numbers instead of integers. This is more than a minor annoyance: it is a fundamental difficulty. No good solution is known for this problem, and we'll see in Chapter 40 that many people believe that no good solution exists. To appreciate the difficulty of the problem, the reader might wish to try solving the case where the values are all 1, the size of the jth item is $\sqrt{j}$ and M is $N/2$.

But when capacities, sizes and values are all integers, we have the fundamental principle that optimal decisions, once made, do not need to be changed. Once we know the best way to pack knapsacks of any size with the first j items, we do not need to reexamine those problems, regardless of what the next items are. Any time this general principle can be made to work, dynamic programming is applicable.

In this algorithm, only a small amount of information about previous optimal decisions needs to be saved. Different dynamic programming applications have widely different requirements in this regard: we'll see other examples below.

Matrix Chain Product

Suppose that the six matrices

$$
\begin{bmatrix} a_{11} & a_{12} \\ a_{21} & a_{22} \\ a_{31} & a_{32} \\ a_{41} & a_{42} \end{bmatrix}
\begin{bmatrix} b_{11} & b_{12} & b_{13} \\ b_{21} & b_{22} & b_{23} \end{bmatrix}
\begin{bmatrix} c_{11} \\ c_{21} \\ c_{31} \end{bmatrix}
\begin{bmatrix} d_{11} & d_{12} \end{bmatrix}
\begin{bmatrix} e_{11} & e_{12} \\ e_{21} & e_{22} \end{bmatrix}
\begin{bmatrix} f_{11} & f_{12} & f_{13} \\ f_{21} & f_{22} & f_{23} \end{bmatrix}
$$

are to be multiplied together. Of course, for the multiplications to be valid, the number of columns in one matrix must be the same as the number of rows in the next. But the total number of scalar multiplications involved depends on the order in which the matrices are multiplied. For example, we could proceed from left to right: multiplying A by B, we get a 4-by-3 matrix after using 24 scalar multiplications. Multiplying this result by C gives a 4-by-1 matrix after 12 more scalar multiplications. Multiplying this result by D gives a 4-by-2 matrix after 8 more scalar multiplications. Continuing in this way, we get a 4-by-3 result after a grand total of 84 scalar multiplications. But if we proceed from right to left instead, we get the same 4-by-3 result with only 69 scalar multiplications.

Many other orders are clearly possible. The order of multiplication can be expressed by parenthesization: for example the left-to-right order described above is the ordering $(((((A*B)*C)*D)*E)*F)$, and the right-to-left order is $(A*(B*(C*(D*(E*F)))))$. Any legal parenthesization will lead to the correct answer, but which leads to the fewest scalar multiplications?

Very substantial savings can be achieved when large matrices are involved: for example, if matrices B, C, and F in the example above were to each have a dimension of 300 where their dimension is 3, then the left-to-right order will require 6024 scalar multiplications but the right-to-left order will use an astronomical $274,200$. (In these calculations we're assuming that the standard method of matrix multiplication is used. Strassen's or some similar method could save some work for large matrices, but the same considerations about the order of multiplications apply. Thus, multiplying a p-by-q matrix by a q-by-r matrix will produce a p-by-r matrix, each entry computed with q multiplications, for a total of pqr multiplications.)

In general, suppose that N matrices are to be multiplied together:

$$M_1 M_2 M_3 \cdots M_N$$

where the matrices satisfy the constraint that M_i has r_i rows and r_{i+1} columns for $1 \leq i < N$. Our task is to find the order of multiplying the matrices that minimizes the total number of multiplications used. Certainly trying all possible orderings is impractical. (The number of orderings is a well-studied combinatorial function called the *Catalan number*: the number of ways to parenthesize N variables is about $4^{N-1}/N\sqrt{\pi N}$.) But it is certainly worthwhile to expend some effort to find a good solution because N is generally quite small compared to the number of multiplications to be done.

As above, the dynamic programming solution to this problem involves working "bottom up," saving computed answers to small partial problems to avoid recomputation. First, there's only one way to multiply M_1 by M_2, M_2 by M_3, ..., M_{N-1} by M_N; we record those costs. Next, we calculate the best way to multiply successive triples, using all the information computed so far. For example, to find the best way to multiply $M_1 M_2 M_3$, first we find the cost of computing $M_1 M_2$ from the table that we saved and then add the cost of multiplying that result by M_3. This total is compared with the cost of first multiplying $M_2 M_3$ then multiplying by M_1, which can be computed in the same way. The smaller of these is saved, and the same procedure followed for all triples. Next, we calculate the best way to multiply successive groups of four, using all the information gained so far. By continuing in this way we eventually find the best way to multiply together all the matrices.

In general, for $1 \leq j \leq N-1$, we can find the minimum cost of computing

$$M_i M_{i+1} \cdots M_{i+j}$$

for $1 \leq i \leq N - j$ by finding, for each k between i and $i + j$, the cost of computing $M_i M_{i+1} \cdots M_{k-1}$ and $M_k M_{k+1} \cdots M_{i+j}$ and then adding the cost of multiplying these results together. Since we always break a group into two smaller groups, the minimum costs for the two groups need only be looked up in a table, not recomputed. In particular, if we maintain an array with entries $cost[l, r]$ giving the minimum cost of computing $M_l M_{l+1} \cdots M_r$, then the cost of the first group above is $cost[i, k-1]$ and the cost of the second group is $cost[k, i+j]$. The cost of the final multiplication is easily determined: $M_i M_{i+1} \cdots M_{k-1}$ is a r_i-by-r_k matrix, and $M_k M_{k+1} \cdots M_{i+j}$ is a r_k-by-r_{i+j+1} matrix, so the cost of multiplying these two is $r_i r_k r_{i+j+1}$. This gives a way to compute $cost[i, i+j]$ for $1 \le i \le N-j$ with j increasing from 1 to $N-1$. When we reach $j = N-1$ (and $i = 1$), then we've found the minimum cost of computing $M_1 M_2 \cdots M_N$, as desired. This leads to the following program:

```
for i:=1 to N do
  for j:=i+1 to N do cost[i,j]:=maxint;
for i:=1 to N do cost[i,i]:=0;
for j:=1 to N−1 do
  for i:=1 to N−j do
    for k:=i+1 to i+j do
      begin
      t:=cost[i,k−1]+cost[k,i+j]+r[i]*r[k]*r[i+j+1];
      if t<cost[i,i+j] then
        begin cost[i,i+j]:=t;  best[i,i+j]:=k end;
      end;
```

As above, we need to keep track of the decisions made in a separate array *best* for later recovery when the actual sequence of multiplications is to be generated.

The following table is derived in a straightforward way from the *cost* and *best* arrays for the sample problem given above:

	B	C	D	E	F
A	24	14	22	26	36
	A B	A B	C D	C D	C D
B		6	10	14	22
		B C	C D	C D	C D
C			6	10	19
			C D	C D	C D
D				4	10
				D E	E F
E					12
					E F

For example, the entry in row A and column F says that 36 scalar multiplications are required to multiply matrices A through F together, and that this can be achieved by multiplying A through C in the optimal way, then multiplying D through F in the optimal way, then multiplying the resulting matrices together. (Only D is actually in the *best* array: the optimal splits are indicated by pairs of letters in the table for clarity.) To find how to multiply A through C in the optimal way, we look in row A and column C, etc. The following program implements this process of extracting the optimal parenthesization from the *cost* and *best* arrays computed by the program above:

```
procedure order(i, j: integer);
  begin
  if i=j then write(name(i)) else
    begin
    write('(');
    order(i, best[i,j]−1); write('*'); order(best[i,j], j);
    write(')')
    end
  end;
```

For our example, the parenthesization computed is $((A*(B*C))*((D*E)*F))$ which, as mentioned above, requires only 36 scalar multiplications. For the example cited earlier with the dimensions of 3 in B, C and F changed to 300, the same parenthesization is optimal, requiring 2412 scalar multiplications.

The triple loop in the dynamic programming code leads to a running time proportional to N^3 and the space required is proportional to N^2, substantially more than we used for the knapsack problem. But this is quite palatable compared to the alternative of trying all $4^{N-1}/N\sqrt{\pi N}$ possibilities.

Optimal Binary Search Trees

In many applications of searching, it is known that the search keys may occur with widely varying frequency. For example, a program which checks the spelling of words in English text is likely to look up words like "and" and "the" far more often than words like "dynamic" and "programming." Similarly, a Pascal compiler is likely to see keywords like "**end**" and "**do**" far more often than "**label**" or "**downto**." If binary tree searching is used, it is clearly advantageous to have the most frequently sought keys near the top of the tree. A dynamic programming algorithm can be used to determine how to arrange the keys in the tree so that the total cost of searching is minimized.

Each node in the following binary search tree on the keys A through G is labeled with an integer which is assumed to be proportional to its frequency of access:

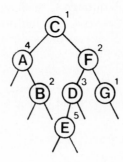

That is, out of every 18 searches in this tree, we expect 4 to be for A, 2 to be for B, 1 to be for C, etc. Each of the 4 searches for A requires two node accesses, each of the 2 searches for B requires 3 node accesses, and so forth. We can compute a measure of the "cost" of the tree by simply multiplying the frequency for each node by its distance to the root and summing. This is the *weighted internal path length* of the tree. For the example tree above, the weighted internal path length is $4*2 + 2*3 + 1*1 + 3*3 + 5*4 + 2*2 + 1*3 = 51$. We would like to find the binary search tree for the given keys with the given frequencies that has the smallest internal path length over all such trees.

This problem is similar to the problem of minimizing weighted external path length that we saw in studying Huffman encoding, but in Huffman encoding it was not necessary to maintain the order of the keys: in the binary search tree, we must preserve the property that all nodes to the left of the root have keys which are less, etc. This requirement makes the problem very similar to the matrix chain multiplication problem treated above: virtually the same program can be used.

Specifically, we assume that we are given a set of search keys $K_1 < K_2 < \cdots < K_N$ and associated frequencies $r_0, r_1, \ldots, r_N$, where r_i is the anticipated frequency of reference to key K_i. We want to find the binary search tree that minimizes the sum, over all keys, of these frequencies times the distance of the key from the root (the cost of accessing the associated node).

We proceed exactly as for the matrix chain problem: we compute, for each j increasing from 1 to $N - 1$, the best way to build a subtree containing $K_i, K_{i+1}, \ldots, K_{i+j}$ for $1 \leq i \leq N - j$. This computation is done by trying each node as the root and using precomputed values to determine the best way to do the subtrees. For each k between i and $i + j$, we want to find the optimal tree containing $K_i, K_{i+1}, \ldots, K_{i+j}$ with K_k at the root. This tree is formed by using the optimal tree for $K_i, K_{i+1}, \ldots, K_{k-1}$ as the left subtree and the optimal tree for $K_{k+1}, K_{k+2}, \ldots, K_{i+j}$ as the right subtree. The internal path length of this tree is the sum of the internal path lengths for the two subtrees plus the sum of the frequencies for all the nodes (since each node is one step further from the root in the new tree). This leads to the following program:

```
for i:=1 to N do
   for j:=i+1 to N+1 do cost[i,j]:=maxint;
for i:=1 to N do cost[i,i]:=f[i];
for i:=1 to N+1 do cost[i,i-1]:=0;
for j:=1 to N-1 do
   for i:=1 to N-j do
      begin
      for k:=i to i+j do
         begin
         t:=cost[i,k-1]+cost[k+1,i+j];
         if t<cost[i,i+j] then
            begin cost[i,i+j]:=t;  best[i,i+j]:=k end;
         end;
      t:=0; for k:=i to i+j do t:=t+f[k];
      cost[i,i+j]:=cost[i,i+j]+t;
      end;
```

Note that the sum of all the frequencies would be added to any cost so it is not needed when looking for the minimum. Also, we must have $cost[i,i-1]=0$ to cover the possibility that a node could just have one son (there was no analog to this in the matrix chain problem).

As before, a short recursive program is required to recover the actual tree from the *best* array computed by the program. For the example given above, the optimal tree computed is

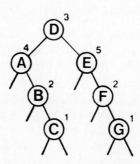

which has a weighted internal path length of 41.

As above, this algorithm requires time proportional to N^3 since it works with a matrix of size N^2 and spends time proportional to N on each entry. It is actually possible in this case to reduce the time requirement to N^2 by taking advantage of the fact that the optimal position for the root of a tree can't be too far from the optimal position for the root of a slightly smaller tree, so that k doesn't have to range over all the values from i to $i + j$ in the program above.

Shortest Paths

In some cases, the dynamic programming formulation of a method to solve a problem produces a familiar algorithm. For example, Warshall's algorithm (given in Chapter 32) for finding the transitive closure of a directed graph follows directly from a dynamic programming formulation. To show this, we'll consider the more general *all-pairs shortest paths* problem: given a graph with vertices $\{1, 2, \ldots, V\}$ determine the shortest distance from each vertex to every other vertex.

Since the problem calls for V^2 numbers as output, the adjacency matrix representation for the graph is obviously appropriate, as in Chapters 31 and 32. Thus we'll assume our input to be a V-by-V array a of edge weights, with $a[i, j] := w$ if there is an edge from vertex i to vertex j of weight w. If $a[i, j] = a[j, i]$ for all i and j then this could represent an undirected graph, otherwise it represents a directed graph. Our task is to find the directed path of minimum weight connecting each pair of vertices. One way to solve this problem is to simply run the shortest path algorithm of Chapter 31 for each vertex, for a total running time proportional to V^3. An even simpler algorithm with the same performance can be derived from a dynamic programming approach.

The dynamic programming algorithm for this problem follows directly from our description of Warshall's algorithm in Chapter 32. We compute, $1 \leq k \leq N$, the shortest path from each vertex to each other vertex *which uses only vertices from* $\{1, 2, \ldots, k\}$. The shortest path from vertex i to vertex j using only vertices from $1, 2, \ldots, k$ is either the shortest path from vertex i to vertex j using only vertices from $1, 2, \ldots, k - 1$ or a path composed of the shortest path from vertex i to vertex k using only vertices from $1, 2, \ldots, k - 1$ and the shortest path from vertex k to vertex j using only vertices from $1, 2, \ldots, k - 1$. This leads immediately to the following program.

```
for y:=1 to V do
    for x:=1 to V do
        if a[x,y]<>maxint div 2 then
            for j:=1 to V do
                if a[x,j]>(a[x,y]+a[y,j])
                    then a[x,j]:=a[x,y]+a[y,j];
```

The value *maxint* **div** *2* is used as a sentinel in matrix positions corresponding to edges not present in the graph. This eliminates the need to test explicitly in the inner loop whether there is an edge from x to j or from y to j. A "small" sentinel value is used so that there will be no overflow.

This is virtually the same program that we used to compute the transitive closure of a directed graph: logical operations have been replaced by arithmetic operations. The following table shows the adjacency matrix before and after this algorithm is run on directed graph example of Chapter 32, with all edge weights set to 1:

```
  A B C D E F G H I J K L M           A B C D E F G H I J K L M
A 0 1       1 1                     A 0 1 2 3 2 1 1       2 3 3 3
B   0                               B   0
C 1   0                             C 1 2 0 4 3 2 2       3 4 4 4
D       0 1                         D       0 2 1
E     1 0                           E     1 0 2
F     1 0                           F     2 1 0
G   1 1   0     1                   G 2 3 1 2 1 2 1 3 0     1 2 2 2
H           1 0 1                   H 3 4 2 3 2 3 2 4 1 0 1 2 3 3 3
I           1 0                     I 4 5 3 4 3 4 3 5 2 1 0 3 4 4 4
J               0 1 1 1             J 4 5 3 4 3 4 3 5 2   0 1 1 1
K                 0                 K                     0
L         1         0 1             L 3 4 2 3 2 3 2 4 1   2 3 0 1
M                   1 0             M 4 5 3 4 3 4 3 5 2   3 4 1 0
```

Thus the shortest path from M to B is of length 5, etc. Note that, for this algorithm, the weight corresponding to the edge between a vertex and itself is 0. Except for this, if we consider nonzero entries as 1 bits, we have exactly the bit matrix produced by the transitive closure algorithm of Chapter 32.

From a dynamic programming standpoint, note that the amount of information saved about small subproblems is nearly the same as the amount of information to be output, so little space is wasted.

One advantage of this algorithm over the shortest paths algorithm of Chapter 31 is that it works properly even if negative edge weights are allowed, as long as there are no cycles of negative weight in the graph (in which case the shortest paths connecting nodes on the cycle are not defined). If a cycle of negative weight is present in the graph, then the algorithm can detect that fact, because in that case $a[i, i]$ will become negative for some i at some point during the algorithm.

Time and Space Requirements

The above examples demonstrate that dynamic programming applications can have quite different time and space requirements depending on the amount of information about small subproblems that must be saved. For the shortest paths algorithm, no extra space was required; for the knapsack problem, space proportional to the size of the knapsack was needed; and for the other problems N^2 space was needed. For each problem, the time required was a factor of N greater than the space required.

The range of possible applicability of dynamic programming is far larger than covered in the examples. From a dynamic programming point of view, divide-and-conquer recursion could be thought of as a special case in which a minimal amount of information about small cases must be computed and stored, and exhaustive search (which we'll examine in Chapter 39) could be thought of as a special case in which a maximal amount of information about small cases must be computed and stored. Dynamic programming is a natural design technique that appears in many guises to solve problems throughout this range.

□

Exercises

1. In the example given for the knapsack problem, the items are sorted by size. Does the algorithm still work properly if they appear in arbitrary order?

2. Modify the knapsack program to take into account another constraint defined by an array $num[1..N]$ which contains the number of items of each type that are available.

3. What would the knapsack program do if one of the values were negative?

4. True or false: If a matrix chain involves a 1-by-k by k-by-1 multiplication, then there is an optimal solution for which that multiplication is last. Defend your answer.

5. Write a program which actually multiplies together N matrices in an optimal way. Assume that the matrices are stored in a three-dimensional array $matrices[1..Nmax, 1..Dmax, 1..Dmax]$, where $Dmax$ is the maximum dimension, with the ith matrix stored in $matrices[i, 1..r[i], 1..r[i+1]]$.

6. Draw the optimal binary search tree for the example in the text, but with all the frequencies increased by 1.

7. Write the program omitted from the text for actually constructing the optimal binary search tree.

8. Suppose that we've computed the optimum binary search tree for some set of keys and frequencies, and say that one frequency is incremented by 1. Write a program to compute the new optimum tree.

9. Why not solve the knapsack problem in the same way as the matrix chain and optimum binary search tree problems, by minimizing, for k from 1 to M, the sum of the best value achievable for a knapsack of size k and the best value achievable for a knapsack of size $M-k$?

10. Extend the program for the shortest paths problem to include a procedure $paths(i, j: integer)$ that will fill an array $path$ with the shortest path from i to j. This procedure should take time proportional to the length of the path each time it is called, using an auxiliary data structure built up by a modified version of the program given in the text.

38. Linear Programming

Many practical problems involve complicated interactions between a number of varying quantities. One example of this is the network flow problem discussed in Chapter 33: the flows in the various pipes in the network must obey physical laws over a rather complicated network. Another example is scheduling various tasks in (say) a manufacturing process in the face of deadlines, priorities, etc. Very often it is possible to develop a precise mathematical formulation which captures the interactions involved and reduces the problem at hand to a more straightforward mathematical problem. This process of deriving a set of mathematical equations whose solution implies the solution of a given practical problem is called *mathematical programming*. In this section, we consider a fundamental variant of mathematical programming, *linear programming*, and an efficient algorithm for solving linear programs, the *simplex* method.

Linear programming and the simplex method are of fundamental importance because a wide variety of important problems are amenable to formulation as linear programs and efficient solution by the simplex method. Better algorithms are known for some specific problems, but few problem-solving techniques are as widely applicable as the process of first formulating the problem as a linear program, then computing the solution using the simplex method.

Research in linear programming has been extensive, and a full understanding of all the issues involved requires mathematical maturity somewhat beyond that assumed for this book. On the other hand, some of the basic ideas are easy to comprehend, and the actual simplex algorithm is not difficult to implement, as we'll see below. As with the fast Fourier transform in Chapter 36, our intent is not to provide a full practical implementation, but rather to learn some of the basic properties of the algorithm and its relationship to other algorithms that we've studied.

497

Linear Programs

Mathematical programs involve a set of *variables* related by a set of mathematical equations called *constraints* and an *objective function* involving the variables that are to be maximized subject to the constraints. If all of the equations involved are simply linear combinations of the variables, we have the special case that we're considering called *linear programming*. The "programming" necessary to solve any particular problem involves choosing the variables and setting up the equations so that a solution to the equations corresponds to a solution to the problem. This is an art that we won't pursue in any further detail, except to look at a few examples. (The "programming" that we'll be interested in involves writing Pascal programs to find solutions to the mathematical equations.)

The following linear program corresponds to the network flow problem that we considered in Chapter 33.

Maximize $x_{AB} + x_{AD}$
subject to the constraints

$$x_{AB} \leq 8 \qquad x_{CD} \leq 2$$
$$x_{AD} \leq 2 \qquad x_{CF} \leq 4$$
$$x_{BC} \leq 6 \qquad x_{EB} \leq 3$$
$$x_{DE} \leq 5 \qquad x_{EF} \leq 5$$
$$x_{AB} + x_{EB} = x_{BC},$$
$$x_{AD} + x_{CD} = x_{DE},$$
$$x_{EF} + x_{EB} = x_{DE},$$
$$x_{CD} + x_{CF} = x_{BC},$$
$$x_{AB}, x_{AC}, x_{BC}, x_{CD}, x_{CF}, x_{DE}, x_{EB}, x_{EF} \geq 0.$$

There is one variable in this linear program corresponding to the flow in each of the pipes. These variables satisfy two types of equations: inequalities, corresponding to capacity constraints on the pipes, and equalities, corresponding to flow constraints at every junction. Thus, for example, the inequality $x_{AB} \leq 8$ says that pipe AB has capacity 8, and the equation $x_{AB} + x_{EB} = x_{BC}$ says that the inflow must equal the outflow at junction B. Note that adding all the equalities together gives the implicit constraint $x_{AB} + x_{AD} = x_{CF} + x_{EF}$ which says that the inflow must equal the outflow for the whole network. Also, of course, all of the flows must be positive.

This is clearly a mathematical formulation of the network flow problem: a solution to this particular mathematical problem is a solution to the particular instance of the network flow problem. The point of this example is not that linear programming will provide a better algorithm for this problem, but rather that linear programming is a quite general technique that can be applied to a variety of problems. For example, if we were to generalize the network flow problem to include costs as well as capacities, or whatever, the linear programming formulation would not look much different, even though the problem might be significantly more difficult to solve directly.

Not only are linear programs richly expressive but also there exists an algorithm for solving them (the simplex algorithm) which has proven to be quite efficient for many problems arising in practice. For some problems (such as network flow) there may be an algorithm specifically oriented to that problem which can perform better than linear programming/simplex; for other problems (including various extensions of network flow), no better algorithms are known. Even if there is a better algorithm, it may be complicated or difficult to implement, while the procedure of developing a linear program and solving it with a simplex library routine is often quite straightforward. This "general-purpose" aspect of the method is quite attractive and has led to its widespread use. The danger in relying upon it too heavily is that it may lead to inefficient solutions for some simple problems (for example, many of those for which we have studied algorithms in this book).

Geometric Interpretation

Linear programs can be cast in a geometric setting. The following linear program is easy to visualize because only two variables are involved.

Maximize $x_1 + x_2$
subject to the constraints

$$-x_1 + x_2 \leq 5,$$
$$x_1 + 4x_2 \leq 45,$$
$$2x_1 + x_2 \leq 27,$$
$$3x_1 - 4x_2 \leq 24,$$
$$x_1, x_2 \geq 0.$$

It corresponds to the following diagram:

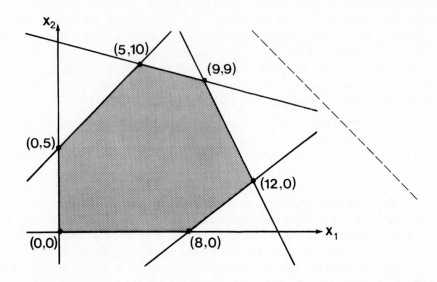

Each inequality defines a halfplane in which any solution to the linear program must lie. For example, $x_1 \geq 0$ means that any solution must lie to the right of the x_2 axis, and $-x_1 + x_2 \leq 5$ means that any solution must lie below and to the right of the line $-x_1 + x_2 = 5$ (which goes through $(0,5)$ and $(5,10)$). Any solution to the linear program must satisfy *all* of these constraints, so the region defined by the intersection of all these halfplanes (shaded in the diagram above) is the set of all possible solutions. To solve the linear program we must find the point within this region which maximizes the objective function.

It is always the case that a region defined by intersecting halfplanes is convex (we've encountered this before, in one of the definitions of the convex hull in Chapter 25). This convex region, called the *simplex*, forms the basis for an algorithm to find the solution to the linear program which maximizes the objective function.

A fundamental property of the simplex, which is exploited by the algorithm, is that the objective function is maximized at one of the vertices of the simplex: thus only these points need to be examined, not all the points inside. To see why this is so for our example, consider the dotted line at the right, which corresponds to the objective function. The objective function can be thought of as defining a line of known slope (in this case -1) and unknown position. We're interested in the point at which the line hits the simplex, as it is moved in from infinity. This point is the solution to the linear program: it satisfies all the inequalities because it is in the simplex, and it maximizes the objective function because no points with larger values were encountered. For our example, the line hits the simplex at (9,9) which maximizes the objective function at 18.

Other objective functions correspond to lines of other slopes, but always the maximum will occur at one of the vertices of the simplex. The algorithm that we'll examine below is a systematic way of moving from vertex to vertex in search of the minimum. In two dimensions, there's not much choice about what to do, but, as we'll see, the simplex is a much more complicated object when more variables are involved.

From the geometric representation, one can also appreciate why mathematical programs involving nonlinear functions are so much more difficult to handle. For example, if the objective function is nonlinear, it could be a curve that could strike the simplex along one of its edges, not at a vertex. If the inequalities are also nonlinear, quite complicated geometric shapes which correspond to the simplex could arise.

Geometric intuition makes it clear that various anomalous situations can arise. For example, suppose that we add the inequality $x_1 \geq 13$ to the linear program in the example above. It is quite clear from the diagram above that in this case the intersection of the half-planes is empty. Such a linear program is called *infeasible*: there are no points which satisfy the inequalities, let alone one which maximizes the objective function. On the other hand the inequality $x_1 \leq 13$ is *redundant*: the simplex is entirely contained within its halfplane, so it is not represented in the simplex. Redundant inequalities do not affect the solution at all, but they need to be dealt with during the search for the solution.

A more serious problem is that the simplex may be an open (unbounded) region, in which case the solution may not be well-defined. This would be the case for our example if the second and third inequalities were deleted. Even if the simplex is unbounded the solution may be well-defined for some objective functions, but an algorithm to find it might have significant difficulty getting around the unbounded region.

It must be emphasized that, though these problems are quite easy to see when we have two variables and a few inequalities, they are very much less apparent for a general problem with many variables and inequalities. Indeed, detection of these anomalous situations is a significant part of the computational burden of solving linear programs.

The same geometric intuition holds for more variables. In 3 dimensions the simplex is a convex 3-dimensional solid defined by the intersection of halfspaces defined by the planes whose equations are given by changing the inequalities to equalities. For example, if we add the inequalities $x_3 \leq 4$ and $x_3 \geq 0$ to the linear program above, the simplex becomes the solid object diagramed below:

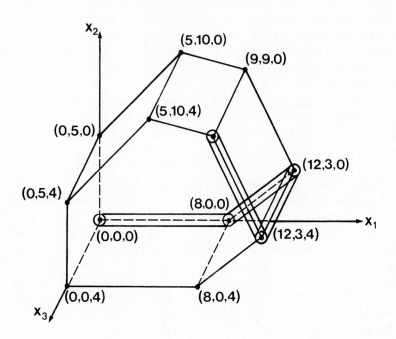

To make the example more three-dimensional, suppose that we change the objective function to $x_1 + x_2 + x_3$. This defines a plane perpendicular to the line $x1 = x2 = x3$. If we move a plane in from infinity along this line, we hit the simplex at the point $(9, 9, 4)$ which is the solution. (Also shown in the diagram is a path along the vertices of the simplex from $(0, 0, 0)$ to the solution, for reference in the description of the algorithm below.)

In n dimensions, we intersect halfspaces defined by $(n-1)$-dimensional hyperplanes to define the n-dimensional simplex, and bring in an $(n-1)$-dimensional hyperplane from infinity to intersect the simplex at the solution point. As mentioned above, we risk oversimplification by concentrating on intuitive two- and three-dimensional situations, but proofs of the facts above involving convexity, intersecting hyperplanes, etc. involve a facility with linear algebra somewhat beyond the scope of this book. Still, the geometric intuition is valuable, since it can help us to understand the fundamental characteristics of the basic method that is used in practice to solve higher-dimensional problems.

The Simplex Method

Simplex is the name commonly used to describe a general approach to solving linear programs by using pivoting, the same fundamental operation used in Gaussian elimination. It turns out that pivoting corresponds in a natural way to the geometric operation of moving from point to point on the simplex, in search of the solution. The several algorithms which are commonly used differ in essential details having to do with the order in which simplex vertices are searched. That is, the well-known "algorithm" for solving this problem could more precisely be described as a generic method which can be refined in any of several different ways. We've encountered this sort of situation before, for example Gaussian elimination or the Ford-Fulkerson algorithm.

First, as the reader surely has noticed, linear programs can take on many different forms. For example, the linear program above for the network flow problem has a mixture of equalities and inequalities, but the geometric examples above use only inequalities. It is convenient to reduce the number of possibilities somewhat by insisting that all linear programs be presented in the same *standard form*, where all the equations are equalities except for an inequality for each variable stating that it is nonnegative. This may seem like a severe restriction, but actually it is not difficult to convert general linear programs to this standard form. For example, the following linear program is the standard form for the three-dimensional example given above:

Maximize $x_1 + x_2 + x_3$
subject to the constraints

$$-x_1 + x_2 + y_1 = 5$$
$$x_1 + 4x_2 + y_2 = 45$$
$$2x_1 + x_2 + y_3 = 27$$
$$3x_1 - 4x_2 + y_4 = 24$$
$$x_3 + y_5 = 4$$
$$x_1, x_2, x_3, y_1, y_2, y_3, y_4, y_5 \geq 0.$$

Each inequality involving more than one variable is converted into an equality by introducing a new variable. The y's are called *slack* variables because they take up the slack allowed by the inequality. Any inequality involving only one variable can be converted to the standard nonnegative constraint simply by renaming the variable. For example, a constraint such as $x_3 \leq -1$ would be handled by replacing x_3 by $-1 - x_3'$ everywhere that it appears.

This formulation makes obvious the parallels between linear programming and simultaneous equations. We have N equations in M unknown variables, all constrained to be positive. In this case, note that there are N slack variables, one for each equation (since we started out with all inequalities).

We assume that $M > N$ which implies that there are many solutions to the equations: the problem is to find the one which maximizes the objective function.

For our example, there is a trivial solution to the equations: take $x_1 = x_2 = x_3 = 0$, then assign appropriate values to the slack variables to satisfy the equalities. This works because $(0,0,0)$ is a point on the simplex. Although this need not be the case in general, to explain the simplex method, we'll restrict attention for now to linear programs where it is known to be the case. This is still a quite large class of linear programs: for example, if all the numbers on the right-hand side of the inequalities in the standard form of the linear program are positive and slack variables all have positive coefficients (as in our example) then there is clearly a solution with all the original variables zero. Later we'll return to the general case.

Given a solution with $M - N$ variables set to 0, it turns out that we can find another solution with the same property by using a familiar operation, *pivoting*. This is essentially the same operation used in Gaussian elimination: an element $a[p, q]$ is chosen in the matrix of coefficients defined by the equations, then the pth row is multiplied by an appropriate scalar and added to all other rows to make the qth column all 0 except for the entry in row q, which is made 1. For example, consider the following matrix, which represents the linear program given above:

$$
\begin{pmatrix}
-1.00 & -1.00 & -1.00 & 0.00 & 0.00 & 0.00 & 0.00 & 0.00 & 0.00 \\
-1.00 & 1.00 & 0.00 & 1.00 & 0.00 & 0.00 & 0.00 & 0.00 & 5.00 \\
1.00 & 4.00 & 0.00 & 0.00 & 1.00 & 0.00 & 0.00 & 0.00 & 45.00 \\
2.00 & 1.00 & 0.00 & 0.00 & 0.00 & 1.00 & 0.00 & 0.00 & 27.00 \\
3.00 & -4.00 & 0.00 & 0.00 & 0.00 & 0.00 & 1.00 & 0.00 & 24.00 \\
0.00 & 0.00 & 1.00 & 0.00 & 0.00 & 0.00 & 0.00 & 1.00 & 4.00
\end{pmatrix}
$$

This $(N+1)$-by-$(M+1)$ matrix contains the coefficients of the linear program in standard form, with the $(M+1)$st column containing the numbers on the right-hand sides of the equations (as in Gaussian elimination), and the 0th row containing the coefficients of the objective function, with the sign reversed. The significance of the 0th row is discussed below; for now we'll treat it just like all of the other rows.

For our example, we'll carry out all computations to two decimal places. Obviously, issues such as computational accuracy and accumulated error are just as important here as they are in Gaussian elimination.

The variables which correspond to a solution are called the *basis* variables and those which are set to 0 to make the solution are called *non-basis* variables. In the matrix, the columns corresponding to basis variables have exactly one 1 with all other values 0, while non-basis variables correspond to columns with more than one nonzero entry.

Now, suppose that we wish to pivot this matrix for $p = 4$ and $q = 1$. That is, an appropriate multiple of the fourth row is added to each of the other rows to make the first column all 0 except for a 1 in row 4. This produces the following result:

$$\begin{pmatrix} 0.00 & -2.33 & -1.00 & 0.00 & 0.00 & 0.00 & 0.33 & 0.00 & 8.00 \\ 0.00 & -0.33 & 0.00 & 1.00 & 0.00 & 0.00 & 0.33 & 0.00 & 13.00 \\ 0.00 & 5.33 & 0.00 & 0.00 & 1.00 & 0.00 & -0.33 & 0.00 & 37.00 \\ 0.00 & 3.67 & 0.00 & 0.00 & 0.00 & 1.00 & -0.67 & 0.00 & 11.00 \\ 1.00 & -1.33 & 0.00 & 0.00 & 0.00 & 0.00 & 0.33 & 0.00 & 8.00 \\ 0.00 & 0.00 & 1.00 & 0.00 & 0.00 & 0.00 & 0.00 & 1.00 & 4.00 \end{pmatrix}$$

This operation removes the 7th column from the basis and adds the 1st column to the basis. Exactly one basis column is removed because exactly one basis column has a 1 in row p.

By definition, we can get a solution to the linear program by setting all the non-basis variables to zero, then using the trivial solution given in the basis. In the solution corresponding to the above matrix, both x_2 and x_3 are zero because they are non-basis variables and $x_1 = 8$, so the matrix corresponds to the point $(8, 0, 0)$ on the simplex. (We're not interested particularly in the values of the slack variables.) Note that the upper right hand corner of the matrix (row 0, column $M + 1$) contains the value of the objective function at this point. This is by design, as we shall soon see.

Now suppose that we perform the pivot operation for $p = 3$ and $q = 2$:

$$\begin{pmatrix} 0.00 & 0.00 & -1.00 & 0.00 & 0.00 & 0.64 & -0.09 & 0.00 & 15.00 \\ 0.00 & 0.00 & 0.00 & 1.00 & 0.00 & 0.09 & 0.27 & 0.00 & 14.00 \\ 0.00 & 0.00 & 0.00 & 0.00 & 1.00 & -1.45 & 0.64 & 0.00 & 21.00 \\ 0.00 & 1.00 & 0.00 & 0.00 & 0.00 & 0.27 & -0.18 & 0.00 & 3.00 \\ 1.00 & 0.00 & 0.00 & 0.00 & 0.00 & 0.36 & 0.09 & 0.00 & 12.00 \\ 0.00 & 0.00 & 1.00 & 0.00 & 0.00 & 0.00 & 0.00 & 1.00 & 4.00 \end{pmatrix}$$

This removes column 6 from the basis and adds column 2. By setting non-basis variables to 0 and solving for basis variables as before, we see that this matrix corresponds to the point $(12, 3, 0)$ on the simplex, for which the objective function has the value 15. Note that the value of the objective function is strictly increasing. Again, this is by design, as we shall soon see.

How do we decide which values of p and q to use for pivoting? This is where row 0 comes in. For each non-basis variable, row 0 contains the amount by which the objective function would increase if that variable were changed from 0 to 1, with the sign reversed. (The sign is reversed so that the standard pivoting operation will maintain row 0, with no changes.) Pivoting using column q amounts to changing the value of the corresponding variable from 0 to some positive value, so we can be sure the objective function will increase if we use any column with a negative entry in row 0.

Now, pivoting on any row with a positive entry for that column will increase the objective function, but we also must make sure that it will result in a matrix corresponding to a point on the simplex. Here the central concern is that one of the entries in column $M+1$ might become negative. This can be forestalled by finding, among the positive elements in column q (not including row 0), the one that gives the smallest value when divided into the $(M+1)$st element in the same row. If we take p to be the index of the row containing this element and pivot, then we can be sure that the objective function will increase and that none of the entries in column $M+1$ will become negative; this is enough to ensure that the resulting matrix corresponds to a point on the simplex.

There are two potential problems with this procedure for finding the pivot row. First, what if there are no positive entries in column q? This is an inconsistent situation: the negative entry in row 0 says that the objective function can be increased, but there is no way to increase it. It turns out that this situation arises if and only if the simplex is unbounded, so the algorithm can terminate and report the problem. A more subtle difficulty arises in the degenerate case when the $(M+1)$st entry in some row (with a positive entry in column q) is 0. Then this row will be chosen, but the objective function will increase by 0. This is not a problem in itself: the problem arises when there are two such rows. Certain natural policies for choosing between such rows lead to *cycling*: an infinite sequence of pivots which do not increase the objective function at all. Again, several possibilities are available for avoiding cycling. One method is to break ties randomly. This makes cycling extremely unlikely (but not mathematically impossible). Another anti-cycling policy is described below.

We have been avoiding difficulties such as cycling in our example to make the description of the method clear, but it must be emphasized that such degenerate cases are quite likely to arise in practice. The generality offered by using linear programming implies that degenerate cases of the general problem will arise in the solution of specific problems.

In our example, we can pivot again with $q = 3$ (because of the -1 in row 0 and column 3) and $p = 5$ (because 1 is the only positive value in column 3). This gives the following matrix:

$$
\begin{pmatrix}
0.00 & 0.00 & 0.00 & 0.00 & 0.00 & 0.64 & -0.09 & 1.00 & 19.00 \\
0.00 & 0.00 & 0.00 & 1.00 & 0.00 & 0.09 & 0.27 & 0.00 & 14.00 \\
0.00 & 0.00 & 0.00 & 0.00 & 1.00 & -1.45 & 0.64 & 0.00 & 21.00 \\
0.00 & 1.00 & 0.00 & 0.00 & 0.00 & 0.27 & -0.18 & 0.00 & 3.00 \\
1.00 & 0.00 & 0.00 & 0.00 & 0.00 & 0.36 & 0.09 & 0.00 & 12.00 \\
0.00 & 0.00 & 1.00 & 0.00 & 0.00 & 0.00 & 0.00 & 1.00 & 4.00
\end{pmatrix}
$$

This corresponds to the point $(12, 3, 4)$ on the simplex, for which the value of the objective function is 19.

In general, there might be several negative entries in row 0, and several different strategies for choosing from among them have been suggested. We have been proceeding according to one of the most popular methods, called the *greatest increment* method: always choose the column with the smallest value in row 0 (largest in absolute value). This does not necessarily lead to the largest increase in the objective function, since scaling according to the row p chosen has to be done. If this column selection policy is combined with the row selection policy of using, in case of ties, the row that will result in the column of lowest index being removed from the basis, then cycling cannot happen. (This anticycling policy is due to R. G. Bland.) Another possibility for column selection is to actually calculate the amount by which the objective function would increase for each column, then use the column which gives the largest result. This is called the *steepest descent* method. Yet another interesting possibility is to choose randomly from among the available columns.

Finally, after one more pivot at $p = 2$ and $q = 7$, we arrive at the solution:

$$
\begin{pmatrix}
0.00 & 0.00 & 0.00 & 0.00 & 0.14 & 0.43 & 0.00 & 1.00 & 22.00 \\
0.00 & 0.00 & 0.00 & 1.00 & -0.43 & 0.71 & 0.00 & 0.00 & 5.00 \\
0.00 & 0.00 & 0.00 & 0.00 & 1.57 & -2.29 & 1.00 & 0.00 & 33.00 \\
0.00 & 1.00 & 0.00 & 0.00 & 0.29 & -0.14 & 0.00 & 0.00 & 9.00 \\
1.00 & 0.00 & 0.00 & 0.00 & -0.14 & 0.57 & 0.00 & 0.00 & 9.00 \\
0.00 & 0.00 & 1.00 & 0.00 & 0.00 & 0.00 & 0.00 & 1.00 & 4.00
\end{pmatrix}
$$

This corresponds to the point $(9, 9, 4)$ on the simplex, which maximizes the objective function at 22. All the entries in row 0 are nonnegative, so any pivot will only serve to decrease the objective function.

The above example outlines the simplex method for solving linear programs. In summary, if we begin with a matrix of coefficients corresponding to a point on the simplex, we can do a series of pivot steps which move to adjacent points on the simplex, always increasing the objective function, until the maximum is reached.

There is one fundamental fact which we have not yet noted but is crucial to the correct operation of this procedure: once we reach a point where no single pivot can improve the objective function (a "local" maximum), then we have reached the "global" maximum. This is the basis for the simplex algorithm. As mentioned above, the proof of this (and many other facts which may seem obvious from the geometric interpretation) in general is quite beyond the scope of this book. But the simplex algorithm for the general case operates in essentially the same manner as for the simple problem traced above.

Implementation

The implementation of the simplex method for the case described above is quite straightforward from the description. First, the requisite pivoting procedure uses code similar to our implementation of Gaussian elimination in Chapter 5:

```
procedure pivot(p, q: integer);
  var j, k: integer;
  begin
  for j:=0 to N do
    for k:=M+1 downto 1 do
      if (j<>p) and (k<>q) then
        a[j, k]:=a[j, k]−a[p, k]*a[j, q]/a[p, q];
  for j:=0 to N do if j<>p then a[j, q]:=0;
  for k:=1 to M+1 do if k<>q then a[p, k]:=a[p, k]/a[p, q];
  a[p, q]:=1
  end;
```

This program adds multiples of row p to each row as necessary to make column q all zero except for a 1 in row q as described above. As in Chapter 5, it is necessary to take care not to change the value of $a[p, q]$ before we're done using it.

In Gaussian elimination, we processed only rows below p in the matrix during forward elimination and only rows above p during backward substitution using the Gauss-Jordan method. A system of N linear equations in N unknowns could be solved by calling $pivot(i, i)$ for i ranging from 1 to N then back down to 1 again.

The simplex algorithm, then, consists simply of finding the values of p and q as described above and calling *pivot*, repeating the process until the optimum is reached or the simplex is determined to be unbounded:

```
repeat
  q:=0; repeat q:=q+1 until (q=M+1) or (a[0, q]<0);
  p:=0; repeat p:=p+1 until (p=N+1) or (a[p, q]>0);
  for i:=p+1 to N do
    if a[i, q]>0 then
      if (a[i, M+1]/a[i, q])<(a[p, M+1]/a[p, q]) then p:=i;
  if (q<M+1) and (p<N+1) then pivot(p, q)
until (q=M+1) or (p=N+1);
```

If the program terminates with $q=M+1$ then an optimal solution has been found: the value achieved for the objective function will be in $a[0, M+1]$ and the values for the variables can be recovered from the basis. If the program terminates with $p=N+1$, then an unbounded situation has been detected.

This program ignores the problem of cycle avoidance. To implement Bland's method, it is necessary to keep track of the column that would leave the basis, were a pivot to be done using row p. This is easily done by setting $outb[p]:=q$ after each pivot. Then the loop to calculate p can be modified to set $p:=i$ also if equality holds in the ratio test and $outb[p]<outb[q]$. Alternatively, the selection of a random element could be implemented by generating a random integer x and replacing each array reference $a[p, q]$ (or $a[i, q]$) by $a[(p+x)\mathbf{mod}(N+1), q]$ (or $a[(i+x)\mathbf{mod}(N+1), q]$). This has the effect of searching through the column q in the same way as before, but starting at a random point instead of the beginning. The same sort of technique could be used to choose a random column (with a negative entry in row 0) to pivot on.

The program and example above treat a simple case that illustrates the principle behind the simplex algorithm but avoids the substantial complications that can arise in actual applications. The main omission is that the program requires that the matrix have a *feasible basis*: a set of rows and columns which can be permuted into the identity matrix. The program starts with the assumption that there is a solution with the $M - N$ variables appearing in the objective function set to zero and that the N-by-N submatrix involving the slack variables has been "solved" to make that submatrix the identity matrix. This is easy to do for the particular type of linear program that we stated (with all inequalities on positive variables), but in general we need to find some point on the simplex. Once we have found one solution, we can make appropriate transformations (mapping that point to the origin) to bring the matrix into the required form, but at the outset we don't even know whether a solution exists. In fact, it has been shown that *detecting* whether a solution exists is as difficult computationally as finding the optimum solution, given that one exists.

Thus it should not be surprising that the technique that is commonly used to detect the existence of a solution is the simplex algorithm! Specifically, we add another set of artificial variables $s_1, s_2, \ldots, s_N$ and add variable s_i to the ith equation. This is done simply by adding N columns to the matrix, filled with the identity matrix. Now, this gives immediately a feasible basis for this new linear program. The trick is to run the above algorithm with the objective function $-s_1-s_2-\cdots-s_N$. If there is a solution to the original linear program, then this objective function can be maximized at 0. If the maximum reached is not zero, then the original linear program is infeasible. If the maximum is zero, then the normal situation is that $s_1, s_2, \ldots, s_N$ all become non-basis variables, so we have computed a feasible basis for the original linear program. In degenerate cases, some of the artificial variables may remain in the basis, so it is necessary to do further pivoting to remove them (without changing the cost).

To summarize, a two-phase process is normally used to solve general linear programs. First, we solve a linear program involving the artificial s variables to get a point on the simplex for our original problem. Then, we dispose of the s variables and reintroduce our original objective function to proceed from this point to the solution.

The analysis of the running time of the simplex method is an extremely complicated problem, and few results are available. No one knows the "best" pivot selection strategy, because there are no results to tell us how many pivot steps to expect, for any reasonable class of problems. It is possible to construct artificial examples for which the running time of the simplex could be very large (an exponential function of the number of variables). However, those who have used the algorithm in practical settings are unanimous in testifying to its efficiency in solving actual problems.

The simple version of the simplex algorithm that we've considered, while quite useful, is merely part of a general and beautiful mathematical framework providing a complete set of tools which can be used to solve a variety of very important practical problems.

Exercises

1. Draw the simplex defined by the inequalities $x_1 \geq 0$, $x_2 \geq 0$, $x_3 \geq 0$, $x_1 + 2x_2 \leq 20$, and $x_1 + x_2 + x_3 \leq 10$.

2. Give the sequence of matrices produced for the example in the text if the pivot column chosen is the largest q for which $a[0, q]$ is negative.

3. Give the sequence of matrices produced for the example in the text for the objective function $x_1 + 5x_2 + x_3$.

4. Describe what happens if the simplex algorithm is run on a matrix with a column of all 0's.

5. Does the simplex algorithm use the same number of steps if the rows of the input matrix are permuted?

6. Give a linear programming formulation of the example in the previous chapter for the knapsack problem.

7. How many pivot steps are required to solve the linear program "Maximize $x_1 + \cdots + x_M$ subject to the constraints $x_1, \ldots, x_M \leq 1$ and $x_1, \ldots, x_M \geq 0$"?

8. Construct a linear program consisting of N inequalities on two variables for which the simplex algorithm requires at least $N/2$ pivots.

9. Give a three-dimensional linear programming problem which illustrates the difference between the greatest increment and steepest descent column selection methods.

10. Modify the implementation given in the text to actually write out the coordinates of the optimal solution point.

39. Exhaustive Search

Some problems involve searching through a vast number of potential solutions to find an answer, and simply do not seem to be amenable to solution by efficient algorithms. In this chapter, we'll examine some characteristics of problems of this sort and some techniques which have proven to be useful for solving them.

To begin, we should reorient our thinking somewhat as to exactly what constitutes an "efficient" algorithm. For most of the applications that we have discussed, we have become conditioned to think that an algorithm must be linear or run in time proportional to something like $N \log N$ or $N^{3/2}$ to be considered efficient. We've generally considered quadratic algorithms to be bad and cubic algorithms to be awful. But for the problems that we'll consider in this and the next chapter, any computer scientist would be absolutely delighted to know a cubic algorithm. In fact, even an N^{50} algorithm would be pleasing (from a theoretical standpoint) because these problems are believed to require *exponential* time.

Suppose that we have an algorithm that takes time proportional to 2^N. If we were to have a computer 1000 times faster than the fastest supercomputer available today, then we could perhaps solve a problem for $N = 50$ in an hour's time under the most generous assumptions about the simplicity of the algorithm. But in two hour's time we could only do $N = 51$, and even in a year's time we could only get to $N = 59$. And even if a new computer were to be developed with a million times the speed, and we were to have a million such computers available, we couldn't get to $N = 100$ in a year's time. Realistically, we have to settle for N on the order of 25 or 30. A "more efficient" algorithm in this situation may be one that could solve a problem for $N = 100$ with a realistic amount of time and money.

The most famous problem of this type is the *traveling salesman problem*: given a set of N cities, find the shortest route connecting them all, with no city visited twice. This problem arises naturally in a number of important applications, so it has been studied quite extensively. We'll use it as an example in this chapter to examine some fundamental techniques. Many advanced methods have been developed for this problem but it is still unthinkable to solve an instance of the problem for $N = 1000$.

The traveling salesman problem is difficult because there seems to be no way to avoid having to check the length of a very large number of possible tours. To check each and every tour is *exhaustive search*: first we'll see how that is done. Then we'll see how to modify that procedure to greatly reduce the number of possibilities checked, by trying to discover incorrect decisions as early as possible in the decision-making process.

As mentioned above, to solve a large traveling salesman problem is unthinkable, even with the very best techniques known. As we'll see in the next chapter, the same is true of many other important practical problems. But what can be done when such problems arise in practice? Some sort of answer is expected (the traveling salesman has to do something): we can't simply ignore the existence of the problem or state that it's too hard to solve. At the end of this chapter, we'll see examples of some methods which have been developed for coping with practical problems which seem to require exhaustive search. In the next chapter, we'll examine in some detail the reasons why no efficient algorithm is likely to be found for many such problems.

Exhaustive Search in Graphs

If the traveling salesman is restricted to travel only between certain pairs of cities (for example, if he is traveling by air), then the problem is directly modeled by a graph: given a weighted (possibly directed) graph, we want to find the shortest simple cycle that connects all the nodes.

This immediately brings to mind another problem that would seem to be easier: given an undirected graph, is there *any* way to connect all the nodes with a simple cycle? That is, starting at some node, can we "visit" all the other nodes and return to the original node, visiting every node in the graph exactly once? This is known as the *Hamilton cycle* problem. In the next chapter, we'll see that it is computationally equivalent to the traveling salesman problem in a strict technical sense.

In Chapters 30–32 we saw a number of methods for systematically visiting all the nodes of a graph. For all of the algorithms in those chapters, it was possible to arrange the computation so that each node is visited just once, and this leads to very efficient algorithms. For the Hamilton cycle problem, such a solution is not apparent: it seems to be necessary to visit each node many times. For the other problems, we were building a tree: when a "dead end" was reached in the search, we could start it up again, working on another part of the tree. For this problem, the tree must have a particular structure (a cycle): if we discover during the search that the tree being built cannot be a cycle, we have to go back and rebuild part of it.

To illustrate some of the issues involved, we'll look at the Hamilton cycle problem and the traveling salesman problem for the example graph from Chapter 31:

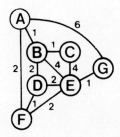

Depth-first search would visit the nodes in this graph in the order A B C E D F G (assuming an adjacency matrix or sorted adjacency list representation). This is not a simple cycle: to find a Hamilton cycle we have to try another way to visit the nodes. It turns out the we can systematically try all possibilities with a simple modification to the *visit* procedure, as follows:

```
procedure visit(k: integer);
  var t: integer;
  begin
  now:=now+1; val[k]:=now;
  for t:=1 to V do
    if a[k, t] then
      if val[t]=0 then visit(t);
  now:=now−1; val[k]:=0
  end;
```

Rather than leaving every node that it touches marked with a nonzero *val* entry, this procedure "cleans up after itself" and leaves *now* and the *val* array exactly as it found them. The only marked nodes are those for which *visit* hasn't completed, which correspond exactly to a simple path of length *now* in the graph, from the initial node to the one currently being visited. To *visit* a node, we simply visit all unmarked adjacent nodes (marked ones would not correspond to a simple path). The recursive procedure checks all simple paths in the graph which start at the initial node.

The following tree shows the order in which paths are checked by the above procedure for the example graph given above. Each node in the tree corresponds to a call of *visit*: thus the descendants of each node are adjacent nodes which are unmarked at the time of the call. Each path in the tree from a node to the root corresponds to a simple path in the graph:

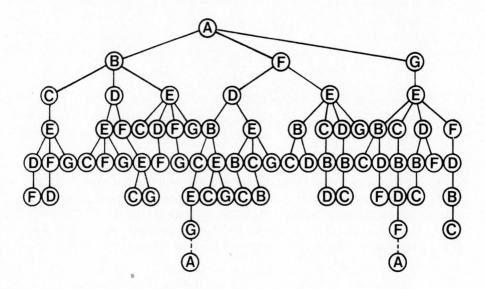

Thus, the first path checked is A B C E D F. At this point all vertices adjacent to F are marked (have non-zero *val* entries), so *visit* for F unmarks F and returns. Then *visit* for D unmarks D and returns. Then *visit* for E tries F which tries D, corresponding to the path A B C E F D. Note carefully that in depth-first search F and D remain marked after they are visited, so that F would *not* be visited from E. The "unmarking" of the nodes makes exhaustive search essentially different from depth-first search, and the reader should be sure to understand the distinction.

As mentioned above, *now* is the current length of the path being tried, and *val*[*k*] is the position of node *k* on that path. Thus we can make the *visit* procedure given above test for the existence of a Hamilton cycle by having it test whether there is an edge from *k* to 1 when *val*[*k*]=*V*. In the example above, there is only one Hamilton cycle, which appears twice in the tree, traversed in both directions. The program can be made to solve the traveling salesman problem by keeping track of the length of the current path in the *val* array, then keeping track of the minimum of the lengths of the Hamilton cycles found.

Backtracking

The time taken by the exhaustive search procedure given above is proportional to the number of calls to *visit*, which is the number of nodes in the exhaustive search tree. For large graphs, this will clearly be very large. For example, if the graph is complete (every node connected to every other node), then there are $V!$ simple cycles, one corresponding to each arrangement of the nodes. (This case is studied in more detail below.) Next we'll examine techniques to greatly reduce the number of possibilities tried. All of these techniques involve adding tests to *visit* to discover that recursive calls should not be made for certain nodes. This corresponds to *pruning* the exhaustive search tree: cutting certain branches and deleting everything connected to them.

One important pruning technique is to remove symmetries. In the above example, this is manifested by the fact that we find each cycle twice, traversed in both directions. In this case, we can ensure that we find each cycle just once by insisting that three particular nodes appear in a particular order. For example, if we insist that node C appear after node A but before node B in the example above, then we don't have to call *visit* for node B unless node C is already on the path. This leads to a drastically smaller tree:

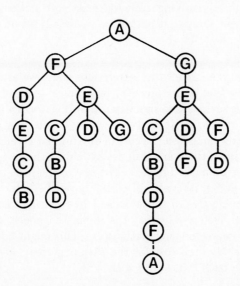

This technique is not always applicable: for example, suppose that we're trying to find the minimum-cost path (not cycle) connecting all the vertices. In the above example, A G E F D B C is a path which connects all the vertices, but it is not a cycle. Now the above technique doesn't apply, since we can't know in advance whether a path will lead to a cycle or not.

Another important pruning technique is to cut off the search as soon as it is determined that it can't possibly be successful. For example, suppose that we're trying to find the minimum cost path in the graph above. Once we've found A F D B C E G, which has cost 11, it's fruitless, for example, to search anywhere further along the path A G E B, since the cost is already 11. This can be implemented simply by making no recursive calls in in *visit* if the cost of the current partial path is greater than the cost of the best full path found so far. Certainly, we can't miss the minimum cost path by adhering to such a policy.

The pruning will be more effective if a low-cost path is found early in the search; one way to make this more likely is to *visit* the nodes adjacent to the current node in order of increasing cost. In fact, we can do even better: often, we can compute a bound on the cost of all full paths that begin with a given partial path. For example, suppose that we have the additional information that all edges in the diagram have a weight of at least 1 (this could be determined by an initial scan through the edges). Then, for example, we know that any full path starting with AG must cost at least 11, so we don't have to search further along that path if we've already found a solution which costs 11.

Each time that we cut off the search at a node, we avoid searching the entire subtree below that node. For very large trees, this is a very substantial savings. Indeed, the savings is so significant that it is worthwhile to do as much as possible within *visit* to avoid making recursive calls. For our example, we can get a much better bound on the cost of any full path which starts with the partial path made up of the marked nodes by adding the cost of the minimum spanning tree of the unmarked nodes. (The rest of the path is a spanning tree for the unmarked nodes; its cost will certainly not be lower than the cost of the minimum spanning tree of those nodes.) In particular, some paths might divide the graph in such a way that the unmarked nodes aren't connected; clearly we stop the search on such paths also. (This might be implemented by returning an artificially high cost for the spanning tree.) For example, there can't be any simple path that starts with ABE.

Drawn below is the search tree that results when all of these rules are applied to the problem of finding the best Hamilton path in the sample graph that we've been considering:

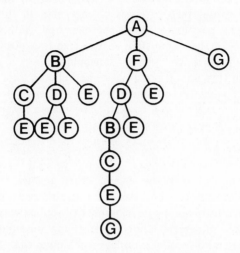

Again the tree is drastically smaller. It is important to note that the savings achieved for this toy problem is only indicative of the situation for larger problems. A cutoff high in the tree can lead to truly significant savings; missing an obvious cutoff can lead to truly significant waste.

The general procedure of solving a problem by systematically generating all possible solutions as described above is called *backtracking*. Whenever we have a situation where partial solutions to a problem can be successively augmented in many ways to produce a complete solution, a recursive implementation like the program above may be appropriate. As above, the process can be described by an exhaustive search tree whose nodes correspond to the partial solutions. Going down in the tree corresponds to forward progress towards creating a more complete solution; going up in the tree corresponds to "backtracking" to some previously generated partial solution, from which point it might be worthwhile to proceed forwards again. The general technique of calculating bounds on partial solutions in order to limit the number of full solutions which need to be examined is sometimes called *branch-and-bound*.

For another example, consider the knapsack problem of the previous chapter, where the values are not necessarily restricted to be integers. For this problem, the partial solutions are clearly some selection of items for the knapsack, and backtracking corresponds to taking an item out to try some other combination. Pruning the search tree by removing symmetries is quite effective for this problem, since the order in which objects are put into the knapsack doesn't affect the cost.

Backtracking and branch-and-bound are quite widely applicable as general problem-solving techniques. For example, they form the basis for many programs which play games such as chess or checkers. In this case, a partial solution is some legal positioning of all the pieces on the board, and the descendant of a node in the exhaustive search tree is a position that can be the result of some legal move. Ideally, it would be best if a program could exhaustively search through all possibilities and choose a move that will lead to a win no matter what the opponent does, but there are normally far too many possibilities to do this, so a backtracking search is typically done with quite sophisticated pruning rules so that only "interesting" positions are examined. Exhaustive search techniques are also used for other applications in artificial intelligence.

In the next chapter we'll see several other problems similar to those we've been studying that can be attacked using these techniques. Solving a particular problem involves the development of sophisticated criteria which can be used to limit the search. For the traveling salesman problem we've given only a few examples of the many techniques that have been tried, and equally sophisticated methods have been developed for other important problems.

However sophisticated the criteria, it is generally true that the running time of backtracking algorithms remains exponential. Roughly, if each node in the search tree has α sons, on the average, and the length of the solution path is N, then we expect the number of nodes in the tree to be proportional to α^N. Different backtracking rules correspond to reducing the value of α, the number of choices to try at each node. It is worthwhile to expend effort to do this because a reduction in α will lead to an increase in the size of the problem that can be solved. For example, an algorithm which runs in time proportional to 1.1^N can solve a problem perhaps eight times a large as one which runs in time proportional to 2^N.

Digression: Permutation Generation

An interesting computational puzzle is to write a program that generates all possible ways of rearranging N distinct items. A simple program for this *permutation generation* problem can be derived directly from the exhaustive search program above because, as noted above, if it is run on a complete graph, then it must try to visit the vertices of that graph in all possible orders.

```
procedure visit(k: integer);
  var t: integer;
  begin
  now:=now+1; val[k]:=now;
  if now=V then writeperm;
  for t:=1 to V do
    if val[t]=0 then visit(t);
  now:=now-1; val[k]:=0
  end;
```

This program is derived from the procedure above by eliminating all reference to the adjacency matrix (since all edges are present in a complete graph). The procedure *writeperm* simply writes out the entries of the *val* array. This is done each time *now*=*V*, corresponding to the discovery of a complete path in the graph. (Actually, the program can be improved somewhat by skipping the **for** loop when *now*=*V*, since at that point is known that all the *val* entries are nonzero.) To print out all permutations of the integers 1 through *N*, we invoke this procedure with the call *visit(0)* with *now* initialized to −1 and the *val* array initialized to 0. This corresponds to introducing a dummy node to the complete graph, and checking all paths in the graph starting with node 0. When invoked in this way for *N=4*, this procedure produces the following output (rearranged here into two columns):

1	2	3	4		2	3	1	4
1	2	4	3		2	4	1	3
1	3	2	4		3	2	1	4
1	4	2	3		4	2	1	3
1	3	4	2		3	4	1	2
1	4	3	2		4	3	1	2
2	1	3	4		2	3	4	1
2	1	4	3		2	4	3	1
3	1	2	4		3	2	4	1
4	1	2	3		4	2	3	1
3	1	4	2		3	4	2	1
4	1	3	2		4	3	2	1

Admittedly, the interpretation of the procedure as generating paths in a complete graph is barely visible. But a direct examination of the procedure reveals that it generates all *N*! permutations of the integers 1 to *N* by first generating all $(N-1)!$ permutations with the 1 in the first position (calling itself recursively to place 2 through N), then generating the $(N-1)!$ permutations with the 1 in the second position, etc.

Now, it would be unthinkable to use this program even for $N = 16$, because $16! > 2^{50}$. Still, it is important to study because it can form the basis for a backtracking program to solve any problem involving reordering a set of elements.

For example, consider the *Euclidean traveling salesman problem*: given a set of N points in the plane, find the shortest tour that connects them all. Since each ordering of the points corresponds to a legal tour, the above program can be made to exhaustively search for the solution to this problem simply by changing it to keep track of the cost of each tour and the minimum of the costs of the full tours, in the same manner as above. Then the same branch-and-bound technique as above can be applied, as well as various backtracking heuristics specific to the Euclidean problem. (For example, it is easy to prove that the optimal tour cannot cross itself, so the search can be cut off on all partial paths that cross themselves.) Different search heuristics might correspond to different ways of ordering the permutations. Such techniques can save an enormous amount of work but always leave an enormous amount of work to be done. It is not at all a simple matter to find an exact solution to the Euclidean traveling salesman problem, even for N as low as 16.

Another reason that permutation generation is of interest is that there are a number of related procedures for generating other combinatorial objects. In some cases, the number of objects generated are not quite so numerous are as permutations, and such procedures can be useful for larger N in practice. An example of this is a procedure to generate all ways of choosing a subset of size k out of a set of N items. For large N and small k, the number of ways of doing this is roughly proportional to N^k. Such a procedure could be used as the basis for a backtracking program to solve the knapsack problem.

Approximation Algorithms

Since finding the shortest tour seems to require so much computation, it is reasonable to consider whether it might be easier to find a tour that is almost as short as the shortest. If we're willing to relax the restriction that we absolutely must have the shortest possible path, then it turns out that we can deal with problems much larger than is possible with the techniques above.

For example, it's relatively easy to find a tour which is longer by at most a factor of two than the optimal tour. The method is based on simply finding the minimum spanning tree: this not only, as mentioned above, provides a lower bound on the length of the tour but also turns out to provide an *upper bound* on the length of the tour, as follows. Consider the tour produced by visiting the nodes of the minimum spanning tree using the following procedure: to process node x, visit x, then visit each son of x, applying this visiting procedure recursively and returning to node x after each son has been visited, ending up at node x. This tour traverses every edge in the spanning tree twice, so its cost is twice the cost of the tree. It is not a simple tour, since a node may be visited many times, but it can be converted to a simple tour simply by deleting all but the first occurrence of each node. Deleting an occurrence of a node corresponds to taking a shortcut past that node: certainly it can't increase the cost of the tour. Thus, we have a simple tour which has a cost less than twice that of the minimum spanning tree. For example, the following diagram shows a minimum spanning tree for our set of sample points (computed as described in Chapter 31), along with a corresponding simple tour.

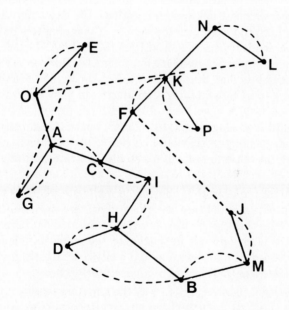

This tour is clearly not the optimum, because it self-intersects. For a large random point set, it seems likely that the tour produced in this way will be close to the optimum, though no analysis has been done to support this conclusion.

Another approach that has been tried is to develop techniques to improve an existing tour in the hope that a short tour can be found by applying such improvements repeatedly. For example, if we have (as above) a *Euclidean* traveling salesman problem where graph distances are distances between points in the plane, then a self-intersecting tour can be improved by removing each intersection as follows. If the line AB intersects the line CD, the situation can be diagramed as at left below, without loss of generality. But it follows immediately that a shorter tour can be formed by deleting AB and CD and adding AD and CB, as diagramed at right:

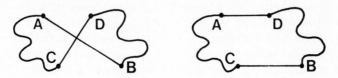

Applying this procedure successively will, given any tour, produce a tour that is no longer and which is not self-intersecting. For example, the procedure applied to the tour produced from the minimum spanning tree in the example above gives the shorter tour AGOENLPKFJMBDHICA. In fact, one of the most effective approaches to producing approximate solutions to the Euclidean traveling salesman problem, developed by S. Lin, is to generalize the procedure above to improve tours by switching around three or more edges in an existing tour. Very good results have been obtained by applying such a procedure successively, until it no longer leads to an improvement, to an initially *random* tour. One might think that it would be better to start with a tour that is already close to the optimum, but Lin's studies indicate that this may not be the case.

The various approaches to producing approximate solutions to the traveling salesman problem which are described above are only indicative of the types of techniques that can be used in order to avoid exhaustive search. The brief descriptions above do not do justice to the many ingenious ideas that have been developed: the formulation and analysis of algorithms of this type is still a quite active area of research in computer science.

One might legitimately question why the traveling salesman problem and the other problems that we have been alluding to *require* exhaustive search. Couldn't there be a clever algorithm that finds the minimal tour as easily and quickly as we can find the minimum spanning tree? In the next chapter we'll see why most computer scientists believe that there is no such algorithm and why approximation algorithms of the type discussed in this section must therefore be studied.

Exercises

1. Which would you prefer to use, an algorithm that requires N^5 steps or one that requires 2^N steps?

2. Does the "maze" graph at the end of Chapter 29 have a Hamilton cycle?

3. Draw the tree describing the operation of the exhaustive search procedure when looking for a Hamilton cycle on the sample graph starting at vertex B instead of vertex A.

4. How long could exhaustive search take to find a Hamilton cycle in a graph where all nodes are connected to exactly two other nodes? Answer the same question for the case where all nodes are connected to exactly three other nodes.

5. How many calls to *visit* are made (as a function of V) by the permutation generation procedure?

6. Derive a nonrecursive permutation generation procedure from the program given.

7. Write a program which determines whether or not two given adjacency matrices represent the same graph, except with different vertex names.

8. Write a program to solve the knapsack problem of Chapter 37 when the sizes can be real numbers.

9. Define another cutoff rule for the Euclidean traveling salesman problem, and show the search tree that it leads to for the first six points of our sample point set.

10. Write a program to count the number of spanning trees of a set of N given points in the plane with no intersecting edges.

11. Solve the Euclidean traveling salesman problem for our sixteen sample points.

40. NP-complete Problems

The algorithms we've studied in this book generally are used to solve practical problems and therefore consume reasonable amounts of resources. The practical utility of most of the algorithms is obvious: for many problems we have the luxury of several efficient algorithms to choose from. Many of the algorithms that we have studied are routinely used to solve actual practical problems. Unfortunately, as pointed out in the previous chapter, many problems arise in practice which do not admit such efficient solutions. What's worse, for a large class of such problems we can't even tell whether or not an efficient solution might exist.

This state of affairs has been a source of extreme frustration for programmers and algorithm designers, who can't find any efficient algorithm for a wide range of practical problems, and for theoreticians, who have been unable to find any reason why these problems should be difficult. A great deal of research has been done in this area and has led to the development of mechanisms by which new problems can be classified as being "as difficult as" old problems in a particular technical sense. Though much of this work is beyond the scope of this book, the central ideas are not difficult to learn. It is certainly useful when faced with a new problem to have some appreciation for the types of problems for which no one knows any efficient algorithm.

Sometimes there is quite a fine line between "easy" and "hard" problems. For example, we saw an efficient algorithm in Chapter 31 for the following problem: "Find the shortest path from vertex x to vertex y in a given weighted graph." But if we ask for the *longest* path (without cycles) from x to y, we have a problem for which no one knows a solution substantially better than checking all possible paths. The fine line is even more striking when we consider similar problems that ask for only "yes-no" answers:

Easy: Is there a path from x to y with weight $\leq M$?
Hard(?):   Is there a path from x to y with weight $\geq M$?

Breadth-first search will lead to a solution for the first problem in linear time, but all known algorithms for the second problem could take exponential time.

We can be much more precise than "could take exponential time," but that will not be necessary for the present discussion. Generally, it is useful to think of an exponential-time algorithm as one which, for some input of size N, takes time proportional to 2^N (at least). (The substance of the results that we're about to discuss is not changed if 2 is replaced by any number $\alpha > 1$.) This means, for example, that an exponential-time algorithm could not be guaranteed to work for all problems of size 100 (say) or greater, because no one could wait for an algorithm to take 2^{100} steps, regardless of the speed of the computer. Exponential growth dwarfs technological changes: a supercomputer may be a trillion times faster than an abacus, but neither can come close to solving a problem that requires 2^{100} steps.

Deterministic and Nondeterministic Polynomial-Time Algorithms

The great disparity in performance between "efficient" algorithms of the type we've been studying and brute-force "exponential" algorithms that check each possibility makes it possible to study the interface between them with a simple formal model.

In this model, the efficiency of an algorithm is a function of the number of bits used to encode the input, using a "reasonable" encoding scheme. (The precise definition of "reasonable" includes all common methods of encoding things for computers: an example of an unreasonable coding scheme is unary, where M bits are used to represent the number M. Rather, we would expect that the number of bits used to represent the number M should be proportional to $\log M$.) We're interested merely in identifying algorithms guaranteed to run in time proportional to some polynomial in the number of bits of input. Any problem which can be solved by such an algorithm is said to belong to

P: the set of all problems which can be solved by deterministic
       algorithms in polynomial time.

By *deterministic* we mean that at any time, whatever the algorithm is doing, there is only one thing that it could do next. This very general notion covers the way that programs run on actual computers. Note that the polynomial is not specified at all and that this definition certainly covers the standard algorithms that we've studied so far. Sorting belongs to P because (for example) insertion sort runs in time proportional to N^2: the existence of $N \log N$ sorting algorithms is not relevant to the present discussion. Also, the time taken by an algorithm obviously depends on the computer used, but it turns out that using a different computer will affect the running time by only a polynomial factor (again, assuming reasonable limits), so that also is not particularly relevant to the present discussion.

Of course, the theoretical results that we're discussing are based on a completely specified model of computation within which the general statements that we're making here can be proved. Our intent is to examine some of the central ideas, not to develop rigorous definitions and theorem statements. The reader may rest assured that any apparent logical flaws are due to the informal nature of the description, not the theory itself.

One "unreasonable" way to extend the power of a computer is to endow it with the power of *nondeterminism*: when an algorithm is faced with a choice of several options, it has the power to "guess" the right one. For the purposes of the discussion below, we can think of an algorithm for a nondeterministic machine as "guessing" the solution to a problem, then verifying that the solution is correct. In Chapter 20, we saw how nondeterminism can be useful as a tool for algorithm design; here we use it as a theoretical device to help classify problems. We have

NP: the set of all problems which can be solved by nondeterministic algorithms in polynomial time.

Obviously, any problem in P is also in NP. But it seems that there should be many other problems in NP: to show that a problem is in NP, we need only find a polynomial-time algorithm to check that a given solution (the guessed solution) is valid. For example, the "yes-no" version of the longest-path problem is in NP. Another example of a problem in NP is the *satisfiability* problem. Given a logical formula of the form

$$(x_1 + x_3 + x_5)^*(x_1 + \overline{x}_2 + x_4)^*(\overline{x}_3 + x_4 + x_5)^*(x_2 + \overline{x}_3 + x_5)$$

where the x_i's represent variables which take on truth values (**true** or **false**), "+" represents **or**, "*" represents **and**, and $\overline{x}$ represents **not**, the satisfiability problem is to determine whether or not there exists an assignment of truth values to the variables that makes the formula **true** ("satisfies" it). We'll see below that this particular problem plays a special role in the theory.

Nondeterminism is such a powerful operation that it seems almost absurd to consider it seriously. Why bother considering an imaginary tool that makes difficult problems seem trivial? The answer is that, powerful as nondeterminism may seem, no one has been able to *prove* that it helps for any particular problem! Put another way, no one has been able to find a single example of a problem which can be proven to be in NP but not in P (or even prove that one exists): we do not know whether or not $P = NP$. This is a quite frustrating situation because many important practical problems belong to NP (they could be solved efficiently on a non-deterministic machine) but may or may not belong to P (we don't know any efficient algorithms for them on a deterministic machine). If we could prove that a problem doesn't belong to P, then we could abandon the search for an efficient solution to it. In the absence of such a proof, there is the lingering possibility that some efficient algorithm has gone undiscovered. In fact, given the current state of our knowledge, it could be the case that there is some efficient algorithm for *every* problem in NP, which would imply that many efficient algorithms have gone undiscovered. Virtually no one believes that $P = NP$, and a considerable amount of effort has gone into proving the contrary, but this remains the outstanding open research problem in computer science.

NP-Completeness

Below we'll look at a list of problems that are known to belong to NP but which might or might not belong to P. That is, they are easy to solve on a non-deterministic machine, but, despite considerable effort, no one has been able to find an efficient algorithm on a conventional machine (or prove that none exists) for any of them. These problems have an additional property that provides convincing evidence that $P \neq NP$: if *any* of the problems can be solved in polynomial time on a deterministic machine, then so can *all* problems in NP (i.e., $P = NP$). That is, the collective failure of all the researchers to find efficient algorithms for all of these problems might be viewed as a collective failure to prove that $P = NP$. Such problems are said to be *NP-complete*. It turns out that a large number of interesting practical problems have this characteristic.

The primary tool used to prove that problems are NP-complete uses the idea of *polynomial reducibility*. We show that any algorithm to solve a new problem in NP can be used to solve some known NP-complete problem by the following process: transform any instance of the known NP-complete problem to an instance of the new problem, solve the problem using the given algorithm, then transform the solution back to a solution of the NP-complete problem. We saw an example of a similar process in Chapter 34, where we reduced bipartite matching to network flow. By "polynomially" reducible, we mean that the transformations can be done in polynomial time: thus the existence of a polynomial-time algorithm for the new problem would imply the existence of a polynomial-time algorithm for the NP-complete problem, and this would (by definition) imply the existence of polynomial-time algorithms for all problems in NP.

The concept of reduction provides a useful mechanism for classifying algorithms. For example, to prove that a problem in NP is NP-complete, we need only show that some known NP-complete problem is polynomially reducible to it: that is, that a polynomial-time algorithm for the new problem could be used to solve the NP-complete problem, and then could, in turn, be used to solve all problems in NP. For an example of reduction, consider the following two problems:

TRAVELING SALESMAN: Given a set of cities, and distances between all pairs, find a tour of all the cities of distance less than M.

HAMILTON CYCLE: Given a graph, find a simple cycle that includes all the vertices.

Suppose that we know the Hamilton cycle problem to be NP-complete and we wish to determine whether or not the traveling salesman problem is also NP-complete. Any algorithm for solving the traveling salesman problem can be used to solve the Hamilton cycle problem, through the following reduction: given an instance of the Hamilton cycle problem (a graph) construct an instance of the traveling salesman problem (a set of cities, with distances between all pairs) as follows: for cities for the traveling salesman use the set of vertices in the graph; for distances between each pair of cities use 1 if there is an edge between the corresponding vertices in the graph, 2 if there is no edge. Then have the algorithm for the traveling salesman problem find a tour of distance less than or equal to N, the number of vertices in the graph. That tour must correspond precisely to a Hamilton cycle. An efficient algorithm for the traveling salesman problem would also be an efficient algorithm for the Hamilton cycle problem. That is, the Hamilton cycle problem reduces to the traveling salesman problem, so the NP-completeness of the Hamilton cycle problem implies the NP-completeness of the traveling salesman problem.

The reduction of the Hamilton cycle problem to the traveling salesman problem is relatively simple because the problems are so similar. Actually, polynomial-time reductions can be quite complicated indeed and can connect problems which seem to be quite dissimilar. For example, it is possible to reduce the satisfiability problem to the Hamilton cycle problem. Without going into details, we can look at a sketch of the proof. We wish to show that if we had a polynomial-time solution to the Hamilton cycle problem, then we could get a polynomial-time solution to the satisfiability problem by polynomial reduction. The proof consists of a detailed method of construction showing how, given an instance of the satisfiability problem (a Boolean formula) to construct (in polynomial time) an instance of the Hamilton cycle problem (a graph) with the property that knowing whether the graph has a Hamilton cycle tells us whether the formula is satisfiable. The graph is built from small components (corresponding to the variables) which can be traversed by a simple path in only one of two ways (corresponding to the truth or falsity of the variables). These small components are attached together as specified by the clauses, using more complicated subgraphs which can be traversed by simple paths corresponding to the truth or falsity of the clauses. It is quite a large step from this brief description to the full construction: the point is to illustrate that polynomial reduction can be applied to quite dissimilar problems.

Thus, if we were to have a polynomial-time algorithm for the traveling salesman problem, then we would have a polynomial-time algorithm for the Hamilton cycle problem, which would also give us a polynomial-time algorithm for the satisfiability problem. Each problem that is proven NP-complete provides another potential basis for proving yet another future problem NP-complete. The proof might be as simple as the reduction given above from the Hamilton cycle problem to the traveling salesman problem, or as complicated as the transformation sketched above from the satisfiability problem to the Hamilton cycle problem, or somewhere in between. Literally thousands of problems have been proven to be NP-complete over the last ten years by transforming one to another in this way.

Cook's Theorem

Reduction uses the NP-completeness of one problem to imply the NP-completeness of another. There is one case where it doesn't apply: how was the *first* problem proven to be NP-complete? This was done by S. A. Cook in 1971. Cook gave a direct proof that satisfiability is NP-complete: that if there is a polynomial time algorithm for satisfiability, then all problems in NP can be solved in polynomial time.

The proof is extremely complicated but the general method can be explained. First, a full mathematical definition of a machine capable of solving any problem in NP is developed. This is a simple model of a general-purpose computer known as a *Turing machine* which can read inputs, perform certain operations, and write outputs. A Turing machine can perform any computation that any other general purpose computer can, using the same amount of time (to within a polynomial factor), and it has the additional advantage that it can be concisely described mathematically. Endowed with the additional power of nondeterminism, a Turing machine can solve any problem in NP. The next step in the proof is to describe each feature of the machine, including the way that instructions are executed, in terms of logical formulas such as appear in the satisfiability problem. In this way a correspondence is established between every problem in NP (which can be expressed as a program on the nondeterministic Turing machine) and some instance of satisfiability (the translation of that program into a logical formula). Now, the solution to the satisfiability problem essentially corresponds to a simulation of the machine running the given program on the given input, so it produces a solution to an instance of the given problem. Further details of this proof are well beyond the scope of this book. Fortunately, only one such proof is really necessary: it is much easier to use reduction to prove NP-completeness.

Some NP-Complete Problems

As mentioned above, literally thousands of diverse problems are known to be NP-complete. In this section, we list a few for purposes of illustrating the wide range of problems that have been studied. Of course, the list begins with *satisfiability* and includes *traveling salesman* and *Hamilton cycle*, as well as *longest path*. The following additional problems are representative:

PARTITION: Given a set of integers, can they be divided into two sets whose sum is equal?

INTEGER LINEAR PROGRAMMING: Given a linear program, is there a solution in integers?

MULTIPROCESSOR SCHEDULING: Given a deadline and a set of tasks of varying length to be performed on two identical processors can the tasks be arranged so that the deadline is met?

VERTEX COVER: Given a graph and an integer N, is there a set of less than N vertices which touches all the edges?

These and many related problems have important natural practical applications, and there has been strong motivation for some time to find good algorithms to solve them. The fact that no good algorithm has been found for any of these problems is surely strong evidence that $P \neq NP$, and most researchers certainly believe this to be the case. (On the other hand, the fact that no one has been able to prove that any of these problem do not belong to P could be construed to comprise a similar body of circumstantial evidence on the other side.) Whether or not $P = NP$, the practical fact is that we have at present no algorithms that are guaranteed to solve any of the NP-complete problems efficiently.

As indicated in the previous chapter, several techniques have been developed to cope with this situation, since some sort of solution to these various problems must be found in practice. One approach is to change the problem and find an "approximation" algorithm that finds not the best solution but a solution that is guaranteed to be close to the best. (Unfortunately, this is sometimes not sufficient to fend off NP-completeness.) Another approach is to rely on "average-time" performance and develop an algorithm that finds the solution in some cases, but doesn't necessarily work in all cases. That is, while it may not be possible to find an algorithm that is guaranteed to work well on all instances of a problem, it may well be possible to solve efficiently virtually all of the instances that arise in practice. A third approach is to work with "efficient" exponential algorithms, using the backtracking techniques described in the previous chapter. Finally, there is quite a large gap between polynomial and exponential time which is not addressed by the theory. What about an algorithm that runs in time proportional to $N^{\log N}$ or $2^{\sqrt{N}}$?

All of the application areas that we've studied in this book are touched by NP-completeness: there are NP-complete problems in numerical applications, in sorting and searching, in string processing, in geometry, and in graph processing. The most important practical contribution of the theory of NP-completeness is that it provides a mechanism to discover whether a new problem from any of these diverse areas is "easy" or "hard." If one can find an efficient algorithm to solve a new problem, then there is no difficulty. If not, a proof that the problem is NP-complete at least gives the information that the development of an efficient algorithm would be a stunning achievement (and suggests that a different approach should perhaps be tried). The scores of efficient algorithms that we've examined in this book are testimony that we have learned a great deal about efficient computational methods since Euclid, but the theory of NP-completeness shows that, indeed, we still have a great deal to learn.

Exercises

1. Write a program to find the longest simple path from x to y in a given weighted graph.

2. Could there be an algorithm which solves an NP-complete problem in an *average* time of $N \log N$, if $P \neq NP$? Explain your answer.

3. Give a nondeterministic polynomial-time algorithm for solving the PARTITION problem.

4. Is there an immediate polynomial-time reduction from the traveling salesman problem on graphs to the Euclidean traveling salesman problem, or vice versa?

5. What would be the significance of a program that could solve the traveling salesman problem in time proportional to 1.1^N?

6. Is the logical formula given in the text satisfiable?

7. Could one of the "algorithm machines" with full parallelism be used to solve an NP-complete problem in polynomial time, if $P \neq NP$? Explain your answer.

8. How does the problem "compute the exact value of 2^N" fit into the P-NP classification scheme?

9. Prove that the problem of finding a Hamilton cycle in a *directed* graph is NP-complete, using the NP-completeness of the Hamilton cycle problem for undirected graphs.

10. Suppose that two problems are known to be NP-complete. Does this imply that there is a polynomial-time reduction from one to the other, if $P \neq NP$?

SOURCES for Advanced Topics

Each of the topics covered in this section is the subject of volumes of reference material. From our introductory treatment, the reader seeking more information should anticipate engaging in serious study; we'll only be able to indicate some basic references here.

The perfect shuffle machine of Chapter 35 is described in the 1968 paper by Stone, which covers many other applications. One place to look for more information on systolic arrays is the chapter by Kung and Leiserson in Mead and Conway's book on VLSI. A good reference for applications and implementation of the FFT is the book by Rabiner and Gold. Further information on dynamic programming (and topics from other chapters) may be found in the book by Hu. Our treatment of linear programming in Chapter 38 is based on the excellent treatment in the book by Papadimitriou and Steiglitz, where all the intuitive arguments are backed up by full mathematical proofs. Further information on exhaustive search techniques may be found in the books by Wells and by Reingold, Nievergelt, and Deo. Finally, the reader interested in more information on NP-completeness may consult the survey article by Lewis and Papadimitriou and the book by Garey and Johnson, which has a full description of various types of NP-completeness and a categorized listing of hundreds of NP-complete problems.

M. R. Garey and D. S. Johnson, *Computers and Intractability: a Guide to the Theory of NP-Completeness*, Freeman, San Francisco, CA, 1979.

T. C. Hu, *Combinatorial Algorithms*, Addison-Wesley, Reading, MA, 1982.

H. R. Lewis and C. H. Papadimitriou, "The efficiency of algorithms," *Scientific American*, 238, 1 (1978).

C. A. Mead and L. C. Conway, *Introduction to VLSI Design*, Addison-Wesley, Reading, MA, 1980.

C. H. Papadimitriou and K. Steiglitz, *Combinatorial Optimization: Algorithms and Complexity*, Prentice-Hall, Englewood Cliffs, NJ, 1982.

E. M. Reingold, J. Nievergelt, and N. Deo, *Combinatorial Algorithms: Theory and Practice*, Prentice-Hall, Englewood Cliffs, NJ, 1982.

L. R. Rabiner and B. Gold, *Digital Signal Processing*, Prentice-Hall, Englewood Cliffs, NJ, 1974.

H. S. Stone, "Parallel processing with the perfect shuffle," *IEEE Transactions on Computing*, **C-20**, 2 (February, 1971).

M. B. Wells, *Elements of Combinatorial Computing*, Pergaman Press, Oxford, 1971.

Index

542

DESIGNS

Cover   *Insertion sort*: Color represents the key value; the ith column (from right to left) shows result of ith insertion.

Page 1   *Relatively prime numbers*: A mark is in positions i, j for which the greatest common divisor of i and j is not 1.

21   *Random points*: A mark is in position i, j with i and j generated by a linear congruential random number generator.

89   *A heap*: Horizontal coordinate is position in heap, vertical coordinate is value.

169   *A binary search tree* laid out in the manner of an H-tree.

239   *Huffman's algorithm before and after*: run on the initial part of the text file for Chapter 22.

305   *One intersecting pair* among a set of random horizontal and vertical lines.

371   *Depth first search on a grid graph*: each node is adjacent to its immediate neighbors; adjacency lists are in random order.

455   *Counting to 2^8*: eight cyclic rotations.

Back   *Random permutation*: Color represents the key value; the ith column (from right to left) shows result of exchanging ith item with one having a random index greater than i.

Heap design inspired by the movie "Sorting out Sorting," R. Baecker, University of Toronto.

Pictures printed by Tom Freeman, using programs from the text.